Biochemistry of Women: Clinical Concepts

Editors:

A. S. Curry
Forensic Science Service
Home Office
London, U.K.

J. V. Hewitt
M.Sc., C.Chem. M.R.I.C.
Area Laboratory
King Edward VII Hospital
Windsor U.K.

CRC Press
Taylor & Francis Group
Boca Raton London New York

CRC Press is an imprint of the
Taylor & Francis Group, an **informa** business

First published 1974 by CRC Press
Taylor & Francis Group
6000 Broken Sound Parkway NW, Suite 300
Boca Raton, FL 33487-2742

Reissued 2018 by CRC Press

A Library of Congress record exists under LC control number: 73090241

Publisher's Note
The publisher has gone to great lengths to ensure the quality of this reprint but points out that some imperfections in the original copies may be apparent.

Disclaimer
The publisher has made every effort to trace copyright holders and welcomes correspondence from those they have been unable to contact.

ISBN 13: 978-1-315-89107-1 (hbk)
ISBN 13: 978-1-351-07017-1 (ebk)

Visit the Taylor & Francis Web site at http://www.taylorandfrancis.com and the
CRC Press Web site at http://www.crcpress.com

INTRODUCTION

This book concerns the clinical chemistry of women – from conception to the grave. It concerns their medical problems and the laboratories' role in elucidating these and in monitoring treatment.

There have been many textbooks dealing with human biochemistry but few concentrate on the differences between male and female; fewer still deal from the laboratories' point of view with those diseases and conditions that affect only women. It is a most important area, for the continuance of the human species depends on the health of the female, and the problems of infertility, teratology, breast cancer, and the effects of the pill are vital. It is also important to keep mother-in-law fit to look after the children, so many will think the section on the menopause equally interesting!

The book represents the current "state of the art" and although in many areas there are sometimes widely divergent opinions, it is good that these should be aired so that problem areas can be revealed and opinions revised when further experimental work is done as a consequence.

We hope those readers who find it good bedside reading will, when the light goes out, find it has enhanced their appreciation of the opposite sex.

THE EDITORS

Dr. Alan S. Curry is Controller of the Home Office Forensic Science Service, U.K. Although primarily a forensic scientist, he has great experience in analytical methodology and toxicology in particular. His interest in clinical chemistry arises from a study of how poisons interfere with metabolic processes, and this had led to an interest in differences in male and female biochemistry. He has over 100 scientific papers to his credit, and as well as being the author of books on forensic toxicology (e.g., *Recent Advances in Forensic and Clinical Toxicology)*, he has also co-edited *Biochemistry of Women: Methods for Clinical Investigation.*

Miss J. V. Hewitt is a clinical chemist who after her degree worked at Manchester Royal Infirmary for 12 years before moving to Windsor. She has extensive experience in routine clinical biochemistry and has brought her special interest in feminine biochemistry to the book. She is responsible for the selection of chapters and authors for the book.

Dr. Curry wishes to make it clear that this work was not done under the aegis of the Home Office nor as any part of his official duties. It has been an entirely private venture.

CONTRIBUTORS

C. M. Andre, Ph.D.
Department of Chemical Pathology
St. Mary's Hospital Medical School
London W2 1PG
England

W. P. Collins, B.Sc., Ph.D.
Department of Obstetrics and Gynaecology
King's College Hospital Medical School
University of London
England

D. Davies, M.A., F.R.C.P.
Manchester Royal Infirmary
Oxford Road
Manchester 13
England

J. C. Gallagher, M.D., M.R.C.P.
M.R.C. Mineral Metabolism Unit
The General Infirmary
Leeds LS1 3EX
England

V.H.T. James, Ph.D., D.Sc., F.R.I.C., F.R.C.Path.
Department of Chemical Pathology
St. Mary's Hospital Medical School
London W2 1PG
England

R. W. Morgan, M.D.
Department of Preventive Medicine
University of Toronto
School of Hygiene
Toronto, Ontario
Canada

S. Naik, M.A., Ch.B., D.A.
Sheffield Supra Regional Endocrine Investigation Centre
The Jessop Hospital for Women
Sheffield S3 7RE
England

J. R. Newton, M.D., M.R.C.O.G.
Department of Obstetrics and Gynaecology
King's College Hospital Medical School
University of London
England

B. E. C. Nordin, M.D., F.R.C.P., D.Sc.
M.R.C. Mineral Metabolism Unit
The General Infirmary
Leeds LS1 3EX
England

D. V. Parke, Ph.D., F.R.I.C., F.I.Biol.
Department of Biochemistry
University of Surrey
Guildford, Surrey GU2 5XH
England

G. W. Pennington, M.D., M.S., M.A., F.R.C.Path.
Sheffield Supra Regional Endocrine Investigation Centre
The Jessop Hospital for Women
Sheffield S3 7RE
England

P. A. Toseland, B.Sc., Ph.D.
Department of Chemical Chemistry
Guy's Hospital
London SE1 9RT
England

D. Tulchinsky, M.D.
Department of Obstetrics and Gynecology
Boston Hospital for Women
221 Longwood Ave.
Boston, Mass. 02115

P. Wilding, B.Sc., Ph.D., F.R.I.C.
Technicon Corporation
Terrytown, New York

TABLE OF CONTENTS

THE OVARIAN CYCLE

W. P. Collins and J. R. Newton

TABLE OF CONTENTS

I. INTRODUCTION

The female anatomy, mentality, and reproductive biochemistry are initiated, maintained, and controlled by hormones. The characteristic glands are the ovaries which perform two complementary and probably interrelated functions. One is the production and release of secondary oocytes and the other is the synthesis and secretion of steroid hormones. Both processes occur in cycles which are related to the life span of the individual. Recent studies have shown that gonadal steroids are secreted throughout life with wide fluctuations in the level of the principal compounds. In contrast, the resumption of gametogenesis and the process of ovulation occur exclusively within the period of reproductive life. This phase is associated with increased levels and regular patterns of hormone secretion and by the occurrence of menstrual cycles.

Numerous studies have shown that steroidogenesis, gametogenesis, and ovulation are under the control of trophic hormones from the anterior pituitary, which in turn is regulated by releasing hormones from the hypothalamus. Furthermore, both hypothalamic and pituitary function are conditioned by neurochemical signals from other parts of the brain and by the feedback effects of various steroids.[1]

A. Approaches to the Study of Ovarian Function

The occurrence of ovulation is frequently inferred from associated changes in hormone levels, as the only positive signs are the isolation of a secondary oocyte or an ensuing pregnancy. The most direct approach to study the endocrine function of the ovary is to monitor the concentrations of secreted products in ovarian venous blood. Unfortunately, this is only possible at operation – a relatively brief moment in time of a process that exhibits cyclical changes over both short and long periods. In view of this limitation, many studies have involved the analysis of metabolites in specimens of urine collected over 24-hr periods. This approach is particularly useful when the endogenous hormone levels are very low, as the extracts may be concentrated. However, it cannot be assumed that there is a consistent relationship between endogenous secretion and urinary excretion; neither does this index reveal acute changes in ovarian activity. The determination of the hormonal content of systemic blood offers the advantage of an approach which is both practical and more direct than urinary assays. Daily, hourly, or more frequent changes can be studied in relation to the menstrual cycle, including the expected time of ovulation. However, at certain times there are formidable difficulties surrounding the interpretation of results. For example, hormonal steroids are produced and secreted by the adrenal cortices as well as both ovaries. Furthermore, the secretions from the adrenal appear to fluctuate over a 24-hr period, so that there are at least two superimposed cycles of steroid secretion. A related problem is that various compounds secreted by one gland may serve as precursors for the other and, in addition, products from either gland may affect hypothalamic-pituitary function and, hence, the secretion of follicle-stimulating hormone (FSH) and luteinizing hormone (LH). Accordingly, it is extremely difficult to evaluate the actual contribution of the ovaries or adrenal cortices to the total secretion of progestogens, androgens, and estrogens. An assessment of the overall production of one of these compounds also presents difficulties as various progestogens may be converted into androgens, and androgens into estrogens in peripheral and growth responsive tissues. Another important approach to the study of ovarian and hormonal function is the calculation of blood production and glandular secretion rates.[2] These values are obtained from the metabolic clearance rates of administered labeled hormones and the mean plasma concentrations of the endogenous substances. The usefulness of this approach adds incentive to attempts to develop accurate methods for the analysis of peripheral venous plasma, but is limited by the inadvisability of exposing young women, who may subsequently conceive, to significant amounts of carbon 14 or tritium.

B. Biological Activity and Chemical Structure of Gonadal Steroids

The multiple interconversions of gonadal steroids in many tissues confuse assessment of their biological activity and role in reproductive biochemistry. Notwithstanding this limitation, however, it is quite clear that estrogens are primarily responsible for the development and maintenance of the sex organs, secondary sexual characteristics, and the menstrual cycle, while the progestogens have an additive effect and are vital for the secretory response of the endometrium and implantation of the fertilized ovum.[3] Some of the glandular and hormonal interrelationships pertaining during the period of active reproductive life are shown in Figure 1. Unfortunately, there is only limited information on the role of androgenic steroids in women. However, it is known that certain naturally occurring C_{19} steroids increase libido[4] and may be involved in the regulation of follicular growth[5] and ovum maturation.

In the early 1930's, it was shown that the naturally occurring gonadal hormones were steroids containing either 21, 19, or 18 carbon atoms. The parent structures of the progestogens, androgens, and estrogens are shown in Figure 2. All naturally occurring steroid hormones have functional groups added to these basic structures which either increase or decrease the inherent activity. The C_{21} steroids all contain methyl groups at the 10th and 13th carbon atoms and an ethyl group at carbon atom 17. The individual members of the group are formed by substitutions of oxo or hydroxyl groups, mainly at carbon atoms 3, 6, 11, 17, and 20, and dehydration of ring A at carbon atoms 4 and 5. With regard to androgens, the basic hydrocarbon (5α androstane) possesses considerable androgenic activity, which is increased by the presence of a hydroxyl group in the β position at carbon 17 and an oxo group at carbon 3 (dihydrotestosterone). Removal of the hydrogen atoms at carbons 4 and 5 gives the structure for testosterone. The naturally occurring

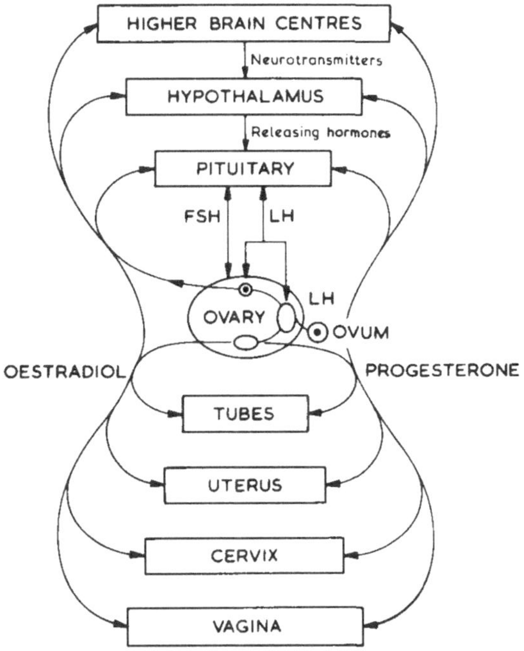

FIGURE 1. A schematic representation of the reproductive cycle in women.

estrogens are characterized by a phenolic A ring and a methyl group at the 13th carbon atom. Specific estrogens with their own characteristic properties are obtained from this parent molecule by substituting oxo or hydroxyl groups at carbon atoms 2, 6, 15, 16, and 17.

According to various estimates, there may be approximately 10^5 naturally occurring steroids based on these structures. To date more than a quarter of a million steroids have been isolated or prepared by partial synthesis. In addition, it is known that several hundred steroids are secreted or formed as metabolic products in the human body. Fortunately, there are only about 60 showing significant biological activity. Even so, this number poses complex analytical problems.

C. Changes in Ovarian Morphology and Function Throughout Life

The time and duration of those periods of life that are associated with marked changes in ovarian morphology are shown in Figure 3. For research purposes, the life span of the ovary may be divided into a number of overlapping periods.

During intrauterine life, the undifferentiated gonads may develop into a recognizable ovary about the 12th week after conception. During childhood, the whole glands increase in size and some endocrine activity may be detected by indirect methods. This phase passes imperceptibly into a relatively short period of sexual maturation, during which time ovarian activity increases, the accessory glands develop, and the secondary sexual characteristics appear. The individual then passes

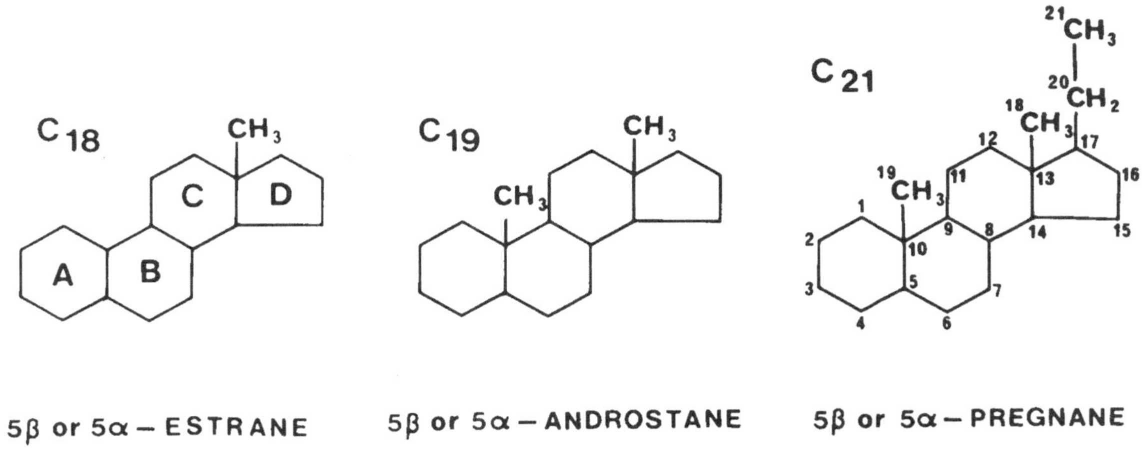

FIGURE 2. The parent structures of gonadal steroids.

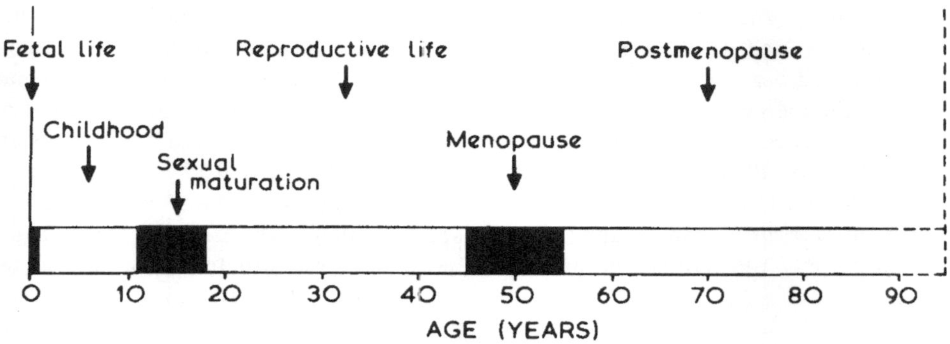

FIGURE 3. Periods of changes in ovarian function.

into a period of active reproductive life which lasts for about 30 years. This phase gradually passes into the menopause, at which time the ovaries start to involute and reproductive capacity gradually ceases. During postmenopausal life, the ovary is thought to retain some of its endocrine capacity, although in a markedly reduced form.

D. Changes in Ovarian Vasculature

An interesting, often forgotten aspect of ovarian function is the marked changes in vascularity and blood flow, which occur throughout the life span of an individual. The studies of Reynolds[6] indicated that spiraling of the ovarian arteries plays an important role in the regulation of ovarian activity. During intrauterine life, this phenomenon has not been observed, but is occasionally seen for a short period after birth – presumably as a consequence of ovarian stimulation by maternal hormones. Throughout childhood, the blood vessels remain poorly developed, but during adolescence they gradually increase in size, complexity, and degree of spiraling. Whether these changes are the cause or result of increased ovarian activity is unknown. During reproductive life, the vasculature adapts to changes in the size and morphology of the ovaries during the course of the menstrual cycle. This adaptation enables the ovarian-stimulating hormones to be evenly distributed throughout the cortex and the products of steroidogenesis to be removed. At the time of the menopause, the number of spiral vessels gradually decreases until the main ovarian artery is virtually the only one remaining. Accordingly, studies on any aspect of ovarian function are incomplete without taking into account the associated changes in blood circulation, flow, and pressure.

II. GAMETOGENESIS AND STEROIDOGENESIS

The process of gametogenesis starts soon after conception, proceeds actively during fetal life, remains quiescent during childhood, and restarts during late pubertal development. The main recognizable stages are summarized in Figure 4. Complete steroidogenesis from precursors containing two carbon atoms is minimal during intrauterine life and childhood. Increased activity is readily discernible during the process of sexual maturation and regular cyclic activity is a characteristic of reproductive life. A great deal of research has been undertaken on the biosynthetic processes which lead to the formation of steroids within the ovary and on the interrelationships between steroidogenesis and ovarian morphology.

A. Fetal Life, Childhood, and Sexual Maturation

Fetal life occupies about 1% of the time from conception to death. During this period the primordial germ cells increase in number and migrate to occupy the undifferentiated gonad.[7,8] This structure contains all the necessary cellular components to develop into either a testis or ovary. The limited amount of evidence that is available suggests that the process of sex differentiation largely consists in the active development of a testis or the passive lack of such a change depending on the presence or absence of the Y chromosome.[9] If the genetic sex is female, the ovary assumes the form of the adult gland and the germ cells become transformed into oogonia. These rapidly increase in number over a short period of time. Thus, the number of germ cells increases from about 1,700 during migration to around 600,000 during the 2nd month of preg-

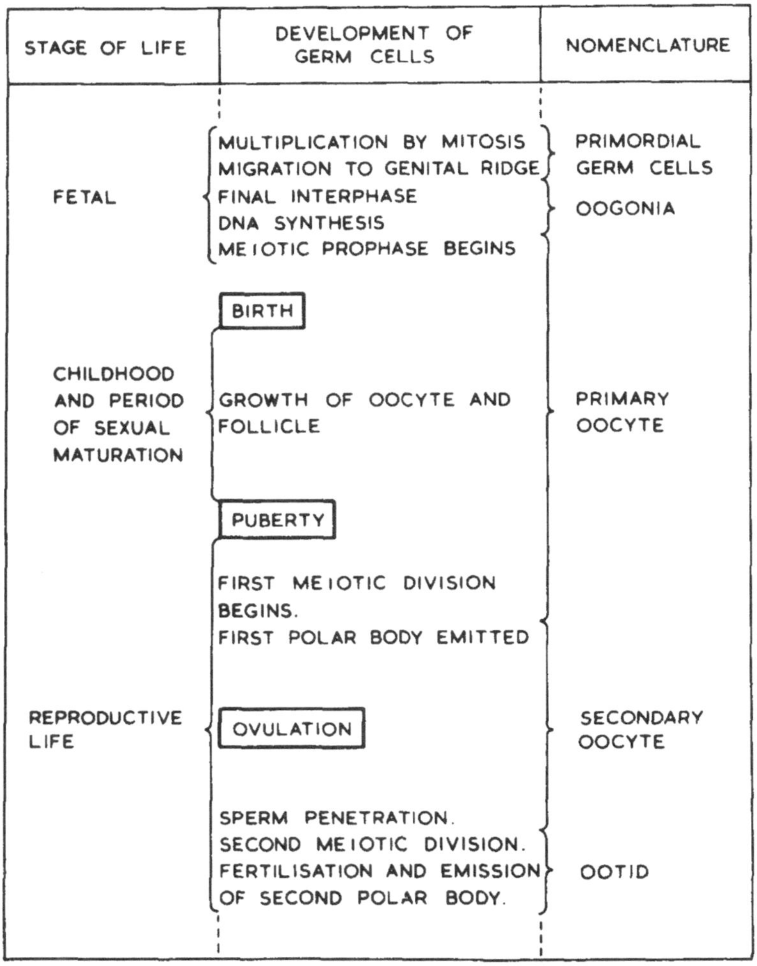

FIGURE 4. Gametogenesis in women.

nancy to almost 7 million at the 5th month. The population then declines rapidly to about 2 million at the time of birth. The oogonia are finally transformed into oocytes, when they enter the prophase of the first of two meiotic divisions. From this time on, the numbers gradually decline to around 250,000 at the start of reproductive life, until the menopause, when few can be detected in histological sections of the ovary. Of the 7 million oocytes that were at one time present in the ovary, only about 500 (i.e., 1 in 14,000) will reach full maturity and be released into the fallopian tube.

The endocrine activity of the fetal ovary is difficult to ascertain. Studies on the incubation of minced tissue with a variety of labeled precursors have indicated that the activity is minimal when compared with the fetal testis; the only enzyme activities readily discernible are 20α, hydrogenation, and 17β, reduction.[10] By contrast, the fetal pituitary gland is capable of synthesizing both FSH and LH as early as 14 weeks of gestation.[11] Whether secretion starts at this time is difficult to determine, owing to the presence of maternal hormones, but one study involving the administration of crude ovine hypothalamic extracts to a 5-week-old infant elicited a prompt rise in serum LH.[12]

During childhood, the ovary increases in size and mature follicles may develop. Stimulation tests with gonadotrophins have indicated that the ovary may respond to the point of ovulation.[18] The results from other studies on the levels of gonadotrophins in patients with gonadal dysgenesis indicate that the prepubertal ovary produces substances which maintain the low levels of FSH and LH.[14]

Although the timing and sequence of physical changes that occur during the process of sexual maturation have been precisely defined, comparatively little is known of the underlying biochemical events. FSH and LH have been measured in urine and plasma of prepubertal females by both bioassay and radioimmunoassay. Notwithstanding the technical difficulties involved, there is general agreement that the level of gonadotrophins rises slowly during the prepubertal years and that there is a better correlation with the stage of pubertal development than with either the bone or chronological age.[15] The relative levels of LH and FSH during this period and throughout life are shown in Figure 5. The values are expressed as a percentage of the maximum to obviate methodological differences and the use of different reference preparations.

With regard to gonadal steroids, the levels of plasma estradiol[15] and urinary estrone, estradiol, and estriol[16] have been studied during the process of sexual maturation. In general, the levels are lower than those observed during reproductive life. The relative levels of progesterone and estradiol in plasma throughout life are shown in Figure 6. The values for progesterone during the period of sexual maturation are thought to be less than 2% of the peak values observed during the luteal phase of the menstrual cycle. From these studies, it would appear that the prepubertal hypothalamus, pituitary, and gonad are all functioning organs and interact with each other continuously. The mechanism of sexual maturation is activated during late childhood and is thought to be associated with a change in sensitivity of the gonadal steroid-brain-

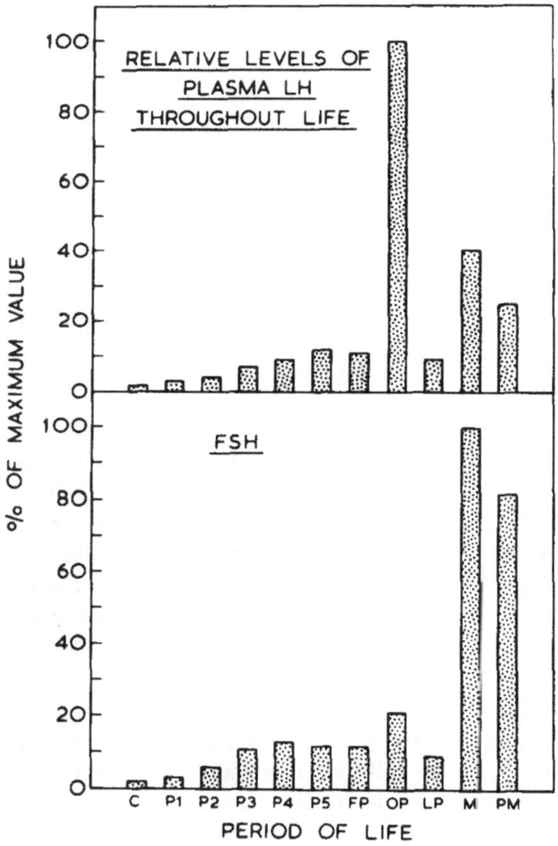

FIGURE 5. The relative levels of LH and FSH throughout life. The abbreviations refer to childhood (C), sexual maturation (P1 – P5), follicular phase of menstrual cycle (FP), ovulatory phase (OP), luted phase (LP), menopause (M), and postmenopause (PM).

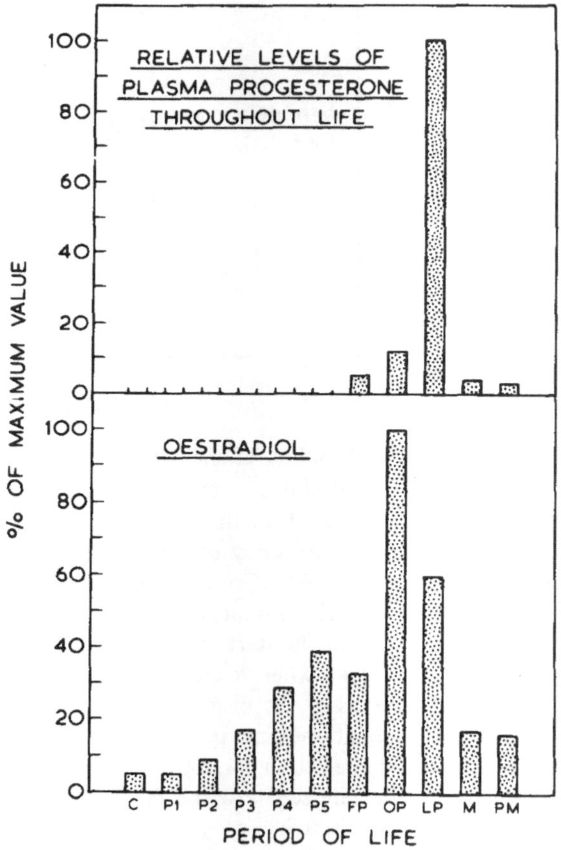

FIGURE 6. The relative levels of the principal gonadal steroids throughout life. The abbreviations refer to childhood (C), sexual maturation (P1 – P5), follicular phase of menstrual cycle (FP), ovulatory phase (OP), luted phase (LP), menopause (M), and postmenopause (PM).

pituitary negative feedback system.[17] Around midpuberty the gonadotrophins start to be released in a cyclical manner and the positive feedback system of estradiol, which causes LH to be released at midcycle, comes into operation.[18] This phase is associated with the resumption of gametogenesis and regular ovulation.

B. Reproductive Life

The period of reproductive life in women occupies approximately 30% of the total life span. During this time, there are more obvious changes in the morphology and related endocrine functions of the ovary. Furthermore, it is only during this period, with the establishment of the correct hormonal balance, that the processes of gametogenesis and follicular growth culminate in ovulation.

Since 1970, the determination of gonadal steroids or their metabolites has been revolutionized by the development of sensitive, accurate, and specific methods based on the principles of competitive protein binding, using either plasma proteins, tissue receptors, or antibodies. Accordingly, a great deal of information has accumulated regarding the levels of many compounds in both plasma and urine. In contrast, comparatively little is known regarding the associated changes in ovarian morphology. For descriptive purposes the menstrual cycle may be divided into four continuous phases,[19] as is illustrated in Figure 7. The length of the preovulatory or follicular phase is variable with a mean value of 14 days. The first part (Phase 1) lasts from 5 to 7 days; during this period a new set of follicles, from 10 to 20, starts to develop. Some rapidly reach a diameter of 1 mm, but the majority become atretic before reaching this size. The earliest signs of follicular growth are a change in shape of the granulosa cells from flat to cuboidal, an increase in mitosis, and formation of the zona pelucida. The next stage (Phase 2) concerns the development of one follicle to the point of ovulation. This process involves a further increase in the number of granulosa cells and the passage of liquid (liquor folliculi) into the spaces between them. As the quantity of liquid increases, the cavity increases to form the antrum. The whole structure is now called a Graafian follicle and the diameter may reach 10 to 15 mm. The oocyte is located at one side of the follicle and is surrounded by the cumulus oophorus and layers of granulosa cells.

The biochemical factors that initiate follicular growth are completely unknown, but circumstantial evidence suggests that the oocyte may be implicated, possibly through the synthesis of RNA. The results from numerous studies on the administration of FSH and LH to women with defective pituitary function have suggested that it is the relative proportion of these two hormones which determines the number of follicles that develops and the proportion that subsequently degenerates.

If ovulation is to follow, the Graafian follicle must undergo further changes (Phase 3). During this period of final maturation, the cells adjacent

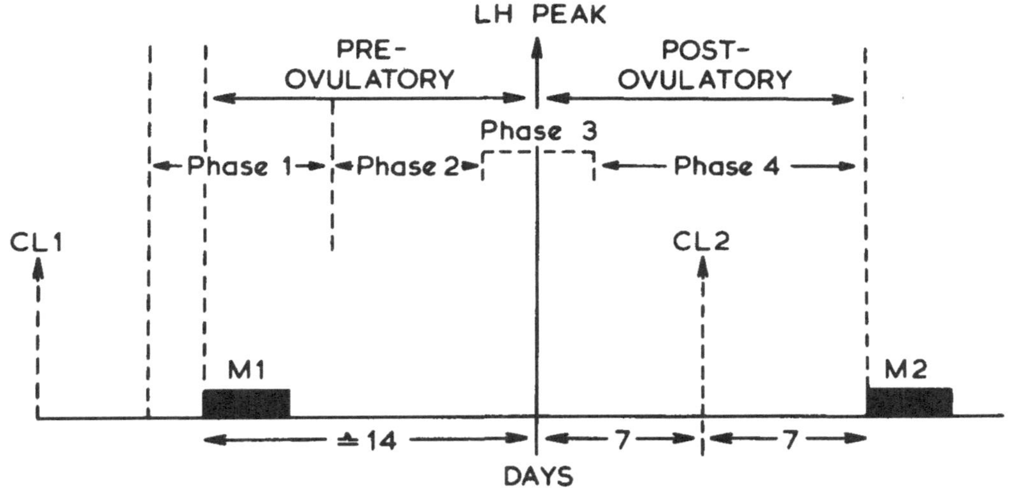

FIGURE 7. A schematic representation of ovarian and menstrual cycles.

to the egg acquire a characteristic columnar shape and the cumulus oophorus gradually becomes detached from the membrana granulosa. There are final increases in the number of granulosa cells and in the volume of follicular fluid. Just prior to release of the oocyte (ovulation), a small area of the wall of the Graafian follicle becomes thin and transparent, forming the "stigma." This area subsequently becomes raised and then tears open, releasing follicular fluid containing the oocyte. The process of ovulation is probably the result of a combination of factors which includes the stimulation of smooth muscle fibers, an increase in intrafollicular pressure, and the action of various proteolytic enzymes on the wall of the follicle.[20]

Dramatic changes occur in the follicle after ovulation (Phase 4). The most obvious is the proliferation of granulosa cells with a concomitant increase in size. In addition, capillary blood vessels from the theca interna invade the granulosa cell mass. The endocrine function of the resultant corpus luteum gradually increases, reaching a maximal secretion of progesterone 6 to 7 days after ovulation. The maximum diameter of the corpus luteum is about 1 cm. The gland has a finite life, depending on whether or not a pregnancy follows. In the absence of a fertilized ovum, the corpus luteum enters a state of regression, characterized by a reduction in the blood flow and a decrease in progesterone biosynthesis. At this time, the next set of follicles is starting to develop and the whole process is repeated.

During early and late follicular growth, there is a corresponding increase in the size of the oocyte. Some 36 hr before ovulation occurs there is a dramatic increase in the level of circulating LH (X 10) and to a lesser extent FSH (X 2). This surge, or possibly the surge in estradiol secretion which precedes it, induces a final increase in the number of granulosa cells and the quantity of follicular fluid. These changes in follicular structure are accompanied by a resumption of meiosis within the primary oocyte.[21] The chromosomes become thicker and enter the stage known as diakinesis. The chromosomes subsequently divide and the oocyte splits into two cells – the secondary oocyte containing most of the cytoplasm and the first polar body which soon degenerates. The secondary oocyte then starts a second meiotic division; it is at this stage that ovulation generally occurs. The continuation of the division and extrusion of a second polar body depend on active penetration of the oocyte by a viable sperm. Unfortunately, the process of gametogenesis and its relationship to pituitary and gonadal hormones is completely unexplained in biochemical terms.

In contrast, a great deal of information has been accrued on the biosynthetic processes which lead to the formation of steroids within the gonad and the significance and interrelationships between the various intermediaries and ovarian morphology.[22] This information has been obtained from numerous studies involving perfusion of the whole gland, the analysis of ovarian venous blood, and the incubation of ovarian components (developing follicles, follicular cysts, corpora lutea, or stromal tissue) with isotopically labeled substrates. A summary of probable transformations involved in the biosynthesis of sterols, which is operative in all compartments of human ovarian tissue, is shown in Figure 8. The basic units are acetate molecules which contain two carbon atoms. These are combined through some 30 enzymatic steps to form cholesterol and a group of cholesterol esters, which may function as a biochemical reserve. However, it should be remembered that steroids may be formed from cholesterol derived from the general circulation. Notwithstanding the source, cholesterol is subsequently degraded in the mitochondria to pregnenolone which contains 21 carbon atoms. Various delta-4-3-ketosteroids are formed from the corresponding delta-5-3β-hydroxysteroids in the microsomes. Some possible sequences and the names of the enzymes are shown in Figure 9. Androstenedione and testosterone are converted by a group of enzymes into estrogens. This process, called aromatization, also occurs in the microsomes. A proposed scheme for this sequence of events is shown in Figure 10. Nineteen hydroxy and 19 aldo derivatives of androstenedione and possibly testosterone are the first and second intermediaries. Removal of the oxygenated carbon atom at position 19 must be effected before the next stage can occur. The simplest mechanism for this step is the simultaneous elimination of the group at carbon 19 together with a hydrogen atom from carbon 1, so forming a 3-keto-1(10),4-diene. The preferential removal of the alpha rather than the beta constituent from carbon 1 is consistent with theoretical considerations which favor diaxial transelimination.[23] This unstable intermediary then aromatizes spontaneously via a heteroenoid structure to the pi bonded form. The phenolic hydroxyl group is formed by the transfer of a proton from the 2β position.

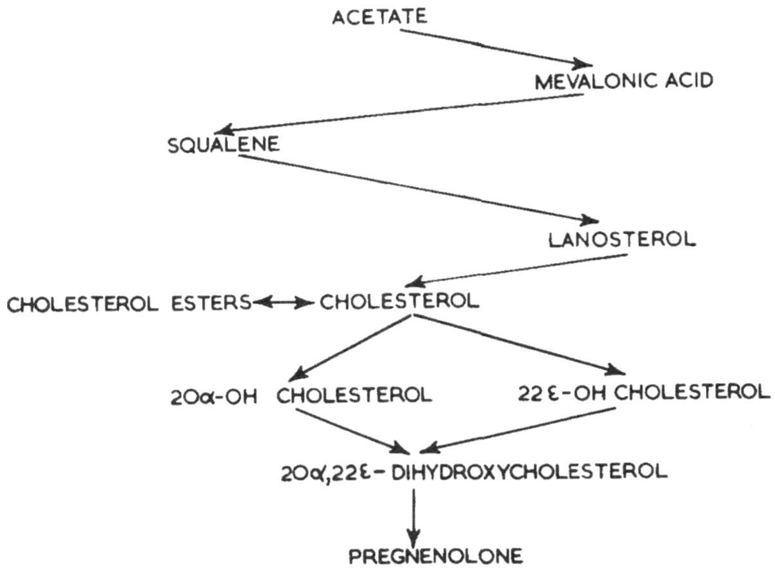

FIGURE 8. Some intermediates in the biosynthesis of pregnenolone.

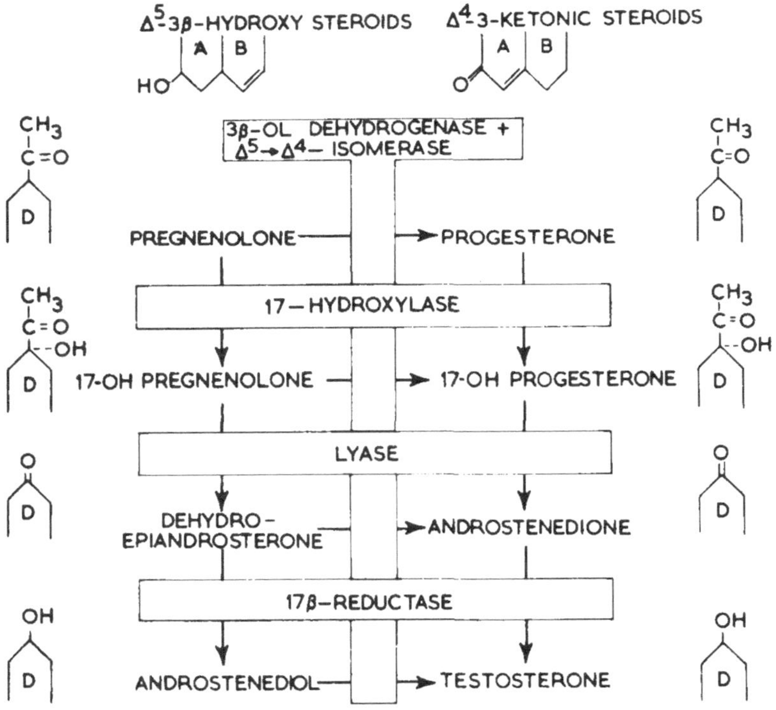

FIGURE 9. Some interconversions of gonadal steroids.

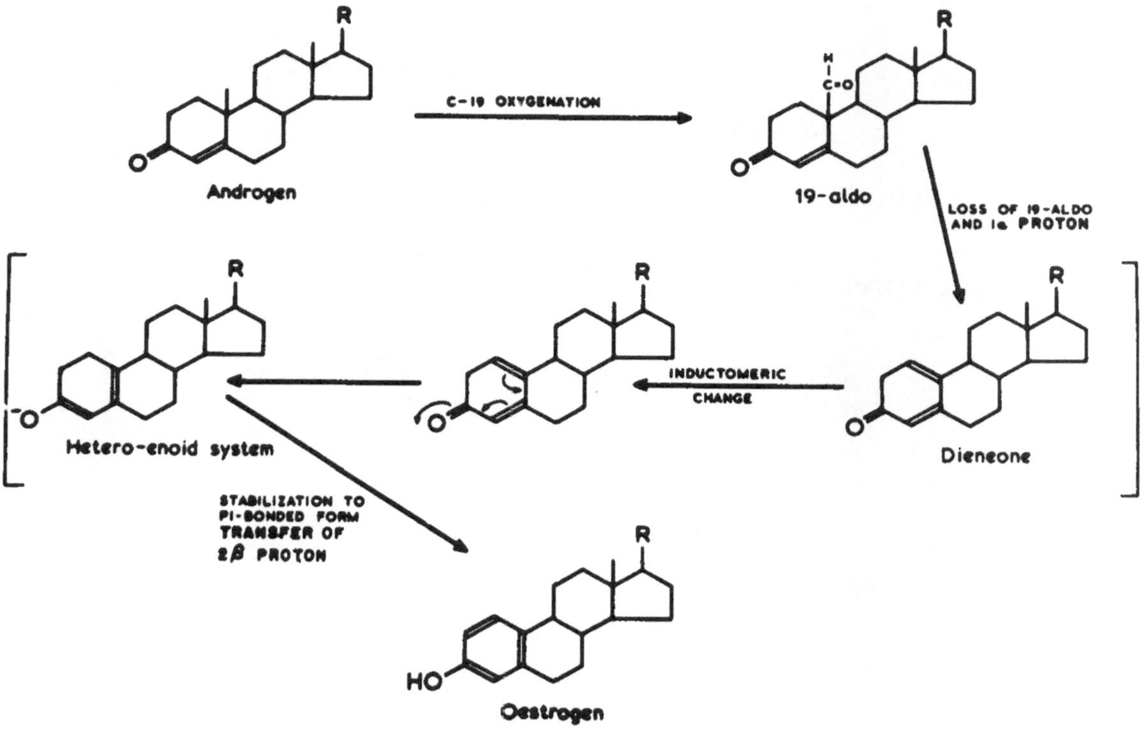

FIGURE 10. A proposed sequence for aromatization.

The enzymes associated with these biosynthetic sequences have been found in all compartments of the ovary, with the possible exception of the oocyte. In addition, their subcellular localization has been determined by histochemical techniques and the high-speed centrifugation of homogenates. However, most studies have been of necessity indirect; few of the enzymes have been isolated and their characteristics studied.

A related problem is the mechanism by which specialized cells within the various ovarian compartments regulate the formation of specific steroids. It is well known, from the results of numerous experimental approaches, that the major product of the developing follicle is estradiol and that of the corpus luteum progesterone. However, the same enzymes are present in both tissues, so some form of intracellular control must be operative. In this connection, the results from various studies have suggested that pregnenolone is a better substrate than progesterone for the production of C_{19} steroids throughout the menstrual cycle in tissue from ovaries which did not contain a corpus luteum.[24,25] In addition, studies with dehydroepiandrosterone have shown the conversion of this substrate to estrogens. The most detailed studies on steroid biosynthesis in follicular tissue have been performed on cyst linings dissected from ovaries after in vivo stimulation with gonadotrophins.[26] Using tissue from the same source, Ryan and Petro[27] have succeeded in separating the thecal and granulosa cells from single follicles. They demonstrated that the granulosa cells were predominantly involved in the production of progesterone and related delta-4-3-ketosteroids, whereas the theca favored the production of androgens and estrogens.

After ovulation, morphological and biochemical luteinization occurs and the overall pattern of steroid biosynthesis appears to change. Huang and Pearlman[28] first demonstrated that the corpus luteum could synthesize progesterone and 20α-dihydroprogesterone from labeled acetate. Subsequent studies with labeled steroids suggested that in this tissue the principal steroid transformations followed the sequence: pregnenolone→ progesterone→17-hydroxyprogesterone→ androstenedione→ estrone and estradiol. Studies on the incubation of aliquots of the same corpus luteum with different substrates have indicated that the levels of delta-5 to delta-4-isomerase and 3β-hydroxy dehydrogenase are relatively high, and the high level of C_{21} delta-4-3 ketosteroids together with the low yields of androstenedione suggest low lyase activity.[29] Incubation studies with androstenedione and testosterone have indicated a high potential for the synthesis of estrone and estradiol.

Steroid biosynthesis in the stroma tissue of the normal human ovary has been demonstrated. The relatively high production of various C_{19} steroids from a number of precursors has led to the suggestion that if the relative weights of the tissue components are taken into consideration, then the stroma may be the principal source of androgenic steroids.[30]

The actual range of hormones secreted by the ovary may be deduced from a comparison of the concentrations of the respective steroids in ovarian and peripheral venous blood. On this basis, there is good evidence that the ovary secretes varying amounts of pregnenolone, progesterone, 20α-dihydroprogesterone, 17-hydroxyprogesterone, dehydroepiandrosterone, androstenedione, testosterone, estrone, and estradiol.[31] An interesting finding is that the concentrations of these steroids in ovarian venous plasma at defined times during the cycle vary more than the corresponding levels in peripheral venous plasma. This variability suggests that the ovaries may secrete at a fluctuating (episodic) rather than at a steady rate,[32] as has been demonstrated for the adrenal cortices and the pituitary.[34]

Attempts have been made to measure the blood production rates of various gonadal steroids.[2] A good estimate of this parameter can be achieved by infusing the labeled hormone until a constant level is reached in peripheral venous blood. From the concentration, a factor known as the metabolic clearance rate (MCR) can be determined according to the equation:

$$\text{MCR (liters/day)} = \frac{\text{Rate of infusion of labeled hormone dpm/min}}{\text{Blood conc. of labeled hormone dpm/ml}} \times 1.44$$

If the mean endogenous hormone level is measured, the blood production rate (P_b) can be calculated according to

$$P_b \text{ (mg/day)} = \text{MCR (liters/day)} \times \text{conc. hormone (mg/liter)}$$

Using this approach, the mean P_b for progesterone during the follicular phase of the cycle has been calculated to be 2.9 mg/24 hr and for the luteal phase 22.1. The corresponding values for estradiol are 0.08 and 0.24 mg/24 hr.

A recent study on the levels of various neutral steroid sulfates in ovarian venous blood indicates that these compounds are neither secreted nor utilized by the ovarian tissue to any significant extent.[35] However, isotopic tracer studies suggest that many other steroids are produced and probably secreted in small amounts.

III. METABOLISM OF GONADAL HORMONES

The application of isotopic tracer techniques to studies on the metabolism of gonadal steroids has led to a better understanding of the problem. For example, the principal metabolites of the ovarian hormones have been identified in urine, plasma, and growth responsive tissues. In addition, reasonable estimates have been made on the dynamics of secretion, the extent of interconversions, and rates of excretion.

A. Progesterone

Numerous studies have shown that labeled progesterone rapidly disappears from the circulation. In 1957, Pearlman[36] presented the concept of a turnover time (T), which was defined as the time taken for the endocrine glands to secrete as much hormone as is present in the circulating blood at any one instant. This value may be calculated as follows:

$$T \text{ (min)} = \frac{C \text{ (}\mu\text{g liter) } V \text{ (liters)}}{r \text{ (mg/day)}} \, 1.44$$

where C is the blood hormone concentration, V the total blood volume, and r the rate of endogenous hormone production.

Subsequently, it was shown that the rate of removal of progesterone from the blood is dependent on the concentration and approximates to a first-order reaction.[37] This work led to the concept of a half-life in blood, rather than to a turnover time. The relationship between the two is

$$\text{Half-life (min)} = \frac{T}{2} \text{ (min)}$$

The half-life may be defined as the time taken for the total blood hormone level to fall by half, should the endogenous secretion cease completely. Various workers have estimated the turnover time and half-life of progesterone using a variety of

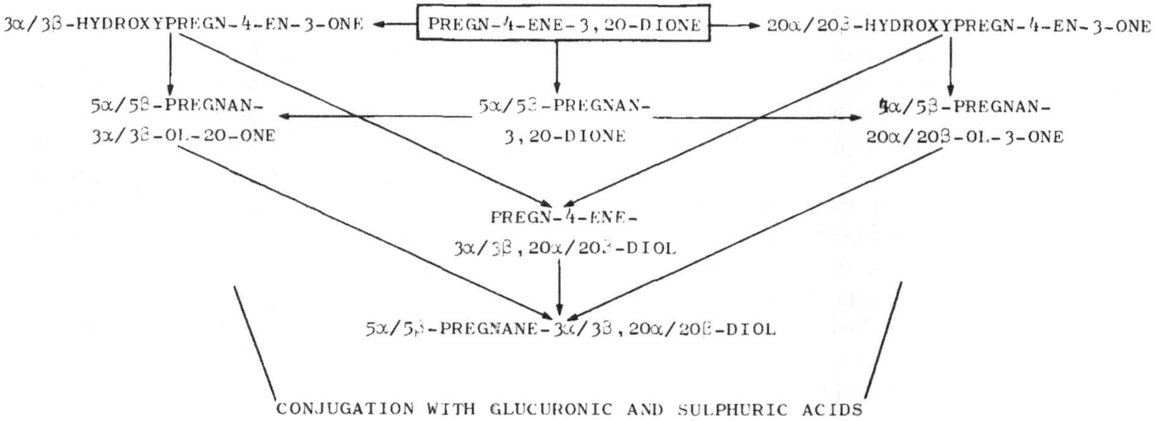

FIGURE 11. Some aspects of the reductive metabolism of progesterone.

techniques and, although there are wide discrepancies in the results, there is general agreement that both values are less than 20 min.

The short half-life appears to be due to a rapid diffusion of plasma progesterone into body tissues, particularly into the subcutaneous fat, and to rapid reduction and conjugation with sulfuric and glucuronic acids in the liver.[38] The principal conversions of progesterone are shown in Figure 11. Either the 20 oxo or 3 oxo group or the double bond may be reduced to give 2 isomeric pregnanediones, 8 possible pregnanolones, and 8 pregnanediols.

Other studies have shown that the excretion of metabolites probably starts immediately after the intravenous injection of labeled progesterone. Thus, approximately 1% of the total dose accumulates in urine during the first 30 min after injection, 9% by 4 hr,[39] 29% by 24 hr, 37% by 48 hr, and 41% by 72 hr.[40] A delayed appearance of metabolites in the feces has been shown.[39] Of the metabolites in urine, approximately 2% are readily extractable with diethyl ether, 62% are released by hydrolysis with β-glucuronidase, and a further 5% by solvolysis.[40] The principal compounds in the glucuronoside fraction are 5β-pregnane-3α,20α-diol (from 5 to 26% of administered progesterone), 5α-pregnane-3α,20α-diol (from 2 to 15%), and 5β-pregnan-3α-ol-20-one (0.5 to 5%).[40] The analysis of peripheral venous plasma has revealed the presence of comparatively small amounts of unconjugated, 5β-pregnan-3α-ol-20-one, 5β-pregnane-3α,20α-diol, and 20α-hydroxypregn-4-en-3-one (20α-dihydroprogesterone). Higher levels of 5β-pregnan-3α-ol-20-one and 5β-pregnane-3α,20α-diol were found in the fraction liberated by hydrolysis with β-glucuronidase.[39]

With regard to growth responsive tissues, it has been shown that pieces of endometrium, removed during both proliferative and secretory phases of the menstrual cycle, have the capacity to transform progesterone into 20α-dihydroprogesterone, 5α-pregnanedione, and other minor products.[41] The order and significance of these changes are unknown, but it has been postulated that they may represent changes in the molecule during the regulation of certain aspects of intermediary metabolism or the formation of local hormones from prehormones. An alternative explanation is that they may represent one mechanism of the cells to remove excessive stimulation.

B. Estradiol

The kinetics and mode of estradiol metabolism have been studied in a similar manner, but the results are less conclusive. The problems involved are associated with the extremely low levels of active compounds in body fluids and tissues and the relative instability of the phenolic A ring. In general, calculations of the turnover time and half-life of estradiol give slightly higher values (from 5 to 25 min) than are obtained for progesterone. This finding may be related to the relatively high affinity of estradiol for specific plasma proteins, a corresponding decrease in the rate of metabolism by the liver, and to a lower excretion of metabolites in the bile. Some aspects of the overall metabolism of estrogens are summarized in Figure 12. The principal compounds are estradiol-17β and estrone, which are

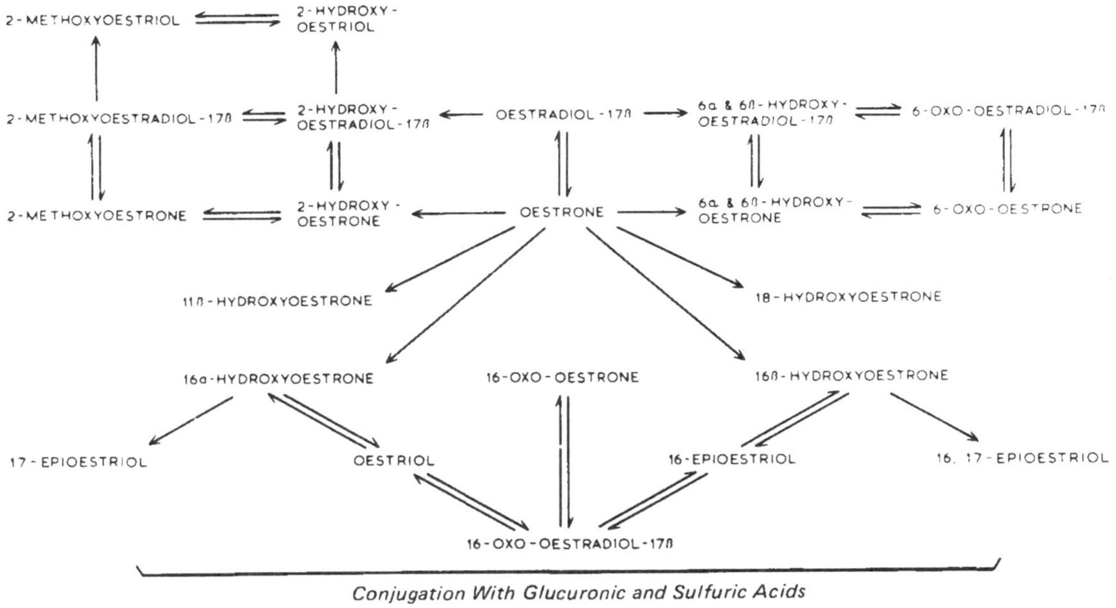

FIGURE 12. The principal interconversions of estrogens.

interconvertible, and estriol. Until 1954, this was virtually all that was known about estrogen metabolism. Since then, at least 15 other metabolites have been identified – mainly conjugated with glucuronic and sulfuric acids in urine from pregnant women.

If tritiated estradiol is administered intravenously, from 52 to 84% of the label may be accounted for by excretion in the urine.[42-44] Of this amount, less than 30% is associated with estrone, estradiol, or estriol, either as the free or conjugated derivatives.[45] Further analysis shows that a high percentage of the total radioactivity in these three compounds is released by hydrolysis with β-glucuronidase.[46] Over 70% of the radioactivity administered as estradiol is associated with unknown polar metabolites or degradation products. If labeled estradiol-17β-glucuronoside is administered, less is metabolized; this finding is associated with a decrease in the number of unknown metabolites.[46]

Other studies have suggested that more estradiol is transformed to estrone (approx. 90%)[47] than estrone to estradiol (5 to 10%).[48] This situation also occurs in endometrial tissue where both in vitro and in vivo studies have demonstrated an intranuclear conversion of estradiol to estrone.[49] This mechanism may represent one method of regulating the level of active hormone in target tissues.

IV. HYPOTHALAMIC-PITUITARY FUNCTION

For many years after the discovery of the gonadotrophic hormones, the anterior pituitary was considered to be responsible for controlling gonadal function. This notion was shaken by the demonstration that gonadotrophin deficiency or precocious puberty could be due to hypothalamic disorders. It is now established beyond doubt that the hypothalamus has a great deal of control over many important functions, including the reproductive cycles.[1]

A. Pituitary Structure and Function

The anterior pituitary gland arises from nonnervous tissue in the roof of the buccal cavity and is in direct contact with the hypothalamus. Different types of cell, identified by their staining characteristics, are thought to be responsible for the secretion of FSH and LH (basophilic cells) and prolactin (acidophilic cells). Further evidence that FSH and LH are contained within a single cell population has been obtained by immunohistological techniques. For example, Phifer and colleagues[50] have demonstrated that the same cells (gonadotroph cells) react with antisera specific to FSH or LH. The hormones are concentrated in granules 200 to 300 μm in diameter and are probably released by the process of reverse pinocytosis.

Numerous studies on the structures of FSH and LH have shown that they are glycoproteins with molecular weights between 25,000 and 40,000.[51,52] Both are composed of at least two subunits (α and β), one of which (α) is common to both gonadotrophins and to TSH.[53] The carbohydrate content is between 9 and 30% for FSH[54] and 15 and 20% for LH.[51] The sialic acid component (5.2% for FSH[53] and 1.4% for LH[56]) is thought to be essential for optimal activity. Prolactin (luteotrophic hormone) has a molecular weight of about 24,000 and is thought to be a single polypeptide chain containing 211 amino acid residues.[57]

In general terms, FSH promotes follicular growth and development. LH has been shown to increase steroidogenesis, initiate the resumption of gametogenesis, and effect ovulation and biochemical luteinization. It has been postulated that under physiological conditions, the observed biological activity is dependent on the relative concentrations of the trophic hormones and the morphological state of the ovary. At present there is no convincing evidence that prolactin has a luteotrophic action in women.[54]

B. Hypothalamus Structure and Function

In humans the hypothalamus is recognizable as a small oval band of grey matter situated directly above the pituitary. The interrelationship between these two structures represents one of the most challenging areas for research in reproductive endocrinology.[58] Early studies on the morphology of both structures established that, in general, nerve fibers from the hypothalamus do not penetrate to the secretory cells of the anterior pituitary. Instead, a vascular connection, the hypophyseal portal systems was observed. Further studies proved that various substances called releasing factors were secreted by hypothalamic neurones into the portal system and were conveyed to the anterior pituitary. Recently, one of the releasing factors (renamed releasing hormone) has been identified as a small polypeptide; the structure has been determined and synthetic material has become available for study.[59] The neurotransmitter substances, which regulate the secretion of releasing hormones, are probably monoamines. The three that have been implicated are dopamine, norepinephrine, and epinephrine.[60] There is strong suggestive evidence from studies on experimental animals that serotonin and melatonin may be inhibitory neurotransmitter substances with regard to the eventual release of LH.

It is now generally accepted that the neural control of LH secretion occurs at two sites, which have different functions. One area is known as the ventromedial nucleus plus the arcuate nucleus and is located in the median eminence. The other area is designated the pre-optic anterior hypothalamic nucleus and is situated above the optic chiasma. The role of these structures in the regulation of ovarian function has been investigated in experimental animals by combinations of surgical differentiation experiments, electrocoagulation, implantation of pituitary tissue, and castration. The results from these studies have shown that lesions involving the ventromedial arcuate nucleus result in gonadal atrophy, whereas lesions of the pre-optic anterior hypothalamic nucleus only abolish ovulation.[61] It is thought that this dual system differentiates in the absence of androgen during a critical period in prenatal or neonatal life.[61] If an active androgen is administered during this period, the male type of hypothalamic-pituitary function develops. During childhood it is thought that various gonadal steroids suppress hypothalamic pituitary function to a relatively low level (Figure 5). The onset of sexual maturation is thought to be associated with a decrease in the sensitivity of the hypothalamus to this negative feedback system. The resultant increase in the levels of circulating FSH and LH stimulates the gonads and the menstrual cycle gradually comes into operation. The ability of the pituitary to release LH following increasing levels of estradiol does not occur until midpubertal development.

V. HORMONAL INTERRELATIONSHIPS

Any attempt to determine the hormonal relationships in the fetus must take into consideration the participation of the placenta and maternal endocrine organs. Accordingly, a complex metabolic situation has been shown to exist between these three anatomical compartments and the interrelationship changes continuously during the course of pregnancy. Throughout childhood there is only a basal secretion of hormones from the gonads and pituitary. After sexual maturation has occurred, ovarian activity settles into regular cyclical activity, as is demonstrated by menstruation.

A. Composite Ovarian Cycle

Prior to 1930, a great deal of confusion existed regarding the relationship of the menstrual cycle in women to the estrus cycle in various animals. This situation arose from the fact that external bleeding signifies a different internal event in the two cycles. In animals vaginal bleeding signifies that ovulation has occurred, whereas in women menstruation marks the end of the functional life of the corpus luteum. Day 1 of bleeding is frequently referred to as the start of a new menstrual cycle.

The changes in level of the principal hormones during the course of a normal cycle (ideal mean) are shown in Figures 13A and B. The mean values for each hormone are expressed as a percentage of the maximum. This method of representation clearly shows that the largest change occurs in the levels of progesterone and 17-hydroxyprogesterone. The maximum level of estradiol occurs prior to the peak of LH. There are only relatively small changes in the level of C_{19} steroids.[62] The significance of the absolute values and relative changes in hormone levels and the correlation of these events with associated alterations in cervical mucus and vaginal cytology have been the subjects of numerous reviews.[5,19,34,63-67]

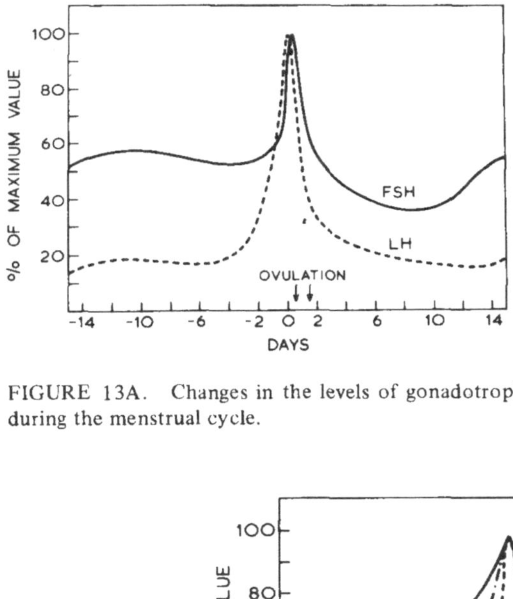

FIGURE 13A. Changes in the levels of gonadotrophins during the menstrual cycle.

B. Acute Changes During Selected Phases

Attempts have been made to relate hormone levels in peripheral venous plasma and urine to the changes in ovarian morphology. Unfortunately, this approach cannot be applied to individuals, but an overall picture may be constructed from the results of group studies.

During Phase 1 (Figure 7), there is a premenstrual rise in the level of FSH which lasts for 3 to 4 days. This change is preceded by a fall in the circulating levels of progesterone and estradiol; during the period of increased FSH secretion, there is little change in gonadal steroid production. Thus, the processes of follicular priming and growth are clearly related to the level of FSH. Subsequent maturation of the follicle and increased secretion of ovarian steroids are associated with a gradual increase in the secretion of LH. It is well known that FSH alone will cause follicular growth when given to women with amenorrhea,

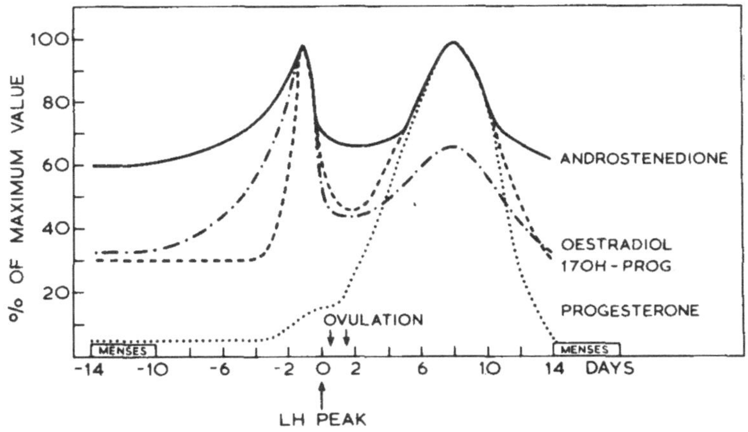

FIGURE 13B. Changes in the levels of various gonadal steroids during the menstrual cycle.

TABLE 1

The Concentration (Mean ± S.D., pg/ml) of Various Ovarian Steroids in Peripheral Venous Plasma from Women

	Days of menstrual cycle		
Steroid	1–10	11–18	19–32
Progesterone	420 ± 190	830 ± 240	7700 ± 1900
17-Hydroxyprogesterone	400 ± 150	810 ± 320	1340 ± 570
Androstenedione	1230 ± 610	1340 ± 590	1430 ± 640
Testosterone	400 ± 140	420 ± 160	380 ± 150
Oestrone	119 ± 46	156 ± 41	156 ± 27
Oestradiol	69 ± 56	130 ± 68	99 ± 54

but ovulation will not occur in these subjects unless LH is administered.

During Phase 2 there is a progressive rise in the level of plasma estradiol, which reaches a peak value just prior to ovulation. At the same time there is a corresponding fall in the level of FSH and either a small rise or no significant change in the level of LH. The rapid rise in estradiol during this part of the cycle may be due to an increase in the sensitivity of the growing follicle to gonadotrophins or to some form of internal regulation possibly involving the oocyte. Evidence for the former mechanism comes from studies on the induction of ovulation in infertile women with constant or decreasing doses of FSH and LH. Some evidence for the latter is provided from experiments on animals involving removal of the oocyte from growing follicles. However, there are alternative explanations to the results from both approaches and more definite information is required before any firm conclusions may be reached.

Until recently, comparatively little was known about the magnitude and duration of changes in the plasma levels of gonadotrophins and ovarian steroids around the time of ovulation (Phase 3). In 1973, Wide and colleagues[34] reported that at midcycle the level of plasma LH increased by a mean factor of 10 and lasted for 16 to 24 hr. There is a concomitant but modest rise in FSH (X 2). Shortly after the start of the surge in plasma LH and before the follicle has ruptured, the level of plasma estradiol starts to fall, while that of progesterone[67] starts to rise (X 3). Ovulation is thought to occur from 16 to 24 hr after the peak of LH in peripheral venous plasma.

The luteal phase of the cycle (Phase 3) is characterized by a vast increase (X 20) in the level of plasma progesterone and to a smaller extent (X 3) in the levels of 17-hydroxyprogesterone, estrone, and estradiol. According to both histological and biochemical criteria, the corpus luteum reaches maximum maturity 6 to 7 days after ovulation. If fertilization and implantation of the ovum do not occur, the corpus luteum starts to regress and the cycle is repeated.

Studies on the diurnal variation of the various hormones in peripheral venous plasma have suggested that there are no significant changes in the levels of FSH, LH,[34] progesterone,[68] and estradiol[69] in either the follicular or luteal phases of the menstrual cycle. A slight decrease in levels during the day has been observed in the concentrations of androstenedione and estrone. More detailed studies over shorter periods of time have indicated episodic changes in the level of plasma LH and to a lesser extent FSH.[34] Similar changes in the level of ovarian hormones have been inferred by indirect methods. The mean and range of values for the principal hormones at various phases of the menstrual cycle are shown in Table 1.

VI. CONTROL MECHANISMS

It is apparent that at least three glands are involved in the correct function of the ovarian and menstrual cycles. Furthermore, there are many functional compartments and cell types within each of these glands. Accordingly, a large number of control mechanisms must be operative at many different levels of organization. In this section, an attempt will be made to summarize some of the information available on the superficial control mechanisms, which are thought to be operative between the gonads, hypothalamus, and pituitary.

TABLE 2

Some Characteristics of Plasma Proteins Which Bind Steroid Hormones

Parameter	Protein fractions HSA	AAG	CBG	SBG
Molecular weight	69000	41000	52000	52000
Concentration				
a. mg/l.	38000	750	36	3.0
b. 10^{-7} X m/l.	5500	180	7	0.5
Association constants at 37°C l./m				
Cortisol	10^3	–	10^7	–
Progesterone	10^4	10^6	10^7	–
Testosterone	10^4	–	10^6	10^8
Oestradiol	10^4	–	10^4	10^7

A. Gonadal Steroids

Estradiol appears to have a negative feedback effect on the secretion of FSH, as is observed during the follicular phase of the menstrual cycle. Other evidence for the relationship comes from the studies of Vande Wiele and colleagues,[5] who infused estradiol at a rate of 100 to 150 μg/24 hr to postmenopausal women and observed a significant depression in the level of FSH within 6 hr. Similarly, the intravenous or intramuscular administration of estradiol-17β in doses from 5 μg to 5 mg caused a decrease in the pretreatment levels of both FSH and LH in oligomenorrheic and postmenopausal women.[70] From these studies, it would appear that estradiol is able to reduce the secretion of both FSH and LH However, Speroff and Vande Wiele[19] have shown a positive feedback of estrogens on LH secretion. This dual action can be explained by the existence of two centers of LH/RH production in the hypothalamus – one for the tonic release of gonadotrophins and the other for the midcycle surge (cyclic center). Thus, during Phase 2 of the cycle, the rising levels of circulating estradiol could cause the rising LH levels and midcycle surge by suppressing one area of the hypothalamus and stimulating another. In this connection, the administration of estradiol benzoate[71] (1 mg intramuscularly) to women with amenorrhea, oligomenorrhea, or regular cycles induced a peak in the level of plasma LH but not FSH within 48 to 72 hr. Other studies on women of reproductive age have shown that the effect of ethinyl estradiol on gonadotrophin levels varies with the time of administration during the menstrual cycle.[72] Further evidence that estradiol acts mainly on the cyclic center of the hypothalamus during the follicular phase of the cycle is the inability of drugs with antidopaminergic activity or adrenergic blocking agents to inhibit the midcycle surge of LH.

The role of progesterone in the regulation of gonadotrophin secretion is not clear. Nillius and Wide[73] have shown that the administration of progesterone (10 or 100 mg intramuscularly) does not result in any significant changes in the level of circulating gonadotrophins. However, if the progesterone is administered after pretreatment of the subjects with ethinyl estradiol (100 μg/day) for 4 weeks, there is a resultant increase in the levels of plasma LH and to a smaller extent FSH. These authors suggested that in low doses progesterone may have a stimulating or facilitating effect on the hypothalamic-pituitary system.

B. Plasma Proteins

Recent studies have suggested that steroids bound to protein in plasma are biologically inactive and that the plasma proteins may be involved in regulating the level of biologically active hormone.[74] At least four groups of proteins in plasma are capable of binding steroids.[75] The relative concentrations of these compounds and their association constants with the principal steroid hormones at 37°C are shown in Table 2. The abbreviations refer to human serum albumin, a_1-glycoprotein, corticosteroid binding globulin, and sex hormone binding globulin. Studies on the relative binding affinities have shown that HSA weakly binds all four steroids, AAG is reasonably specific for progesterone, CBG binds all four, but cortisol and progesterone more strongly, while SBG binds estradiol, testosterone, and other steroids with a 17β-hydroxyl group. In any plasma sample, the steroids are divided among these various protein fractions and a small percentage of the total (from 1 to 5%) remains in solution. Various indirect methods involving ultrafiltration and equilibrium dialysis have been used to assess the amount of apparent free progesterone, testosterone, and estradiol in peripheral venous plasma. Alternatively, the concentrations or the number of binding sites of CBG and SBG have been determined using labeled steroids.[76] Preliminary results from these approaches suggest that the metabolic clearance rates of the respective hormones may be directly related to the level of apparent free hormone in plasma and that the latter index is a better parameter of progestational activity and estrogenicity than measurement of the total hormone concentration.

C. FSH/LH-RH

In 1971, a luteinizing hormone-releasing hormone was isolated and purified from porcine hypothalami.[77] As previously suspected, it proved to be a polypeptide with the amino acid sequence (pyro) Glu-His-Trp-Ser-Tyr-Gly-Leu-Arg-Pro-Gly-NH_2.[78] This structure was synthesized and the resultant material shown to have identical properties to the purified biological extracts. Clinical studies with the synthetic material showed that small doses (25 to 100 μg) administered intravenously would cause a rapid (within 2 min) increase in the level of plasma LH (3- to 16-fold) and to a smaller extent FSH (2- to 3-fold).[79,80] Further studies with this material have shown that there is pronounced variation in the response to 100 μg, depending on the phase of the menstrual cycle. During the late follicular phase the average response was four- to six-fold higher than that obtained in the early and mid-follicular phases. An increased response was also found in the luteal phase of the cycle.[81] These results have been explained in terms of the level of circulating estradiol, but the precise mechanism is not known.[82] In addition, there is considerable doubt as to whether or not there are one or two peptides for the separate release of LH and FSH.[83] Just recently, antisera have been raised to synthetic FSH/LH-RH and used in a method to determine the concentration of gonadotrophin-releasing hormone in urine from men and women.[84] An interesting finding was that the level of urinary FSH/LH-RH increased during the administration of clomiphene.

D. Gonadotrophic Hormones

The precise nature of the relationship between gonadotrophins and optimal ovarian function is not known. However, various studies have indicated that FSH is associated with follicular growth[85] and probably with certain aspects of steroidogenesis during follicular development. LH has been shown to be associated with steroidogenesis, gametogenesis, changes in ovarian vascularity, and luteinization. The time scale for these changes varies considerably. After administration of FSH/LH-RH, peak levels of gonadotrophins in peripheral plasma are obtained after 20 to 30 min. However, maximal increases in the level of plasma estradiol do not occur for a further 6 to 9 hr. The time taken for the resumption of gametogenesis following the midcycle peak of endogenous LH has not been accurately defined, but ovulation occurs after 16 to 24 hr. The time taken for complete luteinization and the optimal synthesis of progesterone is from 7 to 9 days. All of these diverse actions are thought to involve the synthesis of cyclic 3′5′-adenosine monophosphate (C AMP) and possibly various prostaglandins. A scheme depicting the main steps in these processes is shown in Figure 14. The first step involves the selective binding of LH to a receptor protein in the target cell.[56] This interaction results in a stimulation of the enzyme adenyl cyclase, which converts adenosine triphosphate into cyclic 3′5′-adenosine monophosphate (C 3′5′-AMP).[87] The C 3′5′-AMP then increases some form of protein synthesis either through protein kinase, which is involved in the phosphorylation of histones,[88] or through ribosomal activation.[89] The former process involves interaction with DNA and new messenger RNA may be formed (transcription); the latter process increases synthesis of a specific protein at the translational level. The respective roles of calcium ions and prostaglandins in human ovarian function are uncertain. Studies in rodents have shown that indomethacin, a potent inhibitor of prostaglandin synthesis, fails to prevent the resumption of gametogenesis induced by LH, but is effective in preventing ovulation.[90]

The process of luteinization is initiated by LH, but morphological evidence and certain experiments in animals suggest that expulsion of the

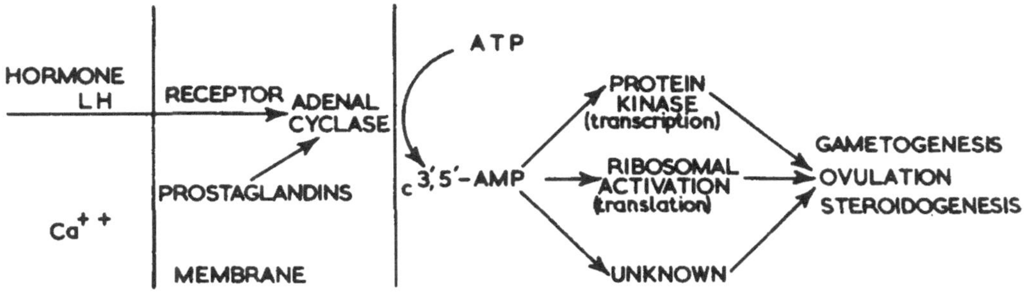

FIGURE 14. Some possible actions of LH on ovarian cells.

oocyte may be a necessary part of the mechanism.[91] During peak activity of the corpus luteum, the plasma levels of LH and FSH are at their lowest. Furthermore, although injections of LH or HCG are able to increase the level of plasma progesterone, it is not possible to extend the life of the corpus luteum by this approach for more than a few days.[92] In subjects with congenital absence of the uterus and tubes, ovarian activity continues in regular cycles, which suggests that luteolytic factors are not being produced in the target organs.[93] Accordingly, the factors that control the function and life span of the corpus luteum in women are completely unknown.

From this account, it is apparent that the regulation of ovarian function throughout life is an extremely complex, ever, changing phenomenon. With the recent advent of isotopic tracer techniques for the determination of the principal compounds involved, the relevant information for superficial analysis is being obtained at an ever increasing rate. However, there are still an infinite number of interrelationships at the cellular and biochemical level to be unraveled. The number of components and variables in the system is already beyond the comprehension of any one individual, and future research must inevitably depend more and more upon computer analysis and simulation.[94,95]

REFERENCES

1. **Schwartz, N. B. and McCormack, C. E.,** Reproduction: gonadal function and its regulation, *Ann. Rev. Physiol.,* 34, 425, 1972.
2. **Baird, D. T., Horton, R., Lancope, C., and Tait, J. F.,** Steroid dynamics under steady-state conditions, *Recent Progr. Horm. Res.,* 25, 611, 1969.
3. **Bradbury, J. T., Brown, W. E., and Gray, L. A.,** Maintenance of the corpus luteum and physiologic action of progesterone, *Recent Progr. Horm. Res.,* 5, 151, 1950.
4. **Dortman, R. I. and Shipley, R. A.,** *Androgens: Biochemistry, Physiology and Clinical Significance,* John Wiley & Sons, New York, and Chpman and Hall, London, 1956.
5. **Vande Wiele, R. L., Bogumil, J., Dyrenfurth, I., Ferin, M., Jewelewicz, R., Wazzen, M., Rizkallah, T., and Mikhail, G. F.,** Mechanisms regulating the menstrual cycle in women, *Recent Progr. Horm. Res.,* 26, 63, 1970.
6. **Reynolds, S. R. M.,** The vasculature of the ovary and ovarian function, *Recent Progr. Horm. Res.,* 5, 65, 1950.
7. **Baker, T. G.,** A quantitative and cytological study of germ cells in human ovaries, *Proc. R. Soc. B.,* 158, 417, 1963.
8. **Witschi, E.,** Migration of the germ cells of human embryos from the yolk sac to the primitive gonadal folds, *Contrib. Embryol.,* 32, 69, 1948.
9. **Greenblatt, R. B.,** Clinical aspects of sexual abnormalities in man, *Recent Progr. Horm. Res.,* 14, 335, 1958.
10. **McNaughton, M. C.,** Endocrinology of the foetus, in *Foetus and Placenta,* Klopper, R. and Diczfalusy, E., Eds., Blackwell, Oxford, 1969.
11. **Kaplan, S. L., Grumback, M. M., and Shepard, T. H.,** Gonadotropins in serum and pituitary of human foetuses and infants, *Pediatr. Res.,* 3, 512, 1969.
12. **Root, A. W., Smith, G. P., Dhariwal, A. P. S., and McCann, S. M.,** Luteinizing hormone releasing activity of crude ovine hypothalamic extract in man, *Nature,* 221, 570, 1969.

13. **Abrams, C. A. L., Grumbach, M. M., Dyrenfurth, I., and Vande Wiele, R. L.,** Ovarian stimulation with human menopausal and chorionic gonadotropins in a prepubertal hypophysectomized female, *J. Clin. Endrocrinol. Metab.*, 27, 467, 1967.
14. **Penny, R., Guyda, H. J., Baghdassarian, A., Johanson, A. J., and Blizzard, R. M.,** Correlation of serum follicular stimulating hormone (FSH) and luteinizing hormone (LH) as measured by radioimmunoassay in disorders of sexual development, *J. Clin. Invest.*, 49, 1847, 1970.
15. **Kelch, R. P., Grumbach, M. M., and Kaplan, S. L.,** Studies on the mechanism of puberty in man, in *Gonadotropins,* Saxena, B. B., Beling, C. G., and Gandy, H. M., Eds., Interscience, New York, 1971.
16. **Loraine, J. A. and Bell, E. T.,** *Hormone Assays and Their Clinical Application,* Livingstone, Edinburgh, 1971, 319.
17. **Kulin, H. E., Grumbach, M. M., and Kaplan, S. L.,** Changing sensitivity of the pubertal gonadal hypothalamic feedback mechanism in man, *Science,* 166, 1012, 1969.
18. **Kulin, H. E., Grumbach, M. M., and Kaplan, S. L.,** Gonadal hypothalamic function in prepubertal and pubertal man: effect of clomiphene citrate on urinary follicular stimulating hormone, luteinizing hormone and plasma testosterone, *Pediatr. Res.*, 6, 162, 1972.
19. **Speroff, L. and Vande Wiele, R. L.,** Regulation of the human menstrual cycle, *Am. J. Obstet. Gynecol.*, 109, 234, 1971.
20. **Rodbard, D.,** Mechanics of ovulation, *J. Clin. Endocrinol.*, 28, 849, 1968.
21. **Baker, T. G.,** Electron microscopy of the primary and secondary oocyte, in *Advances in the Biosciences 6, Schering Symposium on Intrinsic and Extrinsic Factors in Early Mammalian Development,* Raspe, G., Ed., Vieweg-Pergamon, Oxford, 1971.
22. **Inguilla, W. and Forleo, R.,** Studies on steroid metabolism of human ovarian tissue, in *Endocrinologic and Morphologic Correlations of the Ovary,* Inguilla, W. and Greenblatt, R. B., Eds., Charles C Thomas, Springfield, Ill., 1969.
23. **Axelrod, L. and Goldzieher, J.,** Mechanism of biochemical aromatisation of steroids, *J. Clin. Endocrinol. Metab.*, 22, 537, 1962.
24. **Kumari, L. and Goldzieher, J. W.,** In vitro steroidogenesis in normal human ovarian tissue, *Acta Endocrinol.*, 52, 455, 1966.
25. **Aakuaag, A.,** Pathways of the biosynthesis of androstenedione in the human ovary in vitro, *Acta Endrocrinol.*, 60, 517, 1969.
26. **Ryan, K. J. and Smith, O. W.,** Biogenesis of steroid hormones in the human ovary, *Recent Progr. Horm. Res.*, 21, 285, 1965.
27. **Ryan, K. J. and Petro, S.,** Steroid biosynthesis by human ovarian granulosa and thecal cells, *J. Clin. Endocrinol. Metab.*, 26, 46, 1966.
28. **Huang, W. Y. and Pearlman, W. H.,** The corpus luteum and steroid hormone formation, *J. Biol. Chem.*, 238, 1308, 1963.
29. **Forleo, R., Bruni, V., Sbiroli, C., and Ingiulla, W.,** Further studies upon steroid transformations in the human corpus luteum, *Steroids,* 10, 617, 1967.
30. **Savard, K., Marsh, J. M., and Rice, B. F.,** Gonadotropins and ovarian steroidogenesis, *Recent Progr. Horm. Res.*, 21, 285, 1965.
31. **Lloyd, C. W., Lobotsky, J., Baird, D. T., McCracken, J. A., Weisz, J., Pupkin, M., Zanartu, J., and Puga, J.,** Concentration of unconjugated estrogens and androgens and gestragens in ovarian and peripheral venous plasma of women: the normal menstrual cycle, *J. Clin. Endrocrinol. Metab.*, 32, 155, 1971.
32. **Weisz, J. and Lloyd, C. W.,** Some aspects of CNS-pituitary-ovarian interrelationship, in *Gonadotropins,* Saxena, B. B., Beling, C. G., and Gandy, H. M., Eds., Interscience, New York, 1971.
33. **Hellman, L., Nakada, F., Curti, J., Weitzman, E. D., Kream, J., Roffwarg, H., Ellman, S., Fukushima, D. K., and Gallagher, T. F.,** Cortisol is secreted episodically by normal man, *J. Clin. Endocrinol. Metab.*, 30, 411, 1970.
34. **Wide, L., Nillius, S. J., Gemzell, C., and Ross, P.,** Radioimmunosorbent assay of follicle-stimulating hormone and luteinizing hormone in serum and urine from men and women, *Acta Endocrinol.*, Suppl. 174, 1973.
35. **Kalliala, K., Laatikainen, T., Luukkainen, T., and Vihko, R.,** Neutral steroid sulphates in human ovarian vein blood, *J. Clin. Endocrinol.*, 30, 533, 1970.
36. **Pearlman, W. H.,** Circulating steroid hormone levels in relation to steroid hormone production, *Ciba Found. Colloq. Endocrinol.*, 11, 233, 1957.
37. **Short, R. V. and Eton, B.,** Progesterone in blood. III. Progesterone in the peripheral blood of pregnant women, *J. Endocrinol.*, 18, 418, 1959.
38. **Plotz, E. J.,** Studies with labelled progesterone in human pregnancy, in *Progesterone,* Barnes, A. C., Ed., Brook Lodge Press, Augusta, Mich., 1961, 91.
39. **Thijssen, J. H. H. and Zander, J.,** Progesterone-4-^{14}C and its metabolites in the blood after intravenous injection into women, *Acta Endocrinol.*, 51, 563, 1966.
40. **Collins, W. P., Koullapis, E. N., and Sommerville, I. F.,** The effect of chlormadinone acetate on progesterone secretion and metabolism, *Acta Endocrinol.*, 68, 271, 1971.
41. **Collins, W. P., Mansfield, M. P., Bridges, C. E., and Sommerville, I. F.,** Studies on steroid metabolism in human endometrial tissue, *Biochem. J.*, 113, 399, 1969.

42. **Crowley, L. G., Demetriou, J. A., Kotin, P., Donovan, A. J., and Kushinsky, S.,** Excretion patterns of urinary metabolites of estradiol-4-C^{14} in postmenopausal women with benign and malignant disease of the breast, *Cancer Res.*, 25, 371, 1965.
43. **Engel, L. L., Cameron, C. B., Stoffyn, A., Alexander, J. A., and Trofimow, N. D.,** The excretion of urinary metabolites of administered oestradiol-17β-16-C^{14}, *Anal. Biochem.*, 2, 114, 1961.
44. **Eren, S., Reynolds, G. H., Turner, M. E., Schmidt, F. H., Mackay, J. H., Howard, C. M., and Preedy, J. K. R.,** Estrogen metabolism in the human. III. A comparison between females studied during the first and second half of the menstrual cycle, and males, *J. Clin. Endocrinol.*, 27, 1451, 1967.
45. **Fraser, R. C., Cudmore, D. C., Melanson, J., and Morse, W. I.,** Oestrogen metabolism in oligomenorrhoeic hirsute women, *J. Obstet. Gynaecol. Br. Commonw.*, 74, 115, 1967.
46. **Hobkirk, R. and Nilsen, M.,** Identification of some urinary conjugated metabolites of 17β-estradiol-6,7-^{3}H-glucosiduronate in the human female, *Steroids,* 14, 533, 1969.
47. **Fishman, J., Bradbow, H. L., and Gallagher, T. F.,** Oxidative metabolism of estradiol, *J. Biol. Chem.*, 235, 3104, 1960.
48. **Migeon, C. J., Wall, P. E., and Bertrand, J.,** Some aspects of the metabolism of 16-C^{14}-estrone in normal individuals, *J. Clin. Invest.*, 38, 619, 1959.
49. **Siiteri, P. K., Ashby, R., Schwarg, B., and MacDonald, P. C.,** Mechanism of estrogen action studies in the human, *J. Steroid Biochem.*, 3, 459, 1972.
50. **Phifer, R. F., Midgley, A. R., and Spicer, S. S.,** Immunohistologic and histologic evidence that follicle-stimulating hormone and luteinizing hormone are present in the same cell type in the human pars distalis, *J. Clin. Endocrinol. Metab.*, 36, 125, 1973.
51. **Ryan, R. J., Jiang, N. S., and Hanlon, S.,** Some physical and hydrodynamic properties of human FSH and LH, *Recent Prog. Horm. Res.*, 26, 105, 1970.
52. **Papkoff, H. and Li, C. H.,** in *Gonadotrophins and Ovarian Development,* Butt, W. R., Crooke, A. C., and Ryle, M., Eds., E & S Livingstone, Edinburgh, 1971, 138.
53. **Amir, S. M.,** Dissociation of glycoprotein hormones, *Acta Endocrinol.*, 70, 21, 1972.
54. **Barber, S. A., Gray, C. J., Kennedy, J. F., and Butt, W. R.,** Evaluation of human "follicle stimulating" hormone preparations, *J. Endocrinol.*, 45, 275, 1969.
55. **Reichert, L. E., Jr.,** Human FSH: purification, properties, and some structure-function relationships, in *Gonadotrophins,* Sexena, B. B., Beling, C. G., and Gandy, H. M., Eds., Interscience, New York, 1971.
56. **Braunstein, G. D., Reichert, L. E , Jr., Van Hall, E. V., Vaitukaitis, J. L., and Ross, G. T.,** *Biochem. Biophys. Res. Commun.*, **42, 962, 1971.**
57. *Lactogenic Hormone, Ciba Foundation (London) Symposium,* Wolstenholm, G. E. W. and Knight, J., Eds., Churchill Livingstone, Edinburgh, 1972.
58. **Brown-Grant, K. and Cross, B. A.,** Eds., Recent studies on the hypothalamus, *Br. Med. Bull.*, 22(3), 1966.
59. **Schally, A. V., Arimura, A., Bowers, C. Y., Kastin, A. J., Sawano, S., and Redding, T. W.,** Hypothalamic neurohormones regulating anterior pituitary function, *Recent Progr. Horm. Res.*, 24, 497, 1968.
60. **McCann, S. M., Kalra, S. P., Kalra, P. S., Donoso, A. O., Bishop, W., Schneider, H. P. G., Forcett, C. P., and Krulich, L.,** The role of monoamines in the control of gonadotrophin and prolactin secretion, in *Gonadotropins,* Saxena, B. B., Beling, C. G., and Gandy, H. N., Eds., Interscience, New York, 1971, 49.
61. **Gorski, R. A.,** Sexual differentiation of the hypothalamus, in *Neuroendocrinology of Human Reproduction,* Mack, H. C. and Sherman, A. I., Eds., Charles C Thomas, Springfield, Ill., 1971, 60.
62. **Judd, H. C. and Yen, S. S. C.,** Serum androstenedione and testosterone levels during the menstrual cycle, *J. Clin. Endocrinol. Metab.*, 36, 475, 1973.
63. **Ross, G. T., Chargille, C. M., Lipsett, M. B., Rayford, P. L., Marshall, J. R., Strott, C. A., and Rodbard, D.,** Pituitary and gonadal hormones in women during spontaneous and induced ovulatory cycles, *Recent Progr. Horm. Res.*, 26, 1, 1970.
64. **Yen, S. S. C., Vela, P., Rankin, J., and Little, A. S.,** Hormonal relationships in the menstrual cycle, *J.A.M.A.*, 211, 1513, 1970.
65. **Henzl, M. R. and Segre, E. J.,** Physiology of human menstrual cycle and early pregnancy. A review of recent investigations, *Contraception,* 1, 315, 1970.
66. **Mishell, D. R., Nakamura, R. M., Crosignani, P. G., Stone, S., Kharma, K., Nagata, Y., and Thorneycroft, I. H.,** Serum gonadotrophin and steroid patterns during the normal menstrual cycle, *Am. J. Obstet. Gynecol.*, 111, 60, 1971.
67. **Moghissi, K. S., Syner, F. N., and Evans, T. N.,** A composite picture of the menstrual cycle, *Am. J. Obstet. Gynecol.*, 114, 405, 1972.
68. **Runnebaum, B., Bierwirth-V Münstermann, A. M., and Zander, J.,** Circadian variations in plasma progesterone in the luteal phase of the menstrual cycle and during pregnancy, *Acta Endocrinol.*, 69, 731, 1972.
69. **Baird, D. T. and Guevara, A.,** Concentration of unconjugated estrone and estradiol in peripheral plasma in nonpregnant women throughout the menstrual cycle, castrate and postmenopausal women and in men, *J. Clin. Endrocrinol. Metab.*, 29, 149, 1969.
70. **Nillius, S. J. and Wide, L.,** Effects of oestrogen on serum levels of LH and FSH, *Acta Endocrinol.*, 65, 583, 1970.

71. **Nillius, S. J. and Wide, L.,** Introduction of midcycle-like peak of luteinizing hormone in young women by exogenous oestradiol-17β, *J. Obstet. Gynaecol. Br. Commonw.*, 78, 822, 1971.

72. **Chargille, C. M., Vaitukaitis, J. L., Bermudez, J. A., and Ross, G. T.,** A differential effect of ethinyl estradiol upon plasma FSH and LH relating to time of administration in the menstrual cycle, *J. Clin. Endocrinol. Metab.*, 36, 87, 1973.

73. **Nillius, S. J. and Wide, L.,** Effects of progesterone on the serum levels of FSH and LH in postmenopausal women treated with oestrogen, *Acta Endocrinol.*, 67, 362, 1971.

74. **Westphal, U.,** Steroid-protein interactions, in *Monographs on Endocrinology*, Vol. 4, Gross F., Labhart, A., Mann, T., Samuels, L. T., and Zander, J., Eds., Springer-Verlag, Berlin, 1971.

75. **Crepy, O. and Gueriguian, J. L.,** Transport of steroids by proteins, *Res. Steroids*, 4, 61, 1970.

76. **Baulieu, E. E., Raynaud, J. P., and Milgrum, E.,** Measurement of steroid binding proteins, *Acta Endocrinol.*, Suppl. 147, 104, 1970.

77. **Schally, A. V., Arimura, A., Baba, Y., Nair, R. M. G., Matsuo, H., Redding, T. W., Debeljuk, L., and White, W. F.,** *Biochem. Biophys, Res. Commun.*, 43, 243, 1971.

78. **Matsuo, H., Baba, Y., Nair, R. M. G., Arimura, A., and Schally, A. V.,** Structure of the porcine LH- and FSH-releasing hormone. 1. The proposed amino acid sequence, *Biochem. Biophys. Res. Commun.*, 43, 1334, 1971.

79. **Newton, J. R. and Collins, W. P.,** Effect of synthetic luteinizing hormone releasing hormone (LH/FSH-RH) in women with menstrual disorders, *Br. Med. J.*, 3, 271, 1972.

80. **Besser, G. M., McNeilly, A. S., Anderson, D. C., Marshall, J. C., Harsoulis, P., Hall, R., Ormston, B. J., Alexander, L., and Collins, W. P.,** The hormonal responses to synthetic luteinising hormone-follicle stimulating hormone-releasing hormone (LH/FSH-RH) in man, *Br. Med. J.*, 3, 267, 1972.

81. **Nillius, S. J. and Wide, L.,** Variation in LH and FSH response to LH-releasing hormone during the menstrual cycle, *J. Obstet. Gynaecol. Br. Commonw.*, 79, 865, 1972.

82. **Nillius, S. J. and Wide, L.,** The LH-releasing hormone test in 31 women with secondary amenorrhoea, *J. Obstet. Gynaecol. Br. Commonw.*, 79, 874, 1972.

83. **Martini, L., Fraschini, F., and Motta, M.,** Neural control of anterior pituitary functions, *Recent Progr. Horm. Res.*, 24, 439, 1968.

84. **Jeffcoate, S. L., Holland, D. T., Fraser, H. M., and Gunn, A.,** Identification by radioimmunoassay of the LH-RH in human urine, *Nature*, 244, 161, 1973.

85. **Peters, H.,** The development of the mouse ovary from birth to maturity, *Acta Endocrinol.*, 62, 98, 1969.

86. **Coulson, P., Liu, T. C., Morris, P., and Gorski, J.,** Interaction of LH with the ovary, in *Gonadotropins*, Saxena, B. B., Beling, C. G., and Gandy, H. M., Eds., Interscience, New York, 1972, 227.

87. **Robison, G. A., Butcher, R. W., and Sutherland, E. W.,** Eds., *Cyclic-AMP*, Academic Press, New York, 1971.

88. **Kuo, J. R. and Greengard, P.,** Cyclic nucleotide-dependent protein kinases. iv. Widespread occurrence of adenosine 3′5′-monophosphate-dependent protein kinase in various tissues and phyla of the animal kingdom, *Proc. Natl. Acad. Sci.*, 64, 1349, 1969.

89. **Walton, G. M., Gill, G. N., Abrass, I. B., and Garren, L. D.,** Phosphorylation of ribosome-associated protein by an adenosine 3′5′-cyclic monophosphate-dependent protein kinase: location of the microsomal receptor and protein kinase, *Proc. Natl. Acad. Sci.*, 68, 880, 1971.

90. **Tsafriri, A., Lindner, H. R., Zor, U., and Lamprecht, S. A.,** Physiological role of prostaglandins in the induction of ovulation, *Prostaglandins*, 2, 1, 1972.

91. **Nalbandov, A. V.,** Comparative aspects of corpus luteum function, *Biol. Reprod.*, 2, 7, 1970.

92. **Hanson, F. W., Powell, J. E., and Stevens, V. C.,** Effects of H. C. G. and human pituitary LH on steroid secretion and functional life of the human corpus luteum, *J. Clin. Endocrinol. Metab.*, 32, 211, 1971.

93. **Fraser, I. S., Baird, D. T., Hobson, B. M., Michie, E. A., and Hunter, W.,** Cyclic ovarian function in women with congenital absence of the uterus, *J. Clin. Endocrinol. Metab.*, 36, 634, 1973.

94. **Bogumil, R. J., Ferin, M., Rootenberg, J., Speroff, L., and Vande Wiele, R. L.,** Mathematical studies of the human menstrual cycle. 1. Formation of a mathematical model, *J. Clin. Endocrinol. Metab.*, 35, 126, 1972.

95. **Bogumil, R. J., Ferin, M., and Vande Wiele, R. L.,** Mathematical studies of the human menstrual cycle. II. Simulation performance of a model of the human menstrual cycle, *J. Clin. Endocrinol. Metab.*, 35, 144, 1972.

ANDROGEN METABOLISM IN THE HUMAN FEMALE

V. H. T. James and C. M. Andre

TABLE OF CONTENTS

I. INTRODUCTION

One of the difficulties in discussing androgen metabolism in the human female revolves around the problem of finding an adequate definition of the term. Although no entirely satisfactory definition is available, naturally occurring androgens are C_{19} steroids that promote protein anabolism and are necessary for the development of the male reproductive tissues and to maintain their function. No similar, clear role for these compounds exists in the female, although it is possible that the anabolic activity of the weaker androgens may be of minor physiological importance. A further problem arises in attempts to assess the androgenic activity of any particular compound, since this will depend on the assay used. Dissociation of effects is common, and discussion of "androgenicity" requires consideration of the context.

In this chapter, it is convenient to define a group of C_{19} steroids as "androgens" and to consider their overall metabolism. We shall, therefore, discuss the secretion and metabolic interconversions of these steroids and will consider certain abnormalities of these processes, particularly as they relate to hirsutism in women.

The problem of defining more clearly the status of various steroid-producing tissues and evaluating the metabolic processes involved has challenged biochemists and endocrinologists for many years. Initially, the only methods available were those depending upon careful and difficult analyses of urinary steroids, together with maneuvers designed to alter steroid secretion. There are very considerable limitations to the effectiveness of this approach, as will be discussed later, and real progress had to await the introduction of more sophisticated techniques enabling concentrations of androgens to be measured in blood and kinetic investigations of steroid metabolism to be made. None of this work would have been possible without the use of high specific activity radiolabeled steroids, both for microanalytical work and for kinetic studies; the introduction of these materials has revolutionized work in this field, as in every other area of biochemical endocrinology.

II. STRUCTURE AND BIOLOGICAL ACTIVITY OF ANDROGENS

The naturally occurring androgens share, with modifications, the molecular structure of the C_{19} steroid compound androstane. The steroid nucleus has a rigid planar structure and by convention substituents which lie below or above the plane of the molecule are designated α and β, respectively, where stereoisomerism is possible. The modifications in the androstane molecule, that are encountered in various characteristic combinations in androgens, include a carboxyl (C = 0) or hydroxyl (-OH) group at C-3, C-11, and C-17, a double bond either between C-4 and C-5 or C-5 and C-6. Figure 1 shows the structure of various C_{19} steroids which are relevant to this discussion.

Androgenicity is markedly influenced by the orientation of substituent groups, e.g., testo-

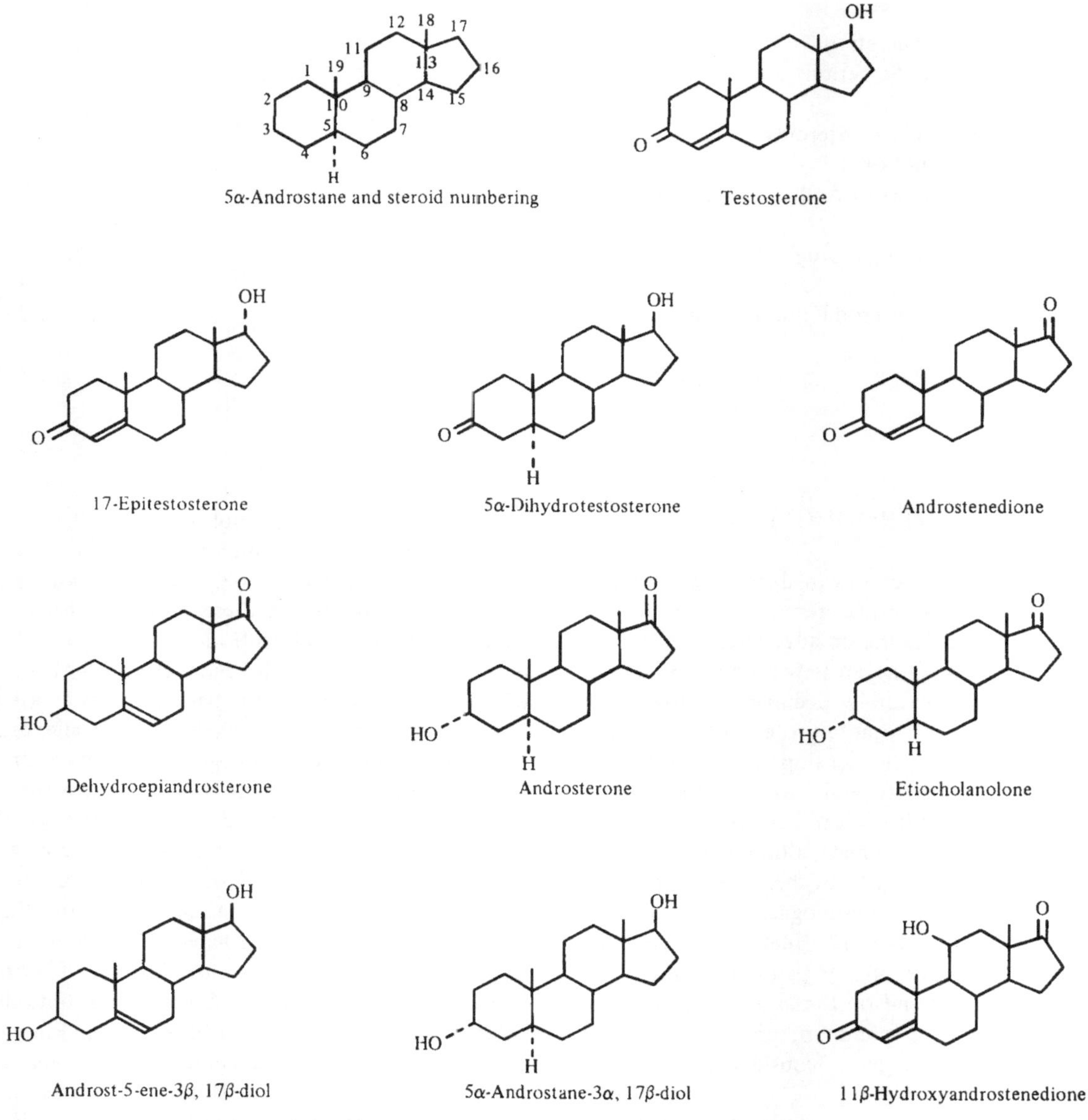

FIGURE 1. Structure of various C_{19} steroids.

sterone, which has an OH group in the 17β configuration, and is considerably more androgenic[1] than 17-epitestosterone, in which the hydroxyl group is in the 17α configuration. Similarly, that the orientation of the hydrogen at C-5 is important as a change from the α to the β configuration in, for example, 5α-hydrotestosterone to 5β-dihydrotestosterone virtually abolishes androgenicity.[2] In quantifying androgenicity, there are, however, additional factors to be considered. The result depends to a great extent upon the tissue used in the bioassay. Compounds that are equipotent in one bioassay may not be so in another. The difference may be due in part to the different metabolic potential of the two systems, since androgenicity probably depends on the interaction of cellular components with a compound with the appropriate molecular configuration, and a compound that lacks the essential structural requirement may still be active if metabolic interconversions in the tissue produce the active principle. Additional factors, such as the route of administration, excipient, and duration of treatment, may all also be of importance in determining the androgenic activity of any particular compound. Table 1 shows the relative biological activity of various androgens.

Information on the biological activity of androgens in women is limited. The evidence is derived mainly from studies of patients in whom excessive androgen production has been demonstrated or where androgens have been administered for therapeutic purposes. Testosterone, dihydrotestosterone, 5α-androstane-3α,17β-diol, and androstenedione are known[1,6-8] to be capable of producing virilizing effects in women (e.g., hair growth, clitoral enlargement, etc.). Dehydroepiandrosterone and androst-5-ene-3β,17β-diol failed to produce such effects when administered to patients with breast cancer in doses of 100 mg three times per week.[9,10] Dehydroepiandrosterone sulfate in plasma is reported to be raised in girls with premature development of sexual hair;[11] in another study it was found to be androgenic and estrogenic in a 3-month-old female with the 21-trisomy syndrome.[12] It would be very useful to have further information about the relative activities of these compounds and to know more about their effects in detail, but such information is obviously difficult to obtain.

III. SOURCES OF C_{19} STEROIDS

In the normal nonpregnant female, the only sources of C_{19} steroids are the adrenal glands (specifically the adrenal cortices) and the ovaries. The situation in pregnancy will be dealt with later.

There are two ways in which androgens are normally produced, both of which are of major importance. These are by direct secretion from an endocrine gland or by indirect production from a precursor, or "pre-hormone."

A. Direct Secretion

There are essentially two methods which can be employed for the study of androgens produced by

TABLE 1

Comparative Biological Activity of Androgens

	Chick's comb inunction with ethanol	Exorbital gland of the castrated rat
Testosterone	100	100
Dihydrotestosterone	228	74
Androst-5-ene-3β,17β-diol	54–155	–
5α-Androstane-3α,17β-diol	–	283
Androstenedione	121	14
Dehydroepiandrosterone	59	< 1
Androsterone	85–139	46
11β-Hydroxyandrostenedione	< 5	–
Reference(s)	3	4, 5

In each bioassay system, testosterone was given a value of 100 and other responses were expressed relative to testosterone.

an endocrine gland. The first is based on in vitro techniques, either perfusion of the isolated gland with blood or an artificial medium or, more commonly, tissue incubations. The latter technique has been extensively used in the investigation of androgen formation by ovarian or adrenal tissues from precursors within the tissue or from added precursors, which are usually labeled with tritium or carbon-14. These investigations have yielded valuable information on the biosynthetic potential of the tissues, the pathways involved, and also the factors which influence steroidogenesis. In this way, it has been possible to show that human ovarian tissues are able to produce androstenedione, dehydroepiandrosterone, and testosterone.[13,14] Similarly, dehydroepiandrosterone and its sulfate, androstenedione, 11β-hydroxyandrostenedione, and testosterone have been shown to be produced by human adrenal tissue.[15,16] There are, however, limitations to the interpretation of these studies, not only because most of them were carried out on tissues bearing pathological stigmata or were obtained from patients suffering from a variety of diseases, but also because great care is needed in extrapolating from in vitro results to the in vivo situation.

The second type of method is based on direct sampling of venous blood from the endocrine gland under study. The concentration of compounds in the blood sample is compared with the levels in peripheral venous blood and a higher level in the effluent blood establishes evidence for secretion. It is assumed, probably with justification, that the concentration of the compounds in question is approximately the same in arterial blood as in peripheral venous blood. In this way, it has been possible to obtain evidence for the secretion of dehydroepiandrosterone and its sulfate,[17] androstenedione,[17] 11β-hydroxyandrostenedione,[18] and testosterone[18] from the adrenals and of dehydroepiandrosterone, testosterone, and androstenedione from the ovary.[19] Some evidence for the secretion of 5-androstene-3β,17β-diol has been obtained.[17] The relative amounts produced, however, vary considerably.

B. Indirect Production

The indirect sources, also of major significance, are the "pre-hormones," which are converted metabolically by various body tissues, to other, and more active hormones. An important example is the conversion of androstenedione to testosterone; this point will be discussed subsequently. Thus, it follows that in considering the total androgen load, both direct secretion and production from pre-hormones must be taken into account:

Androgen load = secretion + production

IV. METABOLISM OF C_{19} STEROIDS

A. Introduction

The metabolism of C_{19} steroids occurs largely in the liver and partly in extrahepatic tissues. There are a number of complex, interrelated pathways; Figure 2 shows major routes which have been established.

The metabolic pattern is complicated by the extensive interconversions which occur between the major secretory products; any attempts to interpret urinary steroid levels in terms of androgen secretion and production must take note of this not inconsiderable problem. It is for this

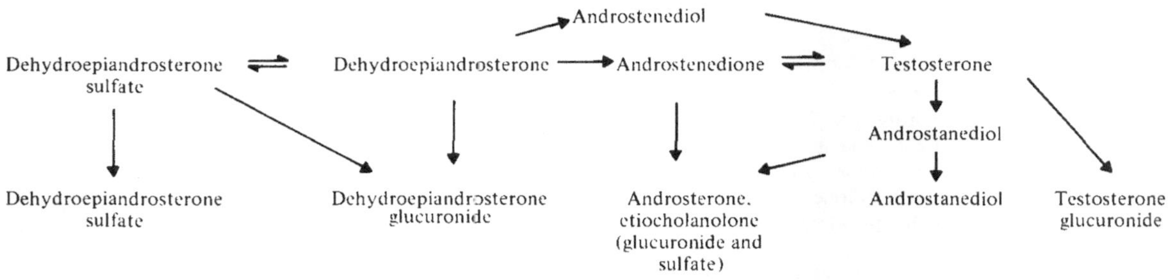

FIGURE 2. Interconversion and metabolism of C_{19} steroids. Compounds in the upper line are present in blood and the interconversions shown occur in plasma or tissues. In the lower line are the metabolic products which are excreted in the urine.

reason that earlier studies on urinary steroid excretion proved relatively unrewarding in trying to further our understanding of the prerenal metabolic processes involved. Thus, while the classical estimation of total 17-oxosteroid excretion (largely dehydroepiandrosterone, androsterone, and etiocholanolone) is of minimal value, even careful quantitative fractionation is not very helpful for this purpose.[20] It has been known for some time[21] that testosterone is excreted in urine as its glucuronide and it was suggested[22] that this might be a useful indicator of testosterone production. However, it is now clear that this straightforward view of testosterone metabolism is invalid since the major part of the urinary testosterone glucuronide is formed in the liver, by direct conjugation of testosterone formed intrahepatically from androstenedione.[23] Thus, the excretion of testosterone glucuronide has no direct relationship to the level of circulating testosterone and urinary testosterone glucuronide is not a simple parameter of testosterone secretion.

Using radiolabeled steroids, it is possible to study the complex quantitative relationships between the various plasma precursors and the urinary metabolites in normal women. Following the intravenous injection of the relevant labeled precursor it has been estimated,[24] from the percentage conversion of the precursor to a particular urinary metabolite and from the blood production rate of the precursor, that approximately 50% of the urinary dehydroepiandrosterone glucuronide originates from plasma dehydroepiandrosterone sulfate and the remaining 50% from plasma dehydroepiandrosterone. With regard to urinary etiocholanolone glucuronide and androsterone glucuronide, approximately 33% originates from plasma dehydroepiandrosterone sulfate, 33% from plasma dehydroepiandrosterone, 20% from plasma androstenedione, and 5% from plasma testosterone. These data show how difficult it is to interpret urinary steroid excretion in terms of steroid production, particularly since the figures quoted are means and may show considerable variation between subjects.

There are many facets to the metabolism of androgens. One of these relates to the metabolic processes, like conjugation with glucuronic or sulfuric acid, which precede the urinary excretion of androgen metabolites. Another concerns the production of active metabolites, as mentioned earlier. These metabolites may only act at the site of production or may merge with the circulating androgen pool as well. Not only may potent androgens like testosterone and dihydrotestosterone be produced from weak androgens like androstenedione,[25,26] but estrogens are also formed in this way. For example, it has been calculated that the main source of estrone in the postmenopausal period is androstenedione.[27] Although of considerable importance, some of the details of these processes are still only poorly understood; this chapter will deal mainly with more direct aspects of androgen metabolism.

B. Parameters of Androgen Metabolism

The proper and complete description of androgen metabolism involves consideration of a relatively large number of factors, all of which are important. Some of the major areas are listed below and will be discussed individually.

1. Urinary excretion of metabolites
2. Plasma androgen levels
3. Metabolic clearance of androgens
4. Binding of androgens to plasma proteins
5. Assessment of androgen production: kinetics and metabolic interconversions

A full description of androgen metabolism would also have to consider the mechanism of action of the steroids at a molecular level, which involves the problem of transport into the cell and the subtle effects which steroids have in regulating protein synthesis.

1. Urinary Excretion of Metabolites

It has been pointed out that the quantitative determination of urinary androgen metabolites can give only a very crude index of overall androgen production. Significant changes in secretion can occur without any marked alteration from normal in the total amount of steroid excreted.[28,29] It is technically a fairly complicated procedure to fractionate the individual steroids and is not a rewarding task. In general, where excessive secretion of C_{19} steroids occurs, this is reflected by an increase in the excretion of urinary steroid metabolites, but these changes may only be minor. Thus, while a relatively small increase in testosterone production may have important clinical implications and major physiological effects, the relatively small change in steroid excretion may be masked against the background level of excretion of the metabolites of other C_{19} steroids. For these reasons, it has proved difficult to interpret urinary steroid patterns, and only relatively gross deviations from normal are of diagnostic assistance. Even then, the technique is too unsophisticated to indicate the nature of the lesion involved.

One may conclude, therefore, that urinary excretion patterns are only useful in indicating total androgen production in a general sense and more specific indicators are necessary.

The normal adult female produces between 5 and 16 mg daily of total urinary 17 oxosteroids.[30] The fractionated components are shown in Table 2.[20,31] Apart from dehydroepiandrosterone, etiocholanolone, and androsterone, the origin of which has been discussed earlier, other 11-oxygenated C_{19} steroids are also present generally in smaller amounts. 11β-Hydroxyetiocholanolone and 11-oxoetiocholanolone derive from cortisol, while 11β-hydroxyandrosterone is essentially a metabolic product of 11β-hydroxyandrostenedione. Thus, the excretion of total 17-oxosteroids, while comprising largely the three 11-deoxysteroids, does reflect the amounts of other steroid metabolic end products which are excreted. The excretion of 17-oxosteroids is very low before puberty. The levels rise at puberty and increase to a maximum around 25 years and then gradually decline with age.[30] There is no consistent change in total 17-oxosteroid excretion during the menstrual cycle,[30] although the levels of androsterone and etiocholanolone have been reported to fluctuate in phase with the biphasic excretion of estrogens.[32] There is also evidence to suggest that the excretion of urinary C_{19} sulfates is higher in the luteal phase than in the follicular phase of the menstrual cycle.[33] The excretion of testosterone glucuronide also varies during the cycle.[34] The most consistent pattern is a rise at midcycle, followed by a secondary rise during the luteal phase. There is also a small sporadic peak during the follicular phase. These findings would suggest an ovarian source for these increases in androgen excretion in the absence of evidence to indicate that the secretion of adrenocortical androgens varies during the menstrual cycle.

2. *Plasma Androgen Levels*

It has been appreciated for many years that the concentration of circulating steroid is likely to be a better parameter of biological activity than is the amount of steroid excreted in the urine. Until

TABLE 2

Range of Urinary Excretion of Individual C_{19} Steroids in Normal Female Subjects

	mg/day
Dehydroepiandrosterone	0.1–3.7
Etiocholanolone	0.6–5.8
Androsterone	0.4–3.6
11β-Hydroxyetiocholanolone	0.1–0.8
11-Oxoetiocholanolone	0.2–1.4
11β-Hydroxyandrosterone	0.2–1.0

recently, the techniques of steroid assay were relatively insensitive and the estimation of plasma androgen levels, particularly in female subjects, posed almost insuperable problems.

The first attempts to assay plasma androgen levels were focused upon testosterone, not unnaturally since it is one of the most active of the naturally occurring androgens. Several successful techniques were eventually developed which are capable of measuring accurately the relatively small amounts of testosterone in female plasma. Nevertheless, these methods require a high degree of technical skill. Techniques employing gas-liquid chromatography, double isotope dilution analysis, and saturation analysis have been developed and applied extensively; more recently, radioimmunoassays using antisera raised against testosterone have been employed almost exclusively because of their superior reliability and sensitivity as compared to other analytical methods. The reader requiring further information is, referred to Chapter 7 in *The Biochemistry of Women: Methods for Clinical Investigation.*

The total level of circulating androgen represents a balance between the rate at which newly formed or secreted androgen enters the circulation and the rate at which it is removed. Thus, the important parameters to consider are the production rate and the metabolic clearance rate (MCR); these three factors are related to each other by the equation:

$$\text{Plasma level} = \frac{\text{production rate (PR)}}{\text{metabolic clearance rate (MCR)}}$$

It is clear that changes in either the PR or the MCR will affect the plasma level of androgen and, equally, knowledge of any two of the three factors will give the third. The normal levels of circulating androgens are shown in Table 3.

TABLE 3

Plasma Level, Metabolic Clearance Rate, and Blood Production Rate of C_{19} Steroids

	Mean plasma level μg/100 ml ± S.D.	Reference	Mean MCR (liters/day)	Mean production rate (mg/24 hr)	Reference
Testosterone	0.035 ± 0.022	18	600	0.23	39
Dihydrotestosterone	0.015 ± 0.005	38	209	0.057	40
Androst-5-ene-3β,17β-diol	0.068 ± 0.024	36	–	–	
5α-Androstane-3α,17β-diol	0.017 ± 0.009	37	–	–	
Androstenedione	0.13 ± 0.08	18	2100	3.3	39
Dehydroepiandrosterone	0.48 ± 0.32	18	1640	7.0	41
Androsterone	0.07 ± 0.03	18	–	–	
11β-Hydroxyandrostenedione	0.18 ± 0.08	18	–	–	
Dehydroepiandrosterone sulfate	60.4 ± 40.3	35	22		42
Androsterone sulfate	13.6 ± 13.1	35			
Androst-5-ene-3β,17β-diol sulfate	11.3 ± 8.2	35			

Complete definition of plasma androgen levels must take into account normal physiological fluctuations, and factors such as age, diurnal rhythms, stage of the menstrual cycle, posture (i.e., recumbent or upright) may all affect the concentration. Figure 3 shows plasma testosterone and androstenedione levels measured through a normal menstrual cycle. Testosterone levels appear to be relatively stable, with only small fluctuations occurring. Androstenedione levels, however, show a marked diurnal change which is related to the fact that the major site of production of this steroid is the adrenal glands, and androstenedione secretion is responsive to adrenocorticotrophin (ACTH), as is cortisol. Thus, the well-known diurnal variation in cortisol levels is paralleled almost exactly by androstenedione, as Figure 4 shows. There is less detailed information available for the levels of dehydroepiandrosterone and dehydroepiandrosterone sulfate, but it has been shown that dehydroepiandrosterone levels fluctuate in parallel with cortisol levels.[43] There is less general agreement concerning dehydroepiandrosterone sulfate.[44]

3. Metabolic Clearance of Androgens

The metabolic clearance rate (MCR) is defined as the volume of plasma which is completely cleared of a compound in unit time. In practice, this is estimated by injecting radiolabeled material and observing the rate at which the labeled compound disappears from the peripheral blood. Alternatively, and more satisfactorily, the labeled material is infused continuously. When an equilibrium state has been achieved, the MCR is calculated from the rate of infusion of the isotope divided by the concentration of isotope in the specific compound per volume of plasma. The labeled compound is assumed to be metabolically indistinguishable from the endogenous compound with which it has been intimately mixed. The theoretical basis for these calculations has been discussed extensively by Tait and Burstein;[45] the reader seeking more detailed information is referred to their paper. Table 3 shows the MCR of various C_{19} steroids.

The MCR, as determined by these techniques, is a measure of the overall metabolism within the body and claims no distinction between various metabolic processes. It should also be remembered that it only describes the rate at which the compound under study is cleared by conversion to another chemical form and is independent of any of the stages subsequent to this step, although these may be of major metabolic significance. It should be also appreciated that the MCR has two major components and it is useful to distinguish between them. The liver is a major site of steroid metabolism and a substantial proportion of the total clearance is hepatic. However, extrahepatic metabolism is also of considerable significance. There is, for example, a marked sex difference in the MCR for testosterone which persists after correction for differences in body surface area.

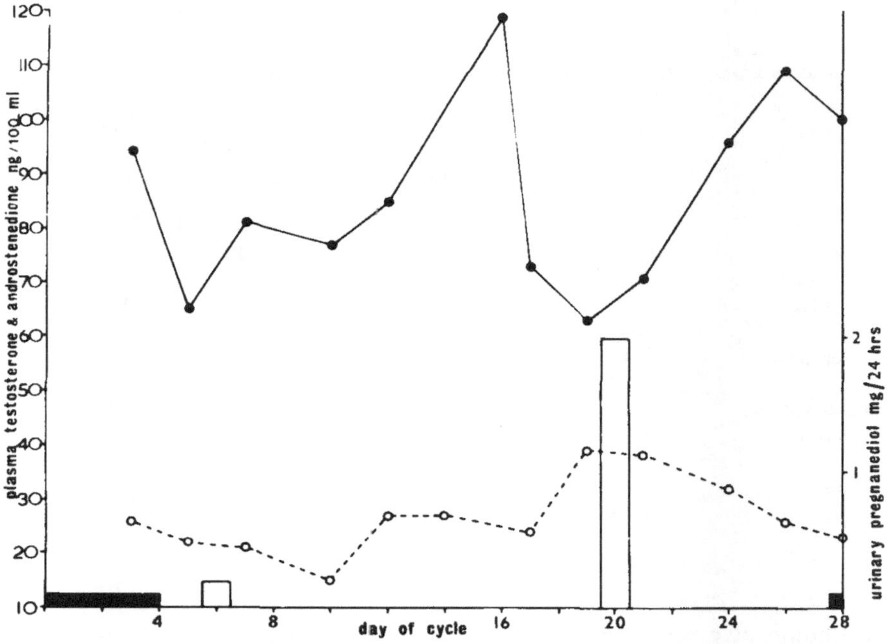

FIGURE 3. Plasma levels of testosterone (o---o) and androstenedione (●-●) through a normal menstrual cycle. Menstrual bleeding (▬) and pregnanediol levels (▯) are also shown.

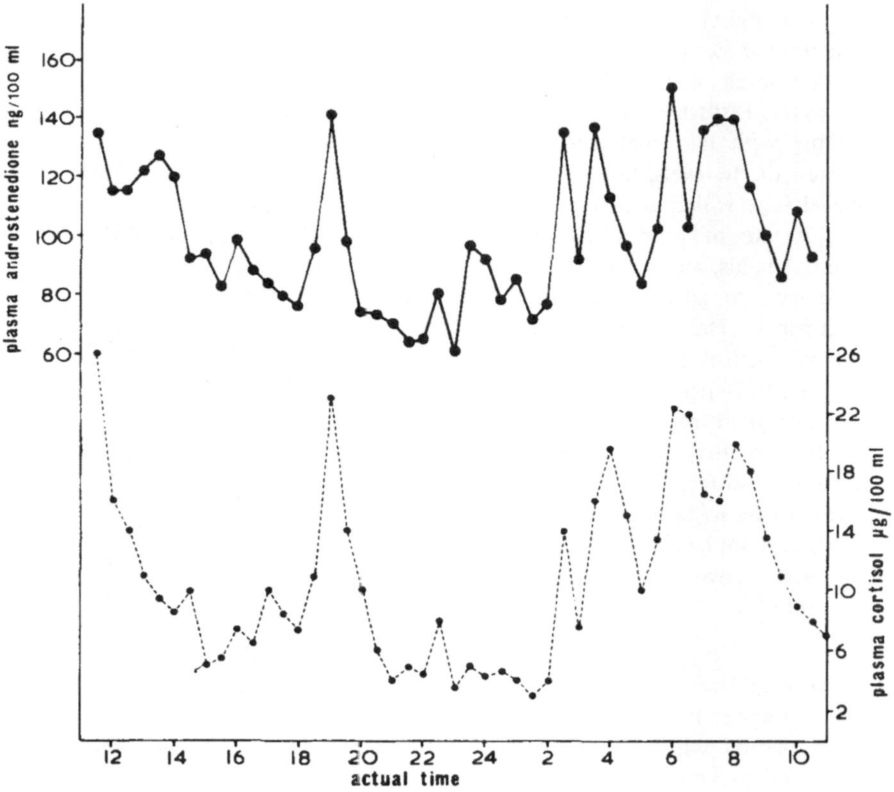

FIGURE 4. Plasma androstenedione and cortisol levels in a normal female subject measured through the 24 hours.

The average MCR in men is 1,200 liters/day and in women 600 liters/day.[46] The reason for this difference is complex. It is partly due to the fact that there is little extrahepatic clearance of testosterone in women, but it is also due to differences in the binding of testosterone by plasma proteins between the sexes, as discussed in another section. In men, there is a larger proportion of circulating testosterone that is freely available for metabolic disposal[47] and there is a good correlation between this fraction and MCR.[47] However, there is evidence to indicate that if the fraction of testosterone available for clearance in women is raised acutely by the administration of testosterone to produce the plasma levels found in men, the MCR under these conditions would still be lower than in men. These findings suggest that there are factors other than protein binding which influence MCR. Thus, it has been shown[47] that after the chronic administration of testosterone to women, the differences observed above disappeared. The evidence indicates that this increase in clearance is likely to be due to an increase in extrahepatic clearance probably through enzyme induction in peripheral tissues. It is apparent then that androgen production itself is able to induce changes in extrahepatic metabolism and that the sex difference that is observed in testosterone clearance is related to the higher secretion of testosterone in the normal male, being mainly testicular in origin.

Although the effect of androgen is important, it should be noted that changes in posture can also alter MCR probably by altering hepatic blood flow. For example, the MCR of testosterone can be decreased by as much as 35% on changing from the recumbent to the upright posture.[48]

Other factors, such as menstrual status, do not appear to influence the MCR of testosterone. In ovariectomized subjects, the MCR of testosterone is significantly lower than in normal subjects, although the controls were in a different age group.[49]

The factors that influence the MCR of other androgens have not been as extensively investigated as they have for testosterone. However, it is known that the MCR of androstenedione is not related to its production rate,[39,49] it is slightly higher during the luteal phase of the menstrual cycle, and unchanged in ovariectomized subjects.[49] The MCR of sulfate conjugates of C_{19} steroids such as dehydroepiandrosterone sulfate is low (Table 3). This may partly be due to the low renal clearances observed for these conjugates,[50] which in turn relates to some extent to the binding to plasma albumin.[51] In contrast, the glucuronides of androsterone and etiocholanolone have high renal clearances.[50]

4. Binding of Androgens to Plasma Proteins

Human plasma contains a number of proteins, some of which bind certain hormones specifically. The best known of these are probably the globulins which bind cortisol (transcortin) and thyroxine (thyroxine binding globulin). During early studies of steroid binding to plasma proteins, it was suspected that testosterone might be bound to a protein other than albumin. Direct evidence for this was subsequently obtained.[52] Further investigations revealed that estradiol was also bound at the same site as testosterone. More recently, it has been found that this protein, a β-globulin, is capable of binding relatively specific steroids with an unhindered 17β-hydroxyl group, largely C_{19} steroids.[53,54] In normal women, a large proportion of the testosterone (about 80%) present in plasma is bound to this protein[55] (referred to as sex hormone binding globulin, SHBG) which has a high association constant for testosterone ($Ka = 7.5 \times 10^8$ l. mol at 25°C) but a limited binding capacity.[47] The bulk of the remaining testosterone is attached largely to albumin ($Ka = 2.4 \times 10^4$ l. mol at 25°C) but also to transcortin.[53] Although virtually all the testosterone in plasma is protein-bound, a small but significant amount is present in the free form, which normally amounts to about 1%.[55] Thus, if the average concentration of testosterone in female plasma is taken to be about 35 ng/100 ml, only 0.35 ng/100 ml is in fact in the free form.

Androgens that lack 17β-hydroxyl groups, such as dehydroepiandrosterone and androstenedione, and the sulfate conjugates of androsterone and dehydroepiandrosterone also circulate bound to protein,[51,56] probably albumin.

It is not possible at present to measure directly the SHBG concentration in plasma. However, several techniques have been developed to measure the concentration of binding sites,[47,57,58] which reflect the concentration of SHBG.

Variations produced by physiological and pathological states in SHBG concentrations have been studied. It has been found that the level of SHBG increases during pregnancy, but returns to normal in the postpartum period.[47,59] There is no variation during the normal menstrual cycle.[59] Before puberty there is no difference between the sexes in levels of SHBG, the values being comparable to those found in adult females. In the newborn of either sex, the levels are in the adult male range, which is significantly lower than in adult females.[47] It is also known that SHBG levels can be altered by estrogens and androgens. Administration of estrogens increases the amount present, possibly by initiating the synthesis of more protein. Conversely, androgens will decrease SHBG levels; this can be demonstrated in women receiving these compounds.[47]

The physiological significance of SHBG is not known. It is unlikely that the solubilization of the relevant plasma steroids is one of its functions, as it has been shown that at physiological concentrations the presence of protein is not essential for the solubility of these steroids in aqueous medium.[60] It may be involved in the secretion and cellular uptake of 17β-hydroxy-C_{19} steroids or act as a reservoir for these potent androgens. There is evidence to suggest that the clearance of testosterone is intimately related to the manner in which it is bound to protein and that it is essentially the free fraction and that bound to albumin which are freely available to metabolic disposal.[47,61] Consequently, the availability of potent androgens to nontarget cells is restricted, whereas target cells, as the evidence suggests, appear to have specific receptors with very high association constants and are not subject to the same restrictions. Second, the rate of clearance by organs such as the liver and kidneys is inhibited, thus limiting the rate of loss of these androgens from the plasma.

The complex interaction between the level of androgen production and its effects on protein binding and cellular metabolism is not as yet understood, but it is likely to be of major importance in a fuller appreciation of normal and abnormal states of androgen secretion.

The investigation of androgen binding to receptors in target cells has largely been carried out with animal tissues; the interested reader is referred to a general review of the topic by Baulieu et al.[62]

5. Assessment of Androgen Production

Measurement of androgen production, which it will be remembered comprises androgen derived from both secretion and metabolic interconversions, is carried out with the help of radiolabeled tracers. The hormone under study, e.g., testosterone, must be available in a stable radiolabeled form and either ^{3}H or ^{14}C material, usually the former, is used. After intravenous infusion of the ^{3}H-testosterone at a known rate for a sufficient time to achieve equilibrium (usually about 3 hr), a sample of blood is removed and the specific radioactivity of the testosterone present in the sample is measured. This procedure enables the MCR to be calculated; from this value, together with the concentration of plasma testosterone which is usually measured before commencing the infusion, it is possible to calculate the production rate of testosterone over the period of the infusion.[45] It should be noted that a condition for the validity of these measurements is that a steady state of steroid production should exist during the period of the study. It is, therefore, essential to take note of diurnal alterations when this technique is employed.

It is also possible to calculate the production rate of a compound by the principle of isotope dilution.[45] A small amount of radiolabeled compound is injected and a specific metabolite of the compound under study is isolated from the urine and its specific activity is measured. The production rate of the compound is then calculated from the dilution principle. However, it is essential with this technique to ensure that the urinary metabolite used is uniquely derived from the compound under study.

Attempts were made to apply this method to testosterone, using urinary testosterone glucuronide. However, it became apparent subsequently that, as discussed earlier, testosterone glucuronide is not a unique metabolite[23] (see Figure 5) and that androstenedione is converted intrahepatically into testosterone and then to testosterone glucuronide, which is excreted in the urine. The dilution technique is invalid and the results obtained are not meaningful.

Table 3 lists the approximate blood production rate of various C_{19} steroids in normal women.

6. Metabolic Interconversions

It is only possible to study metabolic interconversions of C_{19} steriods by using radiolabeled tracers. Each steroid needs to be labeled with a different, recognizable isotope. In practice, since usually only ^{3}H- and ^{14}C-labeled steroids are

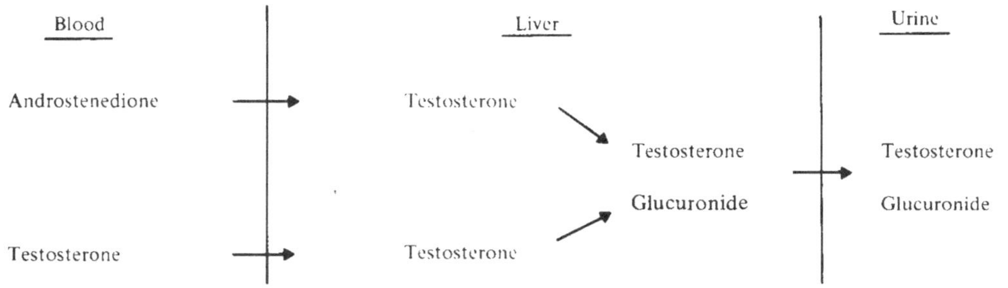

FIGURE 5. Origin of urinary testosterone glucuronide.

available, such studies are limited to pairs of steroids, e.g., testosterone and androstenedione.

The method requires the simultaneous infusion of the radiolabeled steroids until equilibrium is reached.[61] Although this may only be a few hours for some steroids, such as testosterone and androstenedione, other compounds such as dehydroepiandrosterone sulfate take days rather than hours to approach equilibrium with an infused tracer. This, of course, makes accurate studies of these compounds impracticable in human subjects.

After the infused tracers have reached equilibrium, it is necessary to measure the amount of the two tracers in each of the two compounds and then the rate of conversion (ρ) can be calculated. In this way, it is possible to find out what proportion, for example, of the total amount of testosterone produced and entering the bloodstream, arises from androstenedione and vice versa. In the case of testosterone, since it is likely that androstenedione is the only important indirect source of this compound, it is possible to estimate its secretion by difference:

Testosterone secretion = testosterone production – testosterone from androstenedione

These are difficult studies to carry out and they have been made in only a limited number of subjects. However, from such studies it appears that only a small amount (0.7%) of dehydroepiandrosterone is converted to testosterone; thus, taking a production rate of 7 mg/day of dehydroepiandrosterone, about 50 μg of testosterone, at most, arises from this source.[41] However, androstenedione is a major source of testosterone in women; if we assume about 3 mg of androstenedione is produced daily and 5% is converted to testosterone, this gives 150 μg/day, i.e., about 60% of the total.[63] Thus, alterations in the production or conversion of androstenedione can produce major effects on testosterone production in women.

V. STATES OF ALTERED ANDROGEN METABOLISM

A. Pregnancy

During pregnancy, steroid metabolism undergoes a number of complex changes, many of which are not yet well understood.

Several authors have shown that the mean level of testosterone in plasma is progressively increased during pregnancy (see review by Gandy[64]). These elevated levels decrease sharply after parturition, returning to normal within a few days of delivery.[65] The elevated levels of testosterone are apparently due to an increased concentration of testosterone binding globulin[47,59] and the unbound fraction of testosterone is not altered.[55] Plasma dihydrotestosterone levels are increased.[40] Measurements of metabolic clearance rates and production rates of testosterone and dihydrotestosterone in pregnancy have been reported.[40] It was found that the MCR's of testosterone and dihydrotestosterone were diminished, as compared to nonpregnant women, while the production rates of these two compounds are not significantly different from the controls. It has also been reported that the levels of 5α-dihydrotestosterone, 5α-androstane-3α,17β-diol, and androst-5-ene-3β,17β-diol are elevated in pregnancy.[66]

Some authors have reported that androstenedione concentrations are elevated during pregnancy, that this occurs progressively, and that the levels fall after parturition.[64,67] Since androstenedione is not specifically bound to a plasma protein, the increased plasma levels observed imply altered production or metabolism in pregnancy; this additional information is lacking, as are data which would indicate the source of this steroid.

Studies of dehydroepiandrosterone and dehydroepiandrosterone sulfate during pregnancy suggest that in contrast to the C_{19} steroids discussed above, the levels of these two compounds decrease.[64,68,69] Again, the mechanism by which this occurs and its significance have not been elucidated.

B. Androgen Production in Women with Virilism

A considerable amount of the work reviewed in this chapter was initiated because of the interest of several investigators in the clinical problem of female virilization; therefore, it is appropriate to discuss briefly the findings in this group of women and how they compare to normal female subjects.

Increased androgen production, leading to virilism, occurs in association with tumors of the adrenal cortex or the ovary and the biochemical pathology is well recognized. These tumors have been demonstrated to secrete androgens,[70,71] usually testosterone; biosynthetic studies of the tumors have confirmed their ability to produce C_{19} steroids.[16,72] These tumors often result in marked virilization, with the development of hirsutism, clitoral enlargement, and menstrual disturbance or amenorrhea. Although 17-oxosteroid excretion may not necessarily be elevated, plasma testosterone can reach levels which approach or are within the male range and elevated testosterone production rates are observed.[70] These are relatively rare conditions and so not many data are available. A relatively common condition is that of idiopathic hirsutism, which has been investigated extensively. These patients have a much milder degree of virilism. In some cases, this is manifested only by a modest amount of hair growth on the face and the trunk, so that some authors have even questioned whether this may truly be described as "abnormal." Other patients are more severely affected and show a marked degree of hair growth on the face in the regions of the upper lip and sideburns and on the breasts and abdomen, sometimes with development of a male pubic escutcheon. Alterations of menstrual function also occur with oligomenorrhea or amenorrhea. Some of these patients can be subdivided into a group in whom ovarian abnormalities are present. The ovaries contain multiple large cysts with a thickened capsule; this condition is called the Stein-Leventhal syndrome or the polycystic ovary syndrome (PCOS).

It is now becoming clear that these conditions are associated with a relatively mild increase in androgen production which may originate in either the adrenal glands or ovaries or both.

In the patients with idiopathic hirsutism, the assay of urinary 17-oxosteroids shows either results which are within the normal range or which are slightly elevated (Figure 6). In most cases in whom 17-oxosteroid excretion is increased, fractionation of the urinary steroids reveals that the elevation is due to an increase of androsterone and etiocholanolone;[31] in others an increase in the excretion of dehydroepiandrosterone sulfate is seen, either on its own or associated with an increase in the other 17-oxosteroids (see review by Brooksbank[73]). The elevated excretion of androsterone and etiocholanolone implies an increased secretion of C_{19} androgens, such as testosterone, androstenedione, or both; in agreement with this, the excretion of testosterone glucuronide has been shown to be elevated in women with idiopathic hirsutism.[74]

Plasma testosterone levels have been investigated extensively in patients with idiopathic hirsutism. With the more reliable techniques that are now available, various workers have found that there is a modest increase in the plasma testosterone level in some patients, although many have levels within the normal range.[39,75] Androstenedione levels may also be increased or normal[18,39] (see Figure 6). In occasional patients, particularly those with PCOS, higher levels of testosterone are found.[39,75] It has also been shown that the concentration of SHBG is diminished in these patients and that the free testosterone level is relatively elevated compared with normal women.[47,55,76] In addition, measurement of the extrahepatic clearance of testosterone has revealed that, while in normal women this is minimal, in patients with hirsutism, extrahepatic clearance is significantly increased.[46]

The indirect evidence cited above for increased androgen production has been confirmed by more direct investigations. Measurements of testosterone

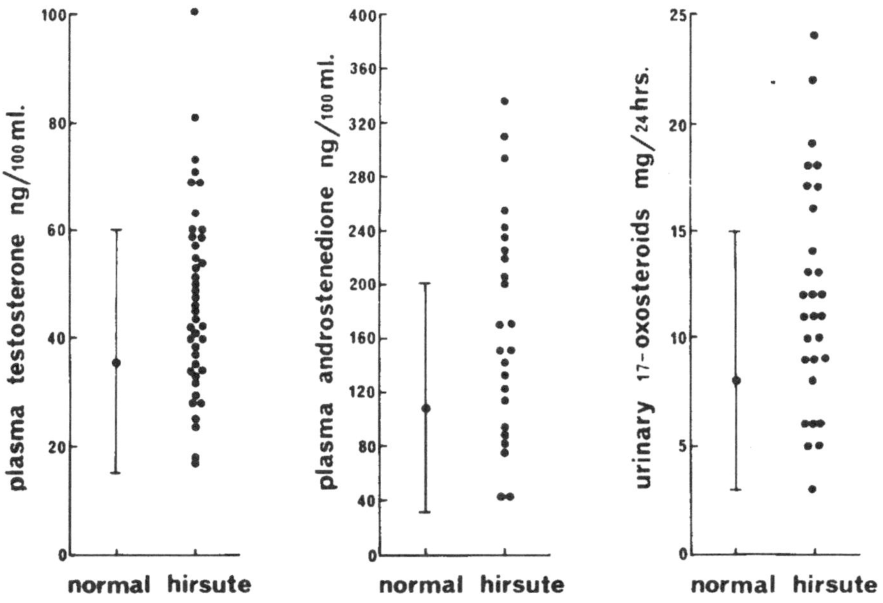

FIGURE 6. Plasma levels of testosterone and androstenedione and urinary 17-oxosteroids in control subjects and in patients with idiopathic hirsutism. The mean and range for the normal subjects are shown.

and androstenedione production rates have shown them to be elevated in patients with hirsutism and PCOS, as compared with normal women.[39] It has also been shown that the fraction (ρ) of the blood androstenedione converted to blood testosterone is similar in patients with PCOS or idiopathic hirsutism and normal subjects.[39] However, in both groups of patients, a smaller percentage of the blood testosterone originated from blood androstenedione (approximately 25% compared to 50% in normal subjects),[39] indicating an increased secretion of testosterone or the synthesis of testosterone from a precursor that does not enter the plasma androstenedione pool.

These studies, of course, cannot reveal the source of the androgens. Attempts have been made to do this by stimulating the adrenal gland with ACTH, suppressing ACTH secretion with glucocorticoids and stimulating the ovary with gonadotrophins, or suppressing gonadotrophin secretion with estrogens while adrenocortical activity remained suppressed. The level of androgens in plasma, urine, or both was measured before and during the tests.[75,77-83] Unfortunately, the interpretation of the results obtained from these investigations has proved extremely difficult, as the tests give inconsistent results in individual cases. It is, therefore, impossible in this way to assess reliably the relative roles of the adrenal cortex and the ovary in the production of androgens in hirsutism or the PCOS.

Recently, direct evidence has been obtained by sampling ovarian and adrenal venous blood by direct cannulation of the appropriate veins. In a remarkable investigation where combined ovarian and adrenal vein catheterization was carried out to determine the sites of androgen overproduction in hirsute women, it was shown that in most of the patients studied there was an increase in the secretion of testosterone or androstenedione or both by the ovaries; in others the adrenal cortex and ovaries were hyperfunctional in this respect.[84] This study also disclosed that dexamethasone, which has been used extensively to inhibit secretion of ACTH, can exert suppressive effects on ovarian steroid secretion, thus rendering invalid the use of this compound as a discriminating agent between ovarian and adrenal secretion.

It is apparent that there is unassailable evidence for increased androgen secretion in many patients with hirsutism and there are also a number of other metabolic abnormalities present which appear to stem from this event. Thus, if plasma androgen levels are increased, plasma SHBG levels are on the whole depressed and the concentration of free testosterone is elevated. In addition, extrahepatic clearance of testosterone is increased. It seems likely that the primary event in these factors is an increase in the secretion of androstenedione, testosterone, or other 17β-hydroxy-C_{19} steroids, giving rise to an increased androgen load. The increase in plasma testosterone levels that is observed is less than might be expected since testosterone clearance is increased. The increased androgen production decreases the concentration of SHBG and free testosterone levels are elevated, leading probably to increased testosterone clearance.

These factors are probably a sufficient explanation for the hirsutism which develops in these patients. However, there is no apparent quantitative correlation between androgen load and hirsutism, as witnessed by the fact that some hirsute patients show no evidence of increased androgen production.[46] This may be due to differences in the responsiveness of hair follicles to androgens. It is well known, for example, that there are differences in the amount of body hair between ethnic groups with identical plasma testosterone levels.[85] The hair growth, on the other hand, may have been initiated by intermittent increases in androgen production and, once initiated, may be maintained by a lower androgen load. Although the androgen load is important, the final determinant of androgen responsiveness rests at the cellular level. For example, in the testicular feminization syndrome (a condition characterized by male genotype and a female phenotype), the testosterone production rate is within the normal male range, but there are no masculinizing effects.[86,87]

It is clear that there is a complex relationship between androgen-dependent events and androgen load. There is evidence to indicate that a measure of this interaction is reflected in the excretion rate of 5α-androstane-3α,17β-diol, a metabolite of testosterone.[88] The hypothesis derives from the observation that testosterone is metabolized to 5α and 5β metabolites in the liver, whereas in extrahepatic tissues such as the skin, the 5α metabolite is mainly produced.[89] Thus, in men, or in women in whom there are clinical signs of probable androgen excess, such as hirsutism or acne vulgaris, there is an increase in the amount of 5α-androstanediol excreted.[88] Taking into account that both in men and hirsute women there is a substantial extrahepatic clearance of testosterone, it is reasonable to assume that the observed increase in the amount of 5α-androstanediol produced reflects the extrahepatic 5α-reductase activity to some extent. As the latter is deemed to be an important metabolic step in androgenic activity at the level of target tissues,[90] the assessment of 5α-androstanediol excretion may prove to be of great importance in evaluating androgenicity.

In conclusion, we may say that although we still do not know why the ovaries or adrenal glands in some women secrete what appear to be excessive amounts of androgen, nevertheless, our understanding of normal and abnormal androgen secretion in the human female has advanced considerably, compared with only a few years ago. It will no doubt continue to do so now that more sophisticated and penetrating techniques are at hand.

This chapter has reviewed briefly and, therefore, selectively one of the major areas of steroid endocrinology. It has not been possible to consider many of the relevant topics in any detail, but this should not be taken to imply that they are of less significance than those which have been discussed. One of the most important areas, namely, the uptake, intracellular transport, and metabolism of androgens is currently attracting intensive study and no picture of androgen metabolism will be complete until this aspect is properly understood. It seems likely that this goal may now be within sight and a discussion of this area will form an integral and major part of future chapters on androgen metabolism.

REFERENCES

1. **Dorfman, R. I. and Shipley, R. A.,** *Androgens,* John Wiley & Sons, New York and Chapman and Hall, Ltd., London, 1956.
2. **Segaloff, A. and Gabbard, R. B.,** Steroid structure and androgenicity, *Endocrinology,* 71, 949, 1962.
3. **Dorfman, R. I. and Dorfman, A. S.,** Assay of androgens administered by inunction to the chick's comb, *Acta Endocrinol.,* Suppl. 74, 1963.
4. **Cavallero, C. and Ofner, P.,** Relative effectiveness of various steroids in an androgen assay using the exorbital lacrimal gland of the castrated rat. II. C_{19}-steroids of the 5α-androstane series, *Acta Endocrinol.,* 55, 131, 1967.
5. **Cavallero, C.,** Relative effectiveness of various steroids in an androgen assay using the exorbital lacrimal gland of the castrated rat. I. Δ^4-3-ketones and Δ^5-3β-hydroxysteroids, *Acta Endocrinol.,* 55, 119, 1967.
6. **Segaloff, A., Horwitt, B. N., Carabasi, R. A., Murison, P. J., and Schlosser, J. V.,** Hormonal therapy in cancer of the breast. VIII. The effect of dihydrotestosterone (androstanolone) on clinical course and hormonal excretion, *Cancer,* 8, 82, 1955.
7. **Segaloff, A., Weeth, J. B., Rongone, E. L., and Murison, P. J.,** Hormonal therapy in cancer of the breast. XVIII. Effect of androstanediol dipropionate (5α-androstan-3α,17β-diol dipropionate) therapy on clinical course and hormonal excretion, *Cancer,* 14, 707, 1961.
8. **Dorfman, R. I.,** Steroid hormones in gynecology, *Obstet. Gynecol. Surv.,* 18, 65, 1963.
9. **Segaloff, A., Gordon, D. L., Bowers, C. Y., Schlosser, J. V., and Murison, P. J.,** Hormonal therapy in cancer of the breast. XI. The effect of dehydroepiandrosterone on clinical course and hormonal excretion, *Cancer,* 10, 1114, 1957.
10. **Segaloff, A., Horwitt, B. N., Gordon, D., Murison, P. J., and Schlosser, J. V.,** Hormonal therapy in cancer of the breast. IV. Effect of androstenediol on clinical course and hormonal excretion, *Cancer,* 5, 1179, 1952.
11. **Rosenfield, R. L.,** Plasma 17-ketosteroids and 17-beta hydroxysteroids in girls with premature development of sexual hair, *J. Pediatr.,* 79, 260, 1971.
12. **Drucker, W. D., Blumberg, J. M., Gandy, H. M., David, R. R., and Verde, A. L.,** Biologic activity of dehydroepiandrosterone sulphate in man, *J. Clin. Endocrinol.,* 35, 48, 1972.
13. **Savard, K., Marsh, J. M., and Rice, B. F.,** Gonadotropins and ovarian steroidogenesis, in *Recent Progress in Hormone Research,* Vol. 21, Pincus, G., Ed., Academic Press, New York, 1965, 285.
14. **Ryan, K. J. and Smith, O. W.,** Biogenesis of steroid hormones in the human ovary, in *Recent Progress in Hormone Research,* Vol. 21, Pincus, G., Ed., Academic Press, New York, 1965, 367.
15. **Grant, J. K.,** The biosynthesis of adrenocortical steroids, *J. Endocrinol.,* 41, 111, 1968.
16. **Griffiths, K. and Cameron, E. H. D.,** Steroid biosynthetic pathways in the human adrenal, in *Advances in Steroid Biochemistry and Pharmacology,* Vol. 2, Briggs, M. H., Ed., Academic Press, New York, 1970, 223.
17. **Wieland, R. G., De Courcy, C., Levy, R. P., Zala, P., and Hirschmann, H.,** C_{19} steroids and some of their precursors in blood from normal human adrenals, *J. Clin. Invest.,* 44, 159, 1965.
18. **Gandy, H. M. and Peterson, R. E.,** Measurement of testosterone and 17-ketosteroids in plasma by the double isotope dilution derivative technique, *J. Clin. Endocrinol.,* 28, 949, 1968.
19. **Baird, D. T.,** Steroids in blood reflecting ovarian function, in *Control of Gonadal Steroid Secretion,* Pfizer Medical Monographs 6, Baird, D. T. and Strong, J. A., Eds., Edinburgh University Press, Edinburgh, 1971, 175.
20. **James, V. H. T.,** The excretion of individual 17-ketosteroids by normal females, *J. Endocrinol.,* 22, 195, 1961.
21. **Schubert, K.,** Steroidstoffwechsel bei Brustkrebs. II. Zur Bestimmung der Testosterone-und Δ^4-Androsten-3, 17-dion-Ausschiedung bei Hormonbehandlung, *Acta Endocrinol.,* 27, 36, 1958.
22. **Korenman, S. G., Wilson, H., and Lipsett, M. B.,** Testosterone production rates in normal adults, *J. Clin. Invest.,* 42, 1753, 1963.
23. **Korenman, S. G. and Lipsett, M. B.,** Is testosterone glucuronide uniquely derived from plasma testosterone? *J. Clin. Invest.,* 43, 2125, 1964.
24. **Hennam, J. F., Collins, W. P., and Sommerville, I. F.,** The metabolism of androgenic steroids in women, *Acta Endocrinol.,* Suppl. 155, 120, 1971.
25. **Ito, T. and Horton, R.,** The source of plasma dihydrotestosterone in man, *J. Clin. Invest.,* 50, 1621, 1971.
26. **Mahoudeau, J. A., Bardin, C. W., and Lipsett, M. B.,** The metabolic clearance rate and origin of plasma dihydrotestosterone in man and its conversion to the 5α-androstanediols, *J. Clin. Invest.,* 50, 1338, 1971.
27. **Grodin, J. M., Siiteri, P. K., and MacDonald, P. C.,** Source of estrogen in postmenopausal women, *J. Clin. Endocrinol.,* 36, 207, 1973.
28. Clinical Staff Conference, Physiologic basis of disorders of androgen metabolism, *Ann. Intern. Med.,* 68, 1328, 1968.
29. **Goldzieher, J. W. and Williams, M. C.,** Urinary testosterone glucuronide as a measure of endogenous testosterone production, *Acta Endocrinol.,* 67, 371, 1971.
30. **Loraine, J. A.,** *The Clinical Application of Hormone Assay,* E & S Livingstone, Edinburgh, 1958.
31. **James, V. H. T., Peart, W. S., and Iles, S. D.,** Steroid excretion in idiopathic hirsutism, *J. Endocrinol.,* 24, 463, 1962.

32. **Adlercreutz, H., Lukkainen, T., and Svanborg, A.,** Simultaneous determination of urinary estrogens, individual 11-deoxy-17-ketosteroids, total 17-ketosteroids, total 17-ketogenic steroids, pregnanediol and plasma cortisol during the menstrual cycle and in early pregnancy, *Ann. Med. Exp. Fenn.,* 45, 277, 1967.
33. **Jänne, O.,** Urinary excretion of mono- and disulphates of C_{19} and C_{21} steroids with reference to the menstrual cycle, *Acta Endocrinol.,* 67, 316, 1971.
34. **Ismail, A. A. A., Harkness, R. A., and Loraine, J. A.,** Some observations on the urinary excretion of testosterone during the normal menstrual cycle, *Acta Endocrinol.,* 58, 685, 1968.
35. **Vihko, R.,** Gas chromatographic mass spectrometric studies on solvolyzable steroids in human peripheral plasma, *Acta Endocrinol.,* Suppl. 109, 1966.
36. **Rosenfield, R. L. and Otto, P.,** Androstenediol levels in human peripheral plasma, *J. Clin. Endocrinol.,* 35, 818, 1972.
37. **Strickland, A. L., Apland, M., and Burton, J.,** Determination of serum testosterone and androstanediol by competitive protein binding, *Steroids,* 21, 27, 1973.
38. **Ito, T. and Horton, R.,** Dihydrotestosterone in human peripheral plasma, *J. Clin. Endocrinol.,* 31, 362, 1970.
39. **Bardin, C. W. and Lipsett, M. B.,** Testosterone and androstenedione blood production rates in normal women and women with idiopathic hirsutism or polycystic ovaries, *J. Clin. Invest.,* 46, 891, 1967.
40. **Saez, J. M., Forest, M. G., Morera, A. M., and Bertrand, J.,** Metabolic clearance rate and blood production rate of testosterone and dihydrotestosterone in normal subjects, during pregnancy, and in hyperthyroidism, *J. Clin. Invest.,* 51, 1226, 1972.
41. **Horton, R. and Tait, J. F.,** In vivo conversion of dehydroisoandrosterone to plasma androstenedione and testosterone in man, *J. Clin. Endocrinol.,* 27, 79, 1967.
42. **Wang, D. Y., Bulbrook, R. D., Sneddon, A., and Hamilton, T.,** The metabolic clearance rates of dehydroepiandrosterone, testosterone and their sulphate esters in man, rat and rabbit, *J. Endocrinol.,* 38, 307, 1967.
43. **Rosenfeld, R. S., Hellman, L., Roffwarg, H., Weitzman, E. D., Fukushima, D. K., and Gallagher, T. F.,** Dehydroepiandrosterone is secreted episodically and synchronously with cortisol by normal man, *J. Clin. Endocrinol.,* 33, 87, 1971.
44. **De Jongh, F. H. and Van der Molen, H. J.,** Determination of dehydroepiandrosterone and dehydroepiandrosterone sulphate in human plasma using electron capture detection of 4-androstene-3,6,17-trione after gas-liquid chromatography, *J. Endocrinol.,* 53, 461, 1972.
45. **Tait, J. F. and Burstein, S.,** In vivo studies of steroid dynamics in man, in *The Hormones,* Vol. 5, Pincus, G., Thiman, K. V., and Astwood, E. B., Eds., Academic Press, London, 1964.
46. **Bardin, C. W. and Mahoudeau, J. A.,** Dynamics of androgen metabolism in women with hirsutism, *Ann. Clin. Res.,* 2, 251, 1970.
47. **Vermeulen, A., Verdonck, L., Van Der Straeten, M., and Orie, N.,** Capacity of the testosterone binding globulin in human plasma and influence of specific binding of testosterone on its metabolic clearance rate, *J. Clin. Endocrinol.,* 29, 1470, 1969.
48. **Lipsett, M. B., Wilson, H., Kirschner, M. A., Korenman, S. G., Fishman, L. M., Sarfaty, G. A., and Bardin, C. W.,** Studies on Leydig cell physiology and pathology; secretion and metabolism of testosterone, in *Recent Progress in Hormone Research,* Vol. 22, Pincus, G., Ed., Academic Press, New York, 1966, 245.
49. **Abraham, G. E., Lobotsky, J., and Lloyd, C. W.,** Metabolism of testosterone and androstenedione in normal and ovariectomised women, *J. Clin. Invest.,* 48, 696, 1969.
50. **Kellie, A. E. and Smith, E. R.,** Renal clearance of 17-oxosteroid conjugates found in human peripheral plasma, *Biochem. J.,* 66, 490, 1957.
51. **Plager, J. E.,** The binding of androsterone sulphate, etiocholanolone sulphate and dehydroepiandrosterone sulphate by human plasma protein, *J. Clin. Invest.,* 44, 1234, 1965.
52. **Mercier-Bodard, C., Alfsen, A., and Baulieu, E. E.,** Sex steroid binding plasma protein (SBP), *Acta Endocrinol.,* Suppl. 147, 204, 1970.
53. **Vermeulen, A. and Verdonck, L.,** Studies on the binding of testosterone to human plasma, *Steroids,* 11, 609, 1968.
54. **Kato, T. and Horton, R.,** Studies of testosterone binding globulin, *J. Clin. Endocrinol.,* 28, 1160, 1968.
55. **Vermeulen, A., Stoica, T., and Verdonck, L.,** The apparent free testosterone concentration, an index of androgenicity, *J. Clin. Endocrinol.,* 33, 759, 1971.
56. **Forest, M. G., Rivarola, M. A., and Migeon, C. J.,** Percentage binding of testosterone, androstenedione and dehydroisoandrosterone in human plasma, *Steroids,* 12, 323, 1968.
57. **Pearlman, W. H.,** Measurement of testosterone binding sites, *Acta Endocrinol.,* Suppl. 147, 225, 1970.
58. **Rosner, W.,** A simplified method for the quantitative determination of testosterone-estradiol binding globulin activity in human plasma, *J. Clin. Endocrinol.,* 34, 985, 1972.
59. **Pearlman, W. H., Crepy, O., and Murphy, M.,** Testosterone binding levels in the serum of women during the normal menstrual cycle, pregnancy and the post-partum period, *J. Clin. Endocrinol.,* 27, 1012, 1967.
60. **Jacobsohn, G. M. and Levenberg, D.,** Solubilities of 17-ketosteroids in water, *Steroids,* 4, 849, 1964.
61. **Baird, D. T., Horton, R., Longcope, C., and Tait, J. F.,** in *Recent Progress in Hormone Research,* Vol. 25, Astwood, E. B., Ed., Academic Press, New York, 1969, 611.

62. **Baulieu, E. E., Alberga, A., Jung, I., Lebeau, M. C., Mercier-Bodard, C., Milgrom, E., Raynaud, J. P., Raynaud-Jammet, C., Rochefort, H., Truong, H., and Robel, P.,** Metabolism and protein binding of sex steroids in target organs: an approach to the mechanism of hormone action, in *Recent Progress in Hormone Research,* Vol. 27, Astwood, E. B., Ed., Academic Press, New York, 1971, 351.
63. **Horton, R. and Tait, J. F.,** Androstenedione production and interconversion rates measured in peripheral blood and studies on the possible site of its conversion to testosterone, *J. Clin. Invest.,* 45, 301, 1966.
64. **Gandy, H. M.,** Androgens, in *Endocrinology of Pregnancy,* Fuchs, F. and Klopper, A., Eds., Harper & Row, New York, 1971.
65. **Demisch, K., Grant, J. K., and Black, W. P.,** Plasma testosterone in late pregnancy and after delivery, *J. Endocrinol.,* 42, 477, 1968.
66. **Murphy, B. E. P.,** Androgen fractionation in normal, pregnant and hirsute women, Program of the 53rd meeting of the Endocrine Society, San Francisco, Abstract No. 202, 1971.
67. **Mizuno, M., Lobotsky, J., Lloyd, C. W., Kobayashi, T., and Murasawa, Y.,** Plasma androstenedione and testosterone during pregnancy and in the newborn, *J. Clin. Endocrinol.,* 28, 1133, 1968.
68. **Rivarola, M. A., Forest, M. G., and Migeon, C. J.,** Testosterone, androstenedione and dehydroepiandrosterone in plasma during pregnancy and at delivery: Concentration and protein binding, *J. Clin. Endocrinol.,* 28, 34, 1968.
69. **Sjovall, K.,** Gas chromatographic determination of steroid sulphates in plasma during pregnancy, *Ann. Clin. Res.,* 2, 393, 1970.
70. **Korenman, S. G., Kirschner, M. A., and Lipsett, M. B.,** Testosterone production in normal and virilised women and in women with the Stein-Leventhal syndrome or idiopathic hirsutism, *J. Clin. Endocrinol.,* 25, 798, 1965.
71. **Rivarola, M. A., Saez, J. M., Jones, H. W., Jr., Jones, G. S., and Migeon, C. J.,** The secretion of androgens by the normal, polycystic and neoplastic ovaries, *Johns Hopkins Med. J.,* 121, 82, 1967.
72. **Prunty, F. T. G.,** Hirsutism, virilism and apparent virilism and their gonadal relationship, *J. Endocrinol.,* 38, 85, 1967.
73. **Brooksbank, B. W. L.,** Endocrinological aspects of hirsutism, *Physiol. Rev.,* 41, 623, 1961.
74. **Ismail, A. A. A., Davidson, D. W., Kirkham, K. E., and Loraine, J. A.,** Studies on sex hormone excretion in normal and hirsute women, *Acta Endocrinol.,* 61, 283, 1969.
75. **Lloyd, C. W., Lobotsky, J., Segre, E. J., Kobayashi, T., Taymor, M. L., and Batt, R. E.,** Plasma testosterone and urinary 17-ketosteroids in women with hirsutism and polycystic ovaries, *J. Clin. Endocrinol.,* 26, 314, 1966.
76. **Rosenfield, R. L.,** Plasma testosterone binding globulin and indexes of the concentration of unbound plasma androgens in normal and hirsute subjects, *J. Clin. Endocrinol.,* 32, 717, 1971.
77. **Bardin, C. W., Hembree, W. C., and Lipsett, M. B.,** Suppression of testosterone and androstenedione production rates with dexamethasone in women with idiopathic hirsutism and polycystic ovaries, *J. Clin. Endocrinol.,* 28, 1300, 1968.
78. **Kirschner, M. A., Bardin, C. W., Hembree, W. C., and Ross, G. T.,** Effect of estrogen administration on androgen production and plasma luteinizing hormone in hirsute women, *J. Clin. Endocrinol.,* 30, 727, 1970.
79. **Ettinger, B., von Werder, K., Thenaers, G. C., and Forsham, P. H.,** Plasma testosterone stimulation – suppression dynamics in hirsute women, *Am. J. Med.,* 51, 170, 1971.
80. **Deller, J. J., Wegienka, L. C., Conte, N. F., Rosner, J. M., and Forsham, P. H.,** Testosterone metabolism in idiopathic hirsutism, *Ann. Intern. Med.,* 63, 369, 1965.
81. **Nichols, T., Nugent, C. A., and Tyler, F. H.,** Glucocorticoid suppression of urinary testosterone excretion in patients with idiopathic hirsutism, *J. Clin. Endocrinol.,* 26, 79, 1966.
82. **Reforzo-Membrives, J., Rocca, D. Z., Enriori, C. L., and Mazza, S. C.,** Adrenal and ovarian tests in female hirsutism, *Acta Endocrinol.,* 55, 685, 1967.
83. **Aakvaag, A., Vogt, J. H., and Fylling, P.,** Plasma and urinary androgens in hirsute women during adrenal and ovarian suppression, *Acta Endocrinol.,* 64, 103, 1970.
84. **Kirschner, M. A. and Jacobs, J. B.,** Combined ovarian and adrenal vein catheterisation to determine the site(s) of androgen overproduction in hirsute women, *J. Clin. Endocrinol.,* 33, 199, 1971.
85. **Kobayashi, T., Lobotsky, J., and Lloyd, C. W.,** Plasma testosterone and urinary 17-ketosteroids in Japanese and Occidentals, *J. Clin. Endocrinol.,* 26, 610, 1966.
86. **Southren, A. L., Tochimoto, S., Carmody, N. C., and Isurugi, K.,** Plasma production rates of testosterone in normal adult men and women and in patients with the syndrome of feminising testes, *J. Clin. Endocrinol.,* 25, 1441, 1965.
87. **Judd, H. L., Hamilton, C. R., Barlow, J. J., Yen, S. S. C., and Kliman, B.,** Androgen and gonadtrophin dynamics in testicular feminisation syndrome, *J. Clin. Endocrinol.,* 34, 229, 1972.
88. **Mauvais-Jarvis, P., Charransol, G., and Bobas-Masson, F.,** Simultaneous determination of urinary androstanediol and testosterone as an evaluation of human androgenicity, *J. Clin. Endocrinol.,* 36, 452, 1973.
89. **Mauvais-Jarvis, P., Bercovici, J. P., Crepy, O., and Gauthier, F.,** Studies on testosterone metabolism in subjects with testicular feminisation syndrome, *J. Clin. Invest.,* 49, 31, 1970.
90. **Wilson, J. D.,** Recent studies on the mechanism of action of testosterone, *N. Engl. J. Med.,* 287, 1284, 1972.

THE BIOCHEMISTRY OF INFERTILITY

D. Davies

TABLE OF CONTENTS

I. INTRODUCTION

The management of infertility was, until relatively recently, largely aimed at seeking lesions that could be corrected surgically, but over the last 20 years the development of increasingly sophisticated and sensitive techniques for hormone assay has led to both better understanding of the processes involved in follicular maturation and ovulation and more efficient methods of diagnosis of abnormalities of these processes. The result of these developments is that the team involved in patient management has increased from the gynecologist assisted by the radiologist and pathologist to include the endocrinologist, biochemist, and most recently of all the geneticist. The advent of drugs and hormone preparations with applications in the therapy of infertility has allowed treatment to be extended to patients who previously could be offered little hope of pregnancy.

The causes of infertility are numerous, and the pursuit of a definitive diagnosis of the etiology is still a lengthy process, requiring persistence and patience in both patient and doctor. In particular the biochemical investigations must frequently be performed in a serial fashion over a considerable period of time to establish the pattern of function, for, as with all dynamic endocrine situations, random sampling of blood or urine for hormone estimation is useless, with specific exceptions. This requires considerable cooperation from the patient and much disturbance of her normal daily routine. The intellectual capacity to cope with relatively complex investigative and therapeutic programs is essential in both the patient and her husband, along with a certain dedication to pursue the matter to its conclusion.

II. PRESENTATION

Patients present with either regular menstrual pattern and infertility after a long marriage or with disturbed or absent menstruation, sometimes with other abnormal features such as hirsutism, galactorrhea, and loss of, or failure of development of, secondary sex characteristics. An initial assessment of the diagnosis may be made on this basis.

A. Regular Menstruation

1. Mechanical Factors

All patients with abnormalities of the reproductive canal that interfere with the process of fertilization may be included in this group. Such factors should be excluded in all cases, whether menstruation is normal or abnormal. During the close questioning of the patient it is necessary to recognize the few individuals who have passed the physiological menopause, who show the progressive loss of fertility that occurs with age, who have been subjected to a radiation menopause, or who even have had surgical oophorectomy or hysterectomy who surprisingly still present to an infertility clinic. It is possible to subdivide patients into several groups.

a. Technique

A surprising number of patients present with problems related to ignorance of the correct method in intercourse, with frigidity, or with premature ejaculation in the husband, and these patients should be sought out at an early stage to prevent extensive and unnecessary investigation. Other patients may be infertile because of ignorance of the correct timing of intercourse in relation to ovulation, and these too should be found and excluded. Finally, it is of value to obtain at least one seminal analysis on the husband at the beginning of the investigation, as it is pointless to continue studies upon the wife if her spouse is infertile, unless his condition can be corrected.

b. Congenital Abnormality

Abnormal development of the uterus, failure of canalization of the fallopian tubes, malformation of the vagina, or absence of any of these organs may lead to obstruction of the reproductive canal and prevent fertilization and conception.

c. Tubal Obstruction

This usually occurs after pelvic infection such as gonorrhea, tuberculosis, or local disease such as appendicitis. The fibrosis and adhesions produced may occlude the fallopian tube at its fimbrial orifice or within its course to the uterus. The tube may also be involved in endometriosis.

d. Disorders of the Endometrium

Endometritis, particularly tuberculous endometritis, is still a significant cause of infertility. Similarly, other causes of abnormal endometrium or menorrhagia may be responsible, and should be excluded by endometrial biopsy and curettage. Rarely it may be found that previous curettage has led to the formation of uterine synechiae causing obstruction of the uterine cavity. Uterine tumors such as fibroids may produce distortion of the uterine cavity sufficient to impede passage of spermatozoa, but more particularly to interfere with nidation of the fertilized ovum.

e. Disorders of the Cervix

Cervicitis is well recognized as a cause of infertility as a result of the abnormal mucus produced. This may be too scanty, with too high a viscosity and too low a pH for survival and capacitation of the spermatozoa.[83]

f. Vaginitis

Infection within the vagina similarly leads to pH and toxic changes which, as with the "hostile cervix," lead to an unsuitable environment for spermatozoa.

g. Other Causes

Under this heading may be included remaining causes of infertility other than anovulation or failure of hormonal production. There has been some evidence in recent years that antibodies against sperm proteins and polypeptides may be responsible for infertility,[39,40,85,115] and there are cases in which no abnormality can be discovered, previously described as having "incompatible" ova and sperm, in which such an antigen-antibody reaction could account for the failure to conceive. Antibodies to spermatozoa have been described in both serum and cervical secretions.

2. Infrequent Ovulation or Anovulation

Some patients may be shown to have a failure to ovulate in many or all their menstrual cycles even when cycles are regular. This may be recognized by a failure of the rise of pregnanediol excretion in the urine, with a monophasic temperature chart showing no evidence of the midcycle rise usually seen in conjunction with ovulation. Similarly, there will be no acute change in the spinnbarkeit of the cervical mucus or in the karyopyknotic index of the lateral vaginal wall smear. If an endometrial biopsy were taken at these times, it would show a persistence of proliferative phase appearances. In those patients with infrequent rather than absent ovulation, the problem of infertility may often be overcome by simple instruction in the keeping of a basal body temperature chart, thus enabling more satisfactory timing of intercourse. This may be the only treatment necessary.

The failure of ovulation in the presence of normal estrogen production in the proliferative phase is almost certainly due to the failure of the midcycle peak of Luteinizing Hormone (LH) release. Treatment will depend on the possibility of stimulating LH release either by Clomiphene or Luteinizing Hormone Releasing Hormone (LHRH), or alternatively supplying an exogenous stimulus for luteinization in the form of Human Chorionic Gonadotrophin (HCG).

The possibility of luteinization with failure or delay of release of the ovum has been postulated in the polycystic ovary syndrome, and although the typical presentation is for irregular menstruation or amenorrhea, occasional patients present with regular menstruation. These patients should be recognized by the presence of other features of the syndrome and the typical steroid excretion changes.

3. Inadequate Luteal Phase

In the normal menstrual cycle the luteal phase following ovulation lasts approximately 14 days, but in some patients the interval is rather less, and estrogen and pregnanediol excretion fall prematurely, leading to menstruation. This may happen regularly in some subjects and is thought to be due to inadequate function of the corpus luteum, which fails prematurely with accompanying failure of progesterone secretion and with endometrial degeneration. Basal body temperature is not sustained and falls before the 8th to 10th day of the secretory phase.

In other patients the length of the luteal phase may be virtually normal, but endometrial biopsy reveals inadequate secretory phase endometrium,[38,50,54] which implies four possible causes.[64] The first is inadequate progesterone production by the corpus luteum. The urinary excretion of pregnanediol, the major urinary excretory product of progesterone, is low. In normal individuals it should exceed 2 mg/24 hr each day between days 19 and 25 of the cycle, and the total cumulative excretion over the whole cycle should exceed 25 mg. The total cycle excretion seems to be a more sensitive index than the daily figure. In patients receiving exogenous FSH to induce ovulation, the occurrence of inadequate luteal phase is well recognized, and under these circumstances the low or shortened phase of pregnanediol excretion is due to an inadequate dosage of LH.[21]

The second cause is inadequate estrogenic preparation of the endometrium prior to ovulation. This may be due to low FSH secretion with inadequate follicular development, which may also prevent development of the corpus luteum.[118] Third, an excess of estrogen in relation to progesterone secretion may lead to similar difficulties. Both these situations are more likely to occur in patients under treatment with human gonadotrophin, but may arise in the spontaneously menstruating patient. Finally, the lack of response of the endometrium to normal hormone production is very rare indeed, and seen usually in the presence of endometritis or neoplastic change.

Although the presence of a diminished urinary pregnanediol excretion and an abnormal basal body temperature record may suggest the presence of an inadequate luteal phase, it may be necessary to confirm this by endometrial biopsy. This should obviously be done as late as possible in the cycle to assess the full effect of hormone secretion, preferably about the 26th day of a 28-day cycle or in relation to the expected date of onset of menstruation, and is best obtained from the uterine fundus.

B. Oligomenorrhea

Menstrual cycles that are reduced in frequency and prolonged in duration are associated with disturbances of hormonal function, which may be due to "inadequate" function of some portion of the hypothalamo-pituitary-ovarian axis or to distinct pathology within the axis.

1. Irregular Periods

Where menstruation occurs irregularly, it may be found on detailed investigation that although ovulation is delayed, the luteal phase may be normal. The onset of the output of FSH, the consequent follicular development, and estrogen secretion may be delayed, leading to a prolonged proliferative phase. In these circumstances the problem may be that of ensuring correct timing of intercourse in order to achieve pregnancy. This would be achieved in most cases by the use of the simple basal body temperature chart.

2. Obesity

In obese patients, particularly if the obesity is of fairly recent onset and rapid development, oligomenorrhea is a common phenomenon and is associated with infertility. FSH secretion may be normal or reduced and the LH peak absent or delayed. The mechanism for this disturbance is far from understood, but appears to be hypothalamic in origin. Weight reduction will usually result in return of normal menstrual habit and fertility.

3. Polycystic Ovary Syndrome

When Stein and Leventhal described the syndrome bearing their name,[117] the main features were given as hirsutism, obesity, and secondary amenorrhea, associated with large polycystic ovaries. However, it has become more apparent that irregular and infrequent menstruation is more typical than amenorrhea. The periods are often painful. Hirsutism is usually present, and develops from the menarche in a progressive fashion. Menstrual cycles are either anovulatory or ovulation may be delayed, possibly by the sclerotic ovarian capsule, leading to infertility.[34,53,95]

In a literature review,[53] infertility was recorded in 74% of reported cases, with a range of incidence from 35 to 94%. In many cases infertility is relative rather than absolute, and pregnancy may be achieved after a considerable period of time or after minimal treatment. The mechanism of the infertility possibly depends on the severity of the hormonal disturbance present. The origin of the disorder has been suggested as arising at one or both of two possible sites. The first is at the hypothalamic level. In experimental animals it has been shown that if electrical hypothalamic injury is induced or injections of androgens are given in newborn rats then polycystic ovaries develop at maturity with interstitial hypertrophy as seen in human polycystic disease.[5] There is loss of the normal cyclical secretion of gonadotrophins, although basal tonic secretion occurs. The pattern of gonadotrophin secretion resembles that described in males with pulsatile secretion over the 24 hr.[10,90] (It is of interest that castration of male animals at birth leads to cyclical release of gonadotrophin.[94]) This persistence of gonadotrophin secretion with both FSH and LH continuing at basal levels leads to the polycystic change. It is known that polycystic ovaries may develop in other situations, e.g., following injections of gonadotrophin,[44] very rarely in adrenal hypersecretory states,[61,87] in virilizing adrenal adenoma,[63] in congenital adrenal hyperplasia,[63,65] and in virilizing ovarian tumors.[18] In these circumstances there is suppression of the normal cyclical surges of pituitary and hypothalamic activity, either by the exogenous gonadotrophin (a fact to be borne in mind during gonadotrophin therapy for infertility) or by the raised levels of androgens, although if the latter reach high values complete gonadotrophin suppression and ovarian atrophy are the rule.

The second possible site of origin is the ovary, which has been shown to be capable of producing the abnormal quantities of androgens found in this condition. There is still a considerable argument as to whether the major source of androgen production is the adrenal cortex or the ovary.[72,73,100] The pathway of estrogen production from pregnenolone is shown in Figure 1. Pregnenolone is formed from acetate within the ovary, as it is in the adrenal cortex, with which it shares the enzyme pathways for the production of androstenedione, which is the major androgen produced by the adrenal. Estrogen production rates have been calculated to be normal for follicular phase activity, but the fact that there is a greater rise in estrogen secretion following gonadotrophin therapy than seen in normal ovaries[51] is thought to be due to chronic hyperstimulation, as may be expected from the continued basal production of FSH and LH. Response to ACTH by an increase in urinary 17-oxosteroids is normal, but injection of HCG causes a rise in urinary 17-oxosteroids as well as the increase in estrogens. This was suggested as representing evidence that

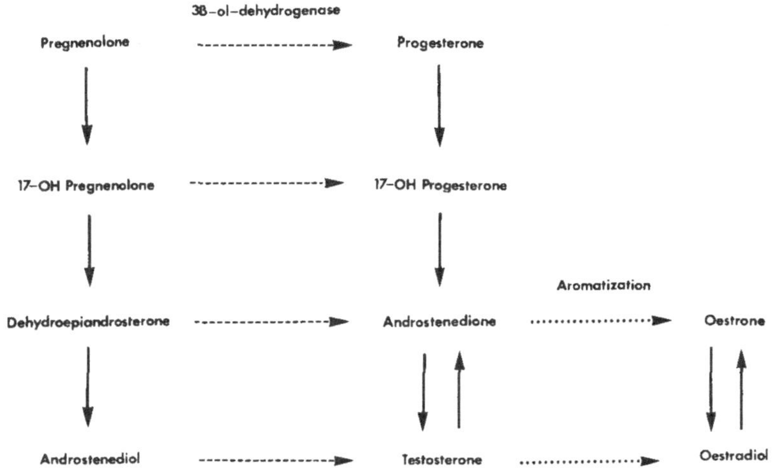

FIGURE 1. Biosynthetic pathway of estrogens.

the primary disorder is within the ovary, but it has been shown that adrenal response to HCG may be obtained in some patients with ovarian disease.[78] Although the administration of dexamethasone will suppress the urinary 17-oxosteroids in many patients, this may represent the suppression of only the adrenal contribution without effect upon the ovarian portion, although this too may be reduced as there is a definite effect on gonadotrophin production from the administration of corticosteroids, due to the close interaction of the hypothalamo-pituitary-adrenal and hypothalamo-pituitary-ovarian systems. The finding of 11-ketopregnentriol in the urine of affected individuals, when this is not usually found in normals, suggests that this is adrenal in origin, as 11-hydroxylation is not known to occur in the ovary.[109,110] The compound is seen in some cases of adrenal hyperplasia, for it is an excretory metabolite of 21-deoxycortisol. Recent ovarian suppression studies using ampicillin are in too early a stage to assess whether the suppression is specifically ovarian or whether the adrenal cortex is also affected.[120]

The ovary, however, is capable of producing the necessary quantities of androgen seen in polycystic ovary syndrome. The theca and granulosa cells produce estrogens in vivo and in vitro by way of dehydroepiandrosterone, androstenedione, and testosterone,[105,106] and the stroma produces largely androgens using the same pathways.[107] In the polycystic ovary two abnormalities of steroid production occur, as assessed from analysis of follicular cyst fluid. In the first type there appears to be an abnormality of aromatization with reduction in estrogen secretion, and an accumulation of androstenedione and testosterone.[3,82,98] This type seems likely to lead to accumulation of testosterone in particular and would be expected to be associated with hirsutism. Certainly, plasma levels of androstenediol and testosterone are increased. In the second type cyst fluid contains an excess of dehydroepiandrosterone and 17-hydroxypregnenolone, but there are reduced amounts of estrogen and of androstenedione.[3,82] This suggests an abnormality of 3β-ol-dehydrogenase, and here there would be an expected reduction in the amount of androgens produced (Figure 1). It has been shown that the deficiency of the 3β-ol-dehydrogenase enzyme may occur concurrently in both ovary and adrenal cortex.[4]

Whatever the basic enzyme defect, the changes in the polycystic ovary are thought to be based initially on the hypothalamic disturbance, with the possibility that the ovarian enzyme changes are induced by chronic hyperstimulation. The sclerocystic changes with thickening of the tunica albuginea are typically found in those individuals with hirsutism, and it is probable that these subjects have a block at the aromatization step. In those cases with a normal tunica, known to be associated with an absence of hirsutism,[113] it may be postulated that a 3β-ol-dehydrogenase defect exists.

The result in the patient is that menstruation is

irregular at best and in a number of individuals amenorrhea exists. The presence of high androgen levels leads to hirsutism and thickening of the tunica albuginea. This results in failure or delay of rupture of the mature follicle in those subjects with menstrual cycles, and so produces infertility. However, in some individual patients the above features appear to be present only to a mild degree, but infertility may be marked, suggesting that other factors are involved in producing the difficulty in conception. Therapy aimed at suppressing the abnormal androgen production and restoring normal cyclical gonadotrophin release seems to offer the best hopes of a successful conception.

4. Ovarian "Inadequacy"

In some individuals there is a history, usually dating back to the menarche, of scanty and irregular periods with vaginal bleeding occurring for 1 to 3 days, often with only scanty loss after the first day, and occurring at intervals of 2 to 6 months. These patients usually present in the "irregular period" group if they present with primary infertility, but they represent a problem group which is worthy of separate consideration if they are ever treated with the contraceptive pill. They have a very high risk of development of amenorrhea. The point at which "inadequacy" occurs is not always clear, and may be at any point in the hypothalamo-pituitary-ovarian axis. Some present the picture of hypothalamic failure and may recover spontaneously after several years; however, a number of fairly rare cases may show high plasma gonadotrophin levels and have the picture of a premature menopause.

5. Lack of Midcycle LH Peak

The patient with scanty periods may present difficulty in assessment of the criteria of ovulation because of the irregular length of the cycle, but in a few individuals careful serial studies have shown the absence of a midcycle peak of LH which would lead to a failure of ovulation. It is possible that the absence of the midcycle rise in LH is the reason for the delay in onset of menstruation, permitting a type of persistent "estrus." Menstruation can be precipitated by the administration of progestogens, showing that estrogen production is adequate to produce endometrial maturation and that oligomenorrhea is due to failure of corpus luteum formation. These studies need to be repeated over several months to be sure that this is a regular phenomenon, as it may occur intermittently with greater or lesser frequency in all women. Treatment here would be aimed at producing a rise in LH levels at a stage when the level of estrogen excretion had reached a point suggesting that adequate follicular maturation had occurred.

C. Secondary Amenorrhea

Patients who present with a history that menstruation has occurred at some stage in their life have demonstrated that the hypothalamo-pituitary-ovarian axis is capable of relatively normal function, and that the uterus is present and sensitive to the hormonal products of the ovary. Therefore, this suggests that the deficiency of the system is either acquired or due to the progression of an abnormality present in the past.

1. Psychogenic Causes

Apart from pregnancy this is probably the most common cause of secondary amenorrhea. It is usually short lived and may last only 2 or 3 months in any young woman exposed to an emotional stress such as a change of job, increase in responsibility, family bereavement, broken engagement, etc. In a few cases it may be more persistent, and amenorrhea may last for many months or even years. In such patients the levels of gonadotrophin are low or at lower proliferative phase levels. Estrogen production is low but compatible with early proliferative phase levels. Vaginal examination reveals a moderate degree of estrogenization and not usually the picture of the dry, poorly estrogenized vagina unless estrogen or gonadotrophin production has failed completely.

The extreme example of this condition is anorexia nervosa, in which the pattern of hypothalamic and hormonal failure is more complex and affected by the failure of adequate nutrition. There are changes in the rate of metabolic destruction and removal of hormones, as well as changes in protein binding and transport. Gonadotrophin levels are very low and in consequence production of estrogens is very low indeed, due to the failure of follicular maturation. When the condition has been present for some time the ovaries become small and rather sclerotic, and there are involutional changes in the uterus.

2. Systemic Disease

Secondary amenorrhea may occur in association with a number of systemic disorders, usually debilitating or metabolic diseases. The typical wasting diseases of tuberculosis, Crohn's disease, ulcerative colitis, or the collagen diseases, rheumatoid arthritis or systemic lupus erythematosus, are not uncommonly associated with amenorrhea. Poorly controlled diabetes mellitus or keto-acidosis may prevent normal menstruation, probably by interfering with normal hypothalamic control of gonadotrophin release. Similarly, malabsorption syndrome and steatorrhea may on occasion produce interruption of normal menstrual cyclical activity.

Endocrine diseases associated in particular with masculinizing tumors of the ovary or adrenal usually disrupt menstrual activity, and amenorrhea occurs in many patients with Cushing's syndrome and in Addison's disease. The excretion of large quantities of metabolites of cortisol in adrenal carcinoma should help to separate the patient with virilization due to this cause from those with secondary hirsutism, deepening voice, and clitoromegaly due to hilus cell tumors or arrhenoblastoma, or granulosa cell tumors and luteomas.

3. Iatrogenic Causes

The common iatrogenic reason for amenorrhea is the exposure of the patient to therapeutic doses of ionizing radiation. It is obvious that exposure of the pelvic area would result in failure of ovarian response to gonadotrophin. However, it is recognized that exposure of other parts of the body may result in amenorrhea, presumably due to inadequate screening of the pelvic organs from the scattered radiation. For example, in the author's personal series, irradiation of thoracic and cervical glands in Hodgkin's disease in a young woman has resulted in amenorrhea due to ovarian failure, with high levels of gonadotrophin in plasma responding very briskly to Clomiphene, but without change in estrogen excretion.

4. Ovarian Failure

Premature menopause due to primary ovarian failure may occur at any age, and such cases are well recognized but fortunately rare in young women. Such patients are identified by the low levels of estrogen and pregnanediol in the urine and high levels of gonadotrophin both in plasma and urine. The patient may complain of hot flushes typical of the menopause. Therapy is unlikely to meet with success. The condition may arise in patients who have previously fallen into the group classed as having "inadequate ovary," and often follows a history of menstrual periods with diminishing vaginal blood loss and of decreasing frequency.

5. Obesity

Secondary amenorrhea may occur in obesity, particularly if severe. There is usually a phase of oligomenorrhea gradually progressing to amenorrhea, which usually responds to adequate dietary restriction and weight loss, although menstruation often reappears after surprisingly little weight loss. Further weight loss is usually needed to restore fertility.

6. Polycystic Ovary Syndrome

In severe cases of the Stein-Leventhal syndrome, secondary amenorrhea is often present, and was one of the triad of symptoms in the original description.[117] Although less common than oligomenorrhea, this condition should not be overlooked in the investigation of secondary amenorrhea, and examination of the urine should reveal the characteristic steroid excretion pattern.

7. "Post-pill" Amenorrhea

A number of patients present with a history of amenorrhea following the contraceptive pill, which may have been taken for only a very short time.[108] There is often a previous history of irregular or scanty periods although a few individuals may have a perfectly normal past menstrual history. There appears to be a failure to reestablish cyclical hypothalamic activity, and although this is often a temporary state, several years may elapse before spontaneous activity is restored. It is essential that pituitary tumors coincidentally arising during contraceptive therapy should be excluded by skull radiography. In those who recover spontaneously, response to Clomiphene may be seen in the later months before activity returns, and one or more courses may precipitate the return of normal hypothalamo-pituitary-ovarian activity. In another group spontaneous activity does not return, and only treatment with gonadotrophin offers any chance of pregnancy, although a few individuals have returned to normal menstrual activity following several courses of therapy. Rarely, galactorrhea may be present.[42]

8. Pituitary Disease

Hypopituitarism results in secondary amenorrhea, where the pituitary disturbance is acquired after puberty. In the case of pituitary tumors such as chromophobe adenoma, the progressive compression of the normal pituitary tissue results in the progressive loss of anterior pituitary hormone production. There is a specific order in which the hormones are affected. Gonadotrophin function appears to be the most sensitive function and is typically lost first. It is therefore not unusual that the first sign of pituitary tumor is the appearance of amenorrhea. Other hormones are lost later, usually in this order: growth hormone, ACTH, and finally TSH. Even in Sheehan's syndrome, the earliest signs are the failure to lactate and the failure of return of menstruation, with atrophy of the secondary sexual characteristics and loss of libido. The loss of other pituitary hormones may only become obvious over several years. In the case of hormone-secreting tumors such as the eosinophile adenoma of acromegaly, the loss of other anterior pituitary hormones is uncommon and a late phenomenon, but once more, gonadotrophin function is the most sensitive.

Finally, there is a group of patients in whom no other abnormality of pituitary function or architecture can be found, and who appears to have a specific deficiency of gonadotrophin. In these individuals the plasma gonadotrophin levels are low, but other pituitary function tests including insulin stress tests, TRH, or metyrapone tests are quite normal. A rise in gonadotrophin production cannot be stimulated by either LHRH or Clomiphene. Gonadotrophin therapy is obviously the treatment of choice here but it is necessary to exercise some caution, as some of these patients are capable of rare bursts of spontaneous cyclical activity, and the author, like others, has had personal experience of patients who presented to the infertility clinic and were found to be in early pregnancy, several years after the last menstrual period.

No mention has been made of lesions of the hypothalamus. Tumors of this area are extremely rare, and at the present time investigation techniques are not sufficiently sensitive to allow the complete diagnosis of functional lesions of the hypothalamus. Releasing factors cannot yet be measured readily in man, but the use of LHRH and TRH may allow hypothalamic lesions to be suspected indirectly. It is possible that many of the patients with apparently isolated gonadotrophin deficiency will eventually be found to have just such a functional disorder of the hypothalamus or its connections.

The presence of galactorrhea may suggest that there is loss of prolactin-release-inhibiting hormone due to hypothalamic disturbances or may be due to prolactin-secreting tumors of the anterior pituitary. Persisting galactorrhea and amenorrhea after pregnancy (Chiari-Frommel syndrome) is well recognized,[17] but in the case of amenorrhea and galactorrhea in the nonparous patient, this may be associated with a pituitary tumor (Forbes-Albright syndrome) or there may be no evidence of tumor involvement (Argonz-del Castillo syndrome).[2,36,111] As prolactin assays are not yet readily available, the diagnosis is largely made on the typical clinical picture. At present the only available treatment to offer much chance of success in restoring fertility is that of gonadotrophin therapy.

D. Primary Amenorrhea

The presentation of patients who have never menstruated at any time leads the investigator to recognize that the abnormality present has been operative since before the time of puberty, and many such patients will not only be infertile but infantile also, with absent secondary sexual development. Such a situation may be due to either genetic or congenital abnormalities or to acquired lesions or tumors developing in childhood.

1. Genetic Disorders

Abnormal chromosome patterns usually produce their particular clinical features, which are recognizable on examination, but chromosome cultures will resolve the diagnosis in those cases not easily recognized. The patient with Turner's syndrome will have an XO chromosome complement. The characteristic appearance of the patient is of a short-statured individual, of female aspect but with sexual infantilism, low hair line, webbing of the neck, shield-shaped chest, wide-carrying angle, and primary amenorrhea which is due to ovarian agenesis. It is usually unnecessary to undertake detailed hormone studies, but if this is done, estrogens and pregnanediol in the urine are low, and urinary and plasma gonadotrophins are increased. Laparoscopy reveals streak ovaries.

Ovarian agenesis may occur as an isolated event, with normal body appearance, but with sexual infantilism. Hormonal excretion patterns are the same as in Turner's syndrome, but chromosome culture reveals a normal chromosome complement. Buccal smear will be chromatin positive, unlike Turner's syndrome in which the buccal smear is chromatin negative. As with Turner's syndrome, there is no hope of fertility in these patients.

The occurrence of chromosome mosaicism may lead to modified sex organ development. A not uncommon combination is that of the XY/XO mosaic. These patients are usually female in appearance, but may develop a testis on one side and an ovary on the other. Often the male organ is undescended, and not uncommonly located in an inguinal hernial sac. The ovary is rarely functional and often abnormal. Tumor development is a danger. In a number of patients virilization may occur at puberty, although it may be difficult to trace the functioning testicular tissue, particularly if it lies within the abdomen. In these patients clitoromegaly may be found, although in a number this has been detected in childhood and may have resulted in removal of the offending organ. There may or may not be a small uterus, with an abnormally narrow vagina or merely a vaginal stump.

Testicular feminization presents in patients who have a normal XY chromosome pattern but are female in appearance. Testes are usually intraabdominal, and external genitalia are female. The vagina is merely a small stump, and no uterus is found. The testicular tissue is functional and the plasma level of testosterone is usually high, although there is no evidence of virilization; indeed, feminization occurs.

One may consider under the heading of genetic disorders resulting in primary amenorrhea the condition of congenital adrenal hyperplasia. Even though there is no typical chromosome abnormality, the condition is genetically determined in that there is a deficiency of an enzyme from the pathway of adrenal production of steroids. Although several such enzyme abnormalities have been described, the most common abnormality is the absence of the 21-hydroxylase enzyme responsible for the conversion of 17-hydroxyprogesterone to 11-desoxycortisol. The result of this abnormality is that intense adrenal activity occurs in order to reach a just-adequate production of cortisol, but this is achieved only at the expense of severe virilization. Amenorrhea is present in those individuals who reach puberty without treatment. In these patients it is likely that the enzyme deficiency is far from complete and the virilization mild, as more severely affected individuals present in childhood; in fact, clitoromegaly may be recognized in babyhood. Growth is stunted and the appearance muscular. Treatment with doses of glucocorticoids sufficient to suppress the abnormal output of steroid will result in the appearance of secondary sex characteristics, menstruation, and, in some, fertility. Some of the patients diagnosed in the early days of treatment have now reached adulthood and had successful pregnancies.

2. X-irradiation

Failure of the onset of ovarian function at puberty may occur as a result of exposure to therapeutic doses of X-rays to the region of the pelvis and abdomen in childhood. Treatment for such lesions as Wilms' tumor or abdominal Hodgkin's disease may result in prepubertal destruction of ovarian tissue. These patients give a history of exposure and are sexually infantile, and hormone studies reveal very low levels of estrogen excretion and high levels of plasma and urinary gonadotrophin. Similarly, exposure of the pituitary area to irradiation, as for optic nerve glioma, may result in failure of gonadotrophin production with or without other evidence of panhypopituitarism. Such individuals have low plasma and urinary gonadotrophin levels which continue after the normal age of puberty. Although feminization may be achieved by giving adequate estrogen therapy, fertility is only likely to respond to human gonadotrophin therapy. Patients with post-irradiation ovarian failure are unresponsive to gonadotrophin therapy, but feminization (although not fertility) may be achieved by adequate estrogen therapy.

3. Hypopituitarism

Patients with panhypopituitarism that begins in childhood are of short stature, showing as well the other features of inadequate production of anterior pituitary hormones. It is unusual for these patients to present with infertility as primary complaint, for they usually present with sexual infantilism and delay of onset of puberty. Although some patients have no evidence of a destructive pituitary lesion, many have evidence of tumors, the most common of which is the cranio-pharyngioma, showing radiological evidence of expansion of the pituitary fossa with calcification above and within it.

Patients with specific deficiency of gonadotrophin present with a characteristic appearance. The body proportions are of eunuchoid type, with growth of the limbs being greater than that of the trunk. On measurement the span is greater than the height, and the heel-to-pubis distance is equal to half the span but greater than half the height, for in normal individuals the pubis is midway between heel and crown and span equals height. Although there is deficiency of gonadotrophin, other pituitary function appears to be normal, and the patients are relatively tall but show sexual infantilism. They will respond to gonadotrophin therapy, and it may be possible to induce ovulation by the use of Human Pituitary Gonadotrophin/Human Menopausal Gonadotrophin-Human Chorionic Gonadotrophin (HPG/HMG-HCG) therapy, but it is preferable to induce sexual maturation and development of uterus and vagina by means of high-dose estrogen therapy in the first place. When withdrawal bleeding has been established and adequate changes of endometrium and vagina have occurred, then gonadotrophin therapy may be instituted.

III. INVESTIGATION

Once the presenting history has been taken and the initial clinical and gynecological examination has been made, it is usually possible to establish a program of investigation that will most clearly demonstrate the etiology of the infertility. The pattern of investigation will vary according to which group of the above classification the patient fits into most accurately. Discussion of the investigation of features listed under the heading of "mechanical" will be avoided, and it will be assumed that as far as possible these problems will have been investigated and excluded by the gynecologist before referral for detailed endocrine and biochemical investigation.

A. Regular Menstruation

In patients with regular menstrual periods, the fact that vaginal bleeding occurs spontaneously suggests that the production of gonadotrophin, the response of the ovary with respect to estrogen production, and the response of the uterus have been adequate to produce enough endometrial changes to permit menstruation. This means that investigation is considerably simplified. It can be divided into three main biochemical studies. All patients should be instructed to keep a basal body temperature chart, for this will help select those patients with anovulation or short luteal phase.

1. Urinary Pregnanediol Excretion

In the normal individual, the urinary excretion of pregnanediol rises after ovulation from its follicular phase value of less than 1 mg/24 hr (of which 0.5 mg is of adrenal origin) to values in excess of 2 mg in 24 hr. Failure of this increase suggests that ovulation has not occurred. Under these circumstances it is necessary to proceed to more detailed study of the estrogen excretion and gonadotrophin production. In patients with inadequate luteal phase the rise occurs as expected, but is not maintained for the usual period of 10 days or more, and menstrual bleeding occurs before the expected 12 to 14 days after ovulation. Usually the interval is less than 8 days, when the excretion of pregnanediol falls, closely followed by the estrogen output, and menstrual bleeding occurs.

Measurement of plasma progesterone levels by competitive protein binding or radioimmunoassay yields the same information.[58,125]

2. Urinary Estrogen Excretion

The excretion of estrogens, measured either as total or fractionated, shows a progressive rise in the proliferative phase of the cycle, reaching levels of 50 to 100 μg/24 hr total estrogen (estriol making up about 50 to 60% and estrone approximately 25 to 30% of this total) at about the time of ovulation.[12,70,81] Following ovulation, estrogen excretion falls sharply for 24 to 48 hr, then rises to a rather more sustained peak in the luteal phase, often of similar amplitude to the preovulatory level. Output then falls prior to menstruation, usually in concert with pregnanediol excretion. In anovulation, the postovulatory trough is not seen and the secondary peak does not occur, the original proliferative phase levels being maintained to some extent through the remainder of the cycle before falling prior to menstruation. In short luteal phase patients, however, serial studies reveal a relatively normal pattern, except for the early fall in the luteal phase, in parallel to the pregnanediol levels.

Measurement of plasma 17-β estradiol by radioimmunoassay yields similar results.[1]

3. Gonadotrophin Production

The pattern of gonadotrophin secretion may be followed by sequential collection of 24-hr urine samples or regular estimation of plasma levels of follicle stimulating hormone (FSH) and luteinizing hormone (LH). Gonadotrophins are released from the hypophysis in a pulsatile fashion throughout the 24-hr period, without a significant nyctohemeral rhythm, with the result that a single sample each day will give a satisfactory pattern for the interpretation of function in the individual case.[67,77,89,129] Typically, the level of FSH rises in the early part of the menstrual cycle, the increase beginning before the onset of menstrual bleeding. There then follows a gradual fall in plasma FSH which continues throughout the remainder of the cycle interrupted only by a surge of secretion coincident with the rise in LH at midcycle. Levels of LH, on the other hand, remain basal through the early part of the cycle, with a sharp peak of secretion at midcycle preceding ovulation. The plasma content of LH then decreases, first quickly, reaching almost basal figures, and then more gradually over the next few days, reaching basal levels before the end of the cycle.[1,70,102] The peak in LH occurs at the peak of estrogen production and precedes the fall and subsequent secondary rise in estrogen with increase in pregnanediol excretion seen after ovulation. The rise in LH precedes the rise of basal body temperature which occurs with ovulation and the rise in pregnanediol.

Urinary excretion of the gonadotrophins shows a similar pattern to the findings in plasma, but the peaks of secretion are less narrow and clear cut. Urinary levels of both FSH and LH are raised over several days at the midcycle peak. If the patient cannot be seen daily or on alternate days for blood sampling, it is equally useful for her to collect daily 24-hr urine samples at home.

The secretion of estrogens by the ovary in the early part of the cycle suggests that there are maturing follicular cells within the ovary responding to the stimulus of FSH from the anterior pituitary. As the ovarian production of estrogen and thus the urinary excretion increases, proliferation of the follicular cells is suggested. However, it is also known that estrogen secretion does not reach normal proportions in the absence of LH secretion.[46,56,106] Thus, a normal degree of estrogen secretion not only means adequate FSH but also at least minimal secretion of LH. In the anovulatory cycle the midcycle peak of LH production may not be seen, but the presence of relatively normal levels of estrogen would suggest that some basal secretion of LH is present, and stimulation with Clomiphene or with LHRH may be adequate to produce release of LH and lead to ovulation. Subnormal levels of estrogen in the presence of adequate levels of radioimmunoassayable FSH in the plasma are almost certainly associated with low levels of LH. In such subjects, therapy with human chorionic gonadotrophin is likely to meet with greater success than attempts to stimulate endogenous LH secretion.

In all patients with suspicion of abnormally low production of gonadotrophin an LHRH or Clomiphene stimulation test should be performed in order to assess the ability of the hypothalamo-pituitary axis to produce LH.[7,68,127,130] In both these tests, some FSH is also released if response occurs. Patients with regular cycles but with lack of ovulation are the most suitable group of all to treat with Clomiphene.

B. Oligomenorrhea

Once more the basal body temperature chart may be of great value in saving time and expensive investigation, in that patients with irregular periods but with ovulation occurring normally, if unpredictably, some 14 days before the onset of the next vaginal bleed, will show the expected rise of temperature at this time. Ovulation may then be confirmed by loss of ferning, change in spinnbarkeit and karyopyknotic index, and if necessary by demonstrating the expected rise in urinary pregnanediol excretion. However, once the patient has become accustomed to the recording of the temperature chart, and has learned to recognize the ovulation rise, the train of investigation is not uncommonly interrupted when pregnancy supervenes, for intercourse can now be more accurately timed. For other types of oligomenorrhea, investigation must be more protracted.

1. Urinary Pregnanediol Excretion

Patients with irregular menstruation but with ovulation will show the characteristic rise in pregnanediol. In some patients with oligomenorrhea, however, ovulation does not occur, and pregnanediol remains at basal levels throughout the cycle. Serial measurement throughout the cycle at, say, 2- or 3-day intervals will clearly separate the two groups.

2. Estrogen Excretion

It is to be expected that estrogen production will be at a low level throughout much of the cycle, rising late and falling again in those subjects who do not ovulate, but showing the typical double peak of estrogen excretion in those who do show normal ovulation. In some individuals the estrogen excretion rises at an earlier stage of the cycle, but often the rise is slow and rather protracted, achieving typical preovulatory levels before the ovum is released in response to the LH stimulus.

3. Urinary 17-oxosteroid Excretion

The investigation of the urinary excretion of 17-oxosteroids and 17-oxogenic steroids (or 17-hydroxycorticosteroids) in cases of oligomenorrhea will help to select those cases with polycystic ovary syndrome. Routine measurement of urinary 17-oxosteroids in an average of 23% of cases reveals a level of excretion in excess of the normal range, which in most laboratories is in the region of 5 to 15 mg/24 hr.[53] In severely affected patients levels over 20 mg are seen, but in most the excretion is in the upper half of the normal range. The significant and almost diagnostic feature is that the 17-oxosteroid excretion is equal to or in excess of the 17-oxogenic steroid excretion, a situation which is the reverse of the normal. The levels of urinary steroid excretion can often be suppressed to normal or low normal values by the exhibition of prednisone in doses of as little as 7.5 mg/day. Urinary testosterone excretion is increased.[28,80]

In severely obese patients with menstrual irregularity, steroid excretion may be increased. It has been necessary to differentiate this from Cushing's syndrome in the occasional patient. However, in obesity the nyctohemeral rhythm of plasma cortisol is preserved, and the suppressibility of urinary 17-oxogenic steroid by 0.5 mg dexamethasone 6 hourly persists, unlike Cushing's syndrome.

4. Plasma Testosterone

In polycystic ovary syndrome, the levels of plasma testosterone and plasma androstenedione are raised above the normal female levels.[66,79] This is a reflection of the increased output of androgens and is also seen from the increased excretion in the urine.[11,28,30,59,60,80]

5. Gonadotrophin Production

Urinary gonadotrophin excretion and plasma gonadotrophin levels reflect a low or normal output of FSH, which rises slowly and often rather late in the cycle, with an absent peak of LH in those patients without ovulation, but with this peak occurring in those who show evidence of ovulation, at about 14 to 15 days before the next menstrual period. The pattern of gonadotrophin production is similar in all those patients with oligomenorrhea, whether due to obesity or polycystic ovary syndrome, but in the latter mean levels may be slightly raised but variable.[131] In those individuals with poorly developed ovarian function, with impending premature menopause, the level of gonadotrophin may be high, with a brisk response to stimulation.[103,112]

In this group of patients with oligomenorrhea, a rise in plasma gonadotrophin levels will be obtained following a suitable period of stimulation with Clomiphene or in response to injections of LHRH. Various dose schedules for the use of Clomiphene have been suggested, up to 2 mg per kilogram body weight daily for 10 to 14 days, but at this dose level there are distinct dangers of cystic ovarian changes, and in the author's own unit, 100 mg daily for 5 days has been an adequate stimulus for most patients. Plasma LH levels rise progressively within the first 72 hr, often, after 24 hr, paralleled by the plasma FSH, and fall within a few days to basal levels. If adequate follicular maturation has occurred, then ovulation will follow, with menstruation occurring, in the absence of pregnancy, within 10 to 14 days.

The LHRH stimulation test is more useful, in that it is a more rapidly performed test. Various doses have been advised by different workers, but a figure between 100 and 400 μg is usually used.[7,127,130] Given intravenously after a resting sample of blood has been taken, it leads to a brisk rise in plasma LH levels (with a lesser rise in FSH levels), reaching a peak at about 30 min after the injection and falling progressively, from 1 hr for the next 12 to 24 hr. Blood samples are usually taken at 30 and 60 min after injection.

C. Secondary Amenorrhea

Secondary amenorrhea is a common way for infertility to present, and investigation may reveal failure at any point on the hypothalamo-pituitary-ovarian axis. The initial pattern of investigation should be based on the clinical picture, e.g., if associated with obesity, hirsutism, a previous history of X-irradiation, or an association with a systemic disease.

1. Pregnanediol Excretion

The excretion of pregnanediol remains low in most patients with secondary amenorrhea for ovulation does not occur spontaneously, except in very rare circumstances. However, all workers in the field have seen patients who have become pregnant with a history of complete amenorrhea for several years.

2. Estrogen Excretion

Without the release of gonadotrophin, or with inability of the ovary to respond to the gonadotrophin stimulus, estrogen excretion in the urine remains persistently low. In patients with psychogenic amenorrhea, in patients with obesity, and in some patients with systemic disease, the level of urinary excretion falls within the range seen in the early proliferative phase of the menstrual cycle, which is sufficient to permit some changes of estrogenization of the vaginal mucosa but insufficient to allow endometrial development adequate to lead to menstrual flow. In ovarian failure or hypopituitarism the levels of estrogen are very low indeed, the vagina is atrophic, and dilatation and curettage rarely produces any material for histological examination. In most such patients the urinary excretion is well under 20 μg and often less than 10 μg total estrogen daily, with estrone figures at less than 3 μg daily.

3. Urinary 17-oxosteroid Excretion

The excretion of the metabolites of the adrenocortical steroids is normal in primary ovarian failure, in psychogenic amenorrhea, and following the contraceptive pill. In obesity the values are normal or slightly raised, and in polycystic ovary patients with amenorrhea the steroid findings are very much as described above, but with a clearly abnormal value for 17-oxosteroids in relation to 17-oxogenic steroid excretion. In those patients with hypopituitarism, 17-oxosteroid output is abnormal if corticotrophin secretion is affected. In some individuals with early pituitary disease there may only be loss of gonadotrophin production, and ACTH secretion may not be involved, although specific deficiency of gonadotrophin secretion may occur in the absence of pituitary disease and may be a manifestation of a hypothalamic disorder of regulation rather than infiltration or tumor. In these cases also ACTH output is usually unimpaired.

In patients with systemic disease, excretion of 17-oxosteroids and 17-oxogenic steroids is not uncommonly abnormal. In those cases with wasting and debilitating diseases the output of adrenal steroid hormones is increased by the stress of the illness, and although the plasma level of cortisol, seen as 11-hydroxycorticosteroids (acid fluorescence) or 17-hydroxycorticosteroids (Porter-Silber chromogens) depending on the method of analysis, is raised, the diurnal rhythm is preserved. The urinary output of steroids is not usually raised to the same degree, although higher than normal basal levels occur. This effect is probably due to reduced metabolic clearance by the liver, reflecting the systemic disturbance of the particular disorder. The increased level of production can be suppressed by exogenous steroid therapy.

In patients receiving corticosteroid drugs as part of the treatment of their disorder, the excretion of 17-oxosteroids and 17-oxogenic steroids is reduced. The greater the dose, the greater that suppression is likely to be. It is uncommon for patients receiving large doses of steroids for a collagen disease or chest disease to present with a complaint of infertility, in view of the severity of their illness. However, it is important to recognize that high doses of steroids may interfere with the production of gonadotrophins by suppression of the secretion of hypothalamic-releasing hormones responsible for gonadotrophin secretion. There is experimental evidence that corticosteroids have an effect on the hypothalamic nuclei responsible for FSHRH and LHRH release as well as the expected effect on corticotrophin releasing hormone secretion.

Patients with adrenocortical disease have abnormalities of urinary steroid excretion consistent with their particular disorder. Patients with Cushing's syndrome show an increased excretion of 17-oxogenic steroids, which is not suppressed by treatment with dexamethasone 0.5 mg 6 hourly for 3 days. The excretion of 17-oxosteroids is often virtually normal in many cases of Cushing's syndrome and only slightly raised in a large proportion of the others. Only a small group have high levels, and in patients with hyperadrenocorticism due to carcinoma of the adrenal cortex the levels of both 17-oxo- and 17-oxogenic steroids are extremely high, very much higher than in typical Cushing's syndrome due to bilateral adrenal hyperplasia or adrenal adenoma. In virilizing tumors the levels of 17-oxosteroids in the urine are high.

4. Testosterone Excretion

Plasma and urinary testosterone levels are raised in those individuals with polycystic ovary syndrome and in virilizing tumors.

5. Gonadotrophin Secretion

The investigation of the production of gonadotrophin in patients with secondary amenorrhea is an essential diagnostic step. Many of the patients are amenorrheic because of the loss of the cyclical peaks of production, although the "tonic" basal secretion still persists. There are two particular situations that differ from this pattern. In primary ovarian failure with premature menopause, the urinary gonadotrophin excretion and the plasma gonadotrophin level are high, as found in menopausal patients in the older age group. Stimulation of gonadotrophin secretion by Clomiphene or LHRH results in a much greater rise in LH than in normal circumstances. The second situation is in cases with hypopituitarism or hypothalamic failure, where the production of gonadotrophin is very low indeed, due either to failure of the pituitary cells producing gonadotrophin or failure of the hypothalamus to produce releasing hormones. Attempts at producing an increase in the gonadotrophin output by stimulation with Clomiphene or LHRH result in a lack of response, with no rise in plasma level. The production of gonadotrophin is well below the basal levels seen in the normal woman and below the levels found in patients with failure of cyclical release of gonadotrophin. It is important to show whether Clomiphene or LHRH can induce a rise in gonadotrophin production. In patients with obesity, polycystic ovary, and in most subjects with systemic disease, such a rise may be obtained in most cases. In patients with "psychogenic" amenorrhea and "post-pill" amenorrhea there may be very little response, and on occasion no response at all, resembling that found in patients with hypothalamic disturbance.[126] In patients with an absent response, particularly those with psychogenic amenorrhea, the return of response to hypothalamic stimuli precedes the return of menstruation and often of fertility, albeit by several months. In a number of patients, a course or courses of Clomiphene will often precipitate the return of menstruation, and regular vaginal bleeding is usually soon followed by the return of ovulation. In many ways this resembles the appearance of the response to Clomiphene and LHRH seen prior to the menarche, although the time scale is rather shorter in most cases of amenorrhea.

6. Pituitary Function

In the patients with secondary amenorrhea who have diminished gonadotrophin output, it is necessary to assess other functions of the anterior pituitary to exclude those cases with hypopituitarism. The advent of radioimmunoassay has rendered this a much more readily available investigation, particularly with the introduction of the commercially available releasing hormones such as Thyrotrophin Releasing Hormone (TRH). It is, however, important to recognize that anterior pituitary function is lost in a progressive fashion with differing sensitivity of the separate hormone-producing cells to the effects of pressure from expanding space-occupying lesions. There is a distinct "pecking order" of function loss, and gonadotrophin production is the most sensitive, followed by growth hormone (GH) secretion, then by corticotrophin (ACTH) and thyrotropin (TSH). It is occasionally possible to follow this sequence quite clearly in the individual patient, particularly in the presence of a slowly developing lesion.

The insulin stress test is the most widely used test of the secretory capacity of the anterior pituitary, but depends on intact hypothalamic function. The dose of 0.1 units per kilogram body weight given intravenously is adequate, except perhaps in the diabetic patient treated with insulin or in acromegaly. This dose usually leads to a degree of hypoglycemia with the blood glucose level of less than 35 mg/100 ml which produces a brisk rise in the plasma level of growth hormone and ACTH (or cortisol). Patients rarely suffer from marked clinical signs of hypoglycemia, providing they are kept recumbent throughout the procedure, but precautions should be taken to see that intravenous glucose solution is available already drawn up in syringes so that severe hypoglycemia may be reversed as quickly as possible should it occur. If symptoms of hypoglycemia are unpleasant and it is necessary to give intravenous 50% glucose, the test is not invalidated, for once the hypoglycemia has occurred the stress stimulus has been achieved and the response will follow. It will be unaffected by the rise in blood glucose produced by the injection. The hormone response to the stress leads to a rise in growth hormone from the fasting level of less than 3 ng/ml by an amount in excess of 6 ng/ml. Plasma cortisol rises from the initial level which depends on the point on the diurnal cycle at which the test is conducted. It is preferable for each unit to establish its own routine in this respect, so that comparable results are obtained in each case. In the author's unit insulin stress tests are routinely carried out at 09.00 with the indwelling cannula inserted an hour earlier to allow growth hormone levels to recover from the stress effect which occurs when this is done. With this routine, plasma cortisol is usually between 15 and 20 μg/100 ml in normal individuals, and rises by at least 10 μg during the test.

Borderline responses, particularly of growth hormone, may be further investigated by the use of β-blockade to augment the response. If 40 mg propanolol are given orally 1 to 2 hr before the test, the rise in growth hormone is markedly increased. This procedure is useful in selecting those patients with partial hypopituitarism from patients who have had a partial response due to some other interfering factor.

In the last few years the TRH test of TSH production has come into widespread use. It was the first direct test of pituitary function, not producing its effect by hypothalamic stimulation but by direct action on the anterior pituitary gland. If 200 μg of TRH are injected i.v., a rise in TSH is seen within 20 min in normal individuals, although a rather slower rise reaching a lower peak is to be expected in patients with hypothalamic disease, suggesting that the pituitary response is sluggish in situations where there is lack of regular stimulation; this is very similar to the situation seen in the adrenal where there is diminished response to ACTH in patients with long-standing hypopituitarism. In patients with panhypopituitarism there is no rise in TSH. There are specific responses seen in patients with primary thyroid disease which need to be mentioned here, in order to allow exclusion of such patients, although they would in most cases present with features other than secondary amenorrhea, except in the rare syndrome of primary hypothyroidism, amenorrhea and galactorrhea in which raised prolactin levels are also found.[6,31,71,101] In patients with primary hypothyroidism the basal level of plasma TSH would be well in excess of 5 mIU/ml and TRH injection would produce a much greater rise than expected in normal subjects. Conversely, patients with thyrotoxicosis have undetectable levels of TSH in peripheral blood, and levels do not rise after stimulation with TRH due to the effects of prolonged suppression of TSH output by the high levels of thyroxine and tri-iodothyronine in peripheral blood.

Other indirect parameters of pituitary function may be sought in patients in whom detailed studies of anterior pituitary polypeptide hormones are not available. Measurement of the 24-hr excretion of 17-oxosteroids and 17-oxogenic or 17-hydroxycorticosteroids will give information concerning the normality of adrenal function. This may be further studied by the measurement of plasma cortisol, by competitive protein binding, radioimmunoassay, or by less expensive but equally reliable techniques as 11-hydroxycorticosteroids or 17-hydroxycorticosteroids. The pattern of the nyctohemeral variation in plasma cortisol levels will show the presence or absence of normal pituitary production of ACTH, assuming the absence of obvious adrenal or systemic disease which would modify the rhythm. Similarly, measurement of serum thyroxine or protein-bound iodine levels enables assessment of thyroid and anterior pituitary TSH function.

In patients with galactorrhea, high prolactin levels may be found.[37,41,43,91,121] As it is only within the last 2 or 3 years that it has been clearly demonstrated that man has prolactin produced by the anterior pituitary, the assay of this hormone is not a widely available investigation.[57] It is likely, however, that the research currently under way in this field will lead to an extension of understanding of the problems of infertility in man, although unlike some animal species no luteotrophic activity has been demonstrated in the human. At the present time most investigators have to rely on the patient as her own bioassay preparation, and in patients with amenorrhea and galactorrhea the suppression of gonadotrophin production by the release of prolactin is to be suspected, either by pituitary tumor or due to the failure of hypothalamic production of prolactin release-inhibiting hormone (PRIH).

D. Primary Amenorrhea

Few patients have primary amenorrhea present with a main complaint of infertility, but more often with complaints of delayed puberty and infantilism. The fact that primary amenorrhea is present narrows the likely field of diagnosis and the range of investigation.

1. Urinary Pregnanediol Excretion

The levels of excretion of pregnanediol are low, as there is no follicular maturation to permit ovulation and luteinization.

2. Estrogen Excretion

Almost by definition the levels of estrogen excretion are low, with delay in or failure of ovarian development.

3. Urinary 17-oxosteroid Excretion

The 24-hr excretion of 17-oxosteroids and 17-oxogenic steroids is abnormal in two particular situations in primary amenorrhea. First, in patients with panhypopituitarism the levels are low, but will rise following the injection of 40 units ACTH b.d. for 3 days. Plasma cortisol levels are also low, and show a poor response to i.m. Synacthen® in a dose of 250 μg, giving a rise of less than 7 μg/100 ml in most patients, from a basal level of less than 10 μg/100 ml and in many subjects less than 5 μg/100 ml. Second, in subjects with congenital adrenal hyperplasia the steroid excretion will be abnormal. The most likely type of this condition to present in the group of patients under consideration is the 21-hydroxylase deficiency. This is the most common type overall, and is typified by the masculinization effects with hirsutism, short stature, advanced bone age, and clitoromegaly. Assay of the urinary levels of 17-oxogenic steroids and 17-oxosteroids reveals that both are increased above the normal adult level, often reaching levels of 40 mg of 17-oxogenic steroids/24 hr. Assay of the output of pregnanetriol, which is the specific urinary metabolite of 17-hydroxy-progesterone, which accumulates in this condition, will show levels above 10 mg/day and often much higher. The normal adult value for urinary pregnanetriol excretion is under 2 mg/24 hr. Plasma cortisol estimation by the 11-hydroxycorticosteroid method is unreliable. The figure obtained is high, but much of this reading is given by 21-desoxycortisol which is present in the plasma of these patients.

In subjects with sexual infantilism the excretion of urinary steroids is below that of the normal adult values, and resembles that in normal prepubertal children. The daily output of 17-oxosteroids is usually less than 5 mg/day, and the 17-oxogenic steroid excretion is of similar magnitude but usually slightly greater than the value of the 17-oxosteroids. Values of 2 or 3 mg/day are not uncommon. Plasma cortisol values are normal.

4. Gonadotrophin Production

In subjects with hypopituitarism the basal levels of gonadotrophin are very low, and no response is seen to LHRH or Clomiphene. The presence of other deficiencies of anterior pituitary function confirms that the lesion is within the pituitary and not due to simple delayed puberty, although such patients are unlikely to present with infertility.

Patients with primary ovarian failure due to maldevelopment or following pelvic irradiation in childhood have high basal levels of gonadotrophin with an increased response to stimulation by LHRH and Clomiphene.

5. Genetic Studies

The buccal smear is a simply performed test from the clinician's point of view, although an experienced cytologist is required to make reliable analyses of the films obtained. The assessment of the proportion of nuclei showing positive sex chromatin is of assistance in finding patients with abnormal chromosome complements. A low number of cells with positive finding of Barr bodies suggest a possible mosaic pattern, while an absence of sex chromatin suggests a possible case of Turner's syndrome with an XO karyotype.

Full chromosome analysis from leukocyte culture is the most certain way of making diagnosis of abnormal genetic states. The most common of these is the XO patient, who presents the classical picture of Turner's syndrome, with short stature, webbed neck, wide carrying angle, and ovarian agenesis. Rather more difficult to recognize is the patient with a mosiac pattern of XX/XO with which ovarian agenesis and streak gonads are found at laparoscopy or laparotomy. In patients with XY/XO mosaic, the picture is of the sexually infantile female, and on laparoscopy and biopsy it is not uncommon to find a streak ovary on one side and a small ovoid testis on the other, sometimes in a small inguinal hernial sac. Recently, the author had care of a patient with

XX/XY/XO mosaicism, with a testis which became functional on one side and a dysgerminoma contralaterally. In none of these patients is fertility possible.

Patients with an apparently female aspect, but with a small vaginal stump and absent or rudimentary internal genitalia, may be found to have an XY karyotype. If measured, the plasma levels of testosterone will be found to be high and within the male range. These patients are examples of testicular feminization, a poorly understood condition in which the tissues react to androgens as if they were estrogen. It is likely that there is an enzyme abnormality in peripheral metabolism of steroids, but this is not yet fully elucidated.

IV. THERAPY

A. Introduction

Having delineated the nature of the basis of the infertility, it is now possible to consider the approach to therapy to be undertaken. This may vary from simple educational advice to prolonged and complex regimes with detailed biochemical monitoring. In all cases with regular or irregular menstruation, the simple discipline of regular keeping of the basal body temperature is imperative. In this way the patient and her spouse may time intercourse rather more successfully in relation to ovulation, although there may be problems in some patients who show a rather gradual temperature rise rather than a sharp increase over a single day. There are further difficulties in that there are suggestions that the temperature may not rise until the day after ovulation, but that it may not be possible to fertilize the ovum 6 to 12 hr after release. If intercourse has taken place in the previous 24 hours then fertilization is likely to occur, but if intercourse has not taken place then it is necessary that it should at the earliest opportunity. Some individuals do have a fairly regular finding of a small trough of temperature preceding ovulation, and with practice these patients may learn to recognize this trough as an indication of impending ovulation and may undertake intercourse at the optimum time. In the situation of human gonadotrophin therapy, the basal body temperature chart is a very necessary adjunct to the monitoring procedures.

It must be appreciated that during any of the treatment regimens discussed it may be found that a second regimen must be combined with the first in order to achieve satisfactory ovulation and luteal phase. For example, ovulation may occur after HCG but a short luteal phase is a possibility. In these circumstances progestogen therapy is indicated. Constant review of the response to each course of therapy is needed, and without such vigilance results will be poor.

1. Short Luteal Phase

The problem here is to supplement the steroid output of the corpus luteum to permit continuing development of the uterine endometrium in the secretory phase and thus allow successful nidation of the ovum. Synthetic progestogens may be given, either as regular doses or as depot injections. Treatment should be instituted immediately following ovulation. Commonly used preparations are medroxyprogesterone, 5 to 10 mg/day, dydrogesterone, 10 to 20 mg/day, or norethisterone, 5 to 10 mg/day. These doses are usually adequate, but, rarely, higher doses may be indicated. With the latter preparation some caution may be needed if prolonged or higher dosage is given, as it is derived from testosterone and retains minimal androgenic activity. Depot preparations such as medroxyprogesterone, 50 to 100 mg, or hydroxyprogesterone, 250 mg, may be given once in each cycle, although a few patients may respond better to a similar dose divided into two weekly parts, the first being given shortly after ovulation. It must be remembered that the corpus luteum may remain inadequate if pregnancy occurs, and regular weekly injections of progestogen will be required throughout the first trimester until placental function is sufficiently established to maintain hormonal secretion, a time when normal luteal function begins to fade. The effect of progestogens may be enhanced by giving a small dose of estrogen 0.01 to 0.05 mg ethinyl estradiol, especially in patients with evidence of lowish estrogen excretion. HCG 2,000 units i.m. on alternate days for 10 days following ovulation may be used as an alternative, but has risks of multiple ovulation.

2. Infrequent Ovulation

The lack of the midcycle peak of LH in the presence of normal estrogen secretion leads to failure of ovulation, while suggesting that FSH production is relatively normal. It should be pointed out that the estrogen response to FSH stimulation does not reach normal proportions unless some basal LH secretion occurs also. Under these circumstances two avenues of treatment are available. First, the missing peak of LH may be provided by the use of injections of HCG. On occasion, the dose needed may be as low as 1,500 i.u. but most workers use doses of 5,000 to 12,000 i.u. per injection. It is the author's own experience that 10,000 i.u. is more reliable than lower doses, although many do respond to 5,000 units. The timing of the injection should be based on levels of adequate urinary excretion of estrogens or other evidence of follicular maturation such as ferning, karyopyknotic index, or spinnbarkeit. The latter parameters are indirect measures of estrogen secretion, insofar as they represent the tissue response to the hormone. Less accurate than the urinary excretion levels, they may be used to provide confirmatory evidence, or when laboratory facilities required for the estrogen estimations are not readily available. To make serial estrogen estimations worthwhile, the answers must be available on the same day on which the samples are received, for the HCG must be given at the optimum time. Once the 24-hr excretion of total estrogen is in excess of 50 μg, the HCG may be given. Ovulation usually occurs within 24 hr, the basal body temperature rises, and the urinary estrogen excretion falls sharply before rising again through the luteal phase, in parallel with the rising output of pregnanediol. At the same time there is loss of ferning of cervical mucus, which becomes scanty and of high viscosity. Menstruation occurs, in the absence of pregnancy, some 10 to 14 days after the injection of HCG.

A second approach is to attempt to stimulate endogenous release of LH by the use of Clomiphene. This drug was introduced for the antiestrogenic effect seen in animal studies, also seen in man,[104,123] but the main result in man is the stimulation of the hypothalamus leading to the release of gonadotrophin, in particular LH.[84,99] In clinical practice, the ideal case for treatment with Clomiphene is the patient in whom FSH production and the resulting estrogen secretion is normal, but who has failure of the cyclical surge of LH. Ovulation may be induced in the patient with anovulatory cycles but is less likely to be successful in the patient who is amenorrheic unless there is basal function, and follicular maturation has occurred to an adequate degree.

Once again it is necessary to monitor the rise in estrogen excretion which indicates the stage of follicular maturation, if the best chance of ensuring ovulation is to be obtained. It has been the practice in many centers, in patients with regular but anovulatory cycles, to give treatment immediately following the previous menstrual cycle, which means that the course of therapy will begin on the 5th day of the cycle in most patients. Treatment is usually given for 5 days with tablets taken once daily. The dose commences at 50 mg daily, increasing to 100 or even 150 mg in subsequent courses if a response is not obtained. The peak level of LH is usually obtained at about the 5th day of treatment and is usually slightly more sustained than the normal physiological peak seen in normal individuals. Thus, in individuals with normal 28-day cycles, to start therapy on the 5th day is not always likely to produce the peak of LH secretion at the optimum time for ovulation, but will almost certainly be 3 or 4 days early. In theory, a more suitable starting date for treatment would be on the 8th day of a regular cycle. It will be recognized that treatment has a greater chance of success if detailed studies of the normal pattern of estrogen excretion are undertaken before the treatment schedule is started. In this way the optimum period for treatment may be discovered, particularly if it is also possible to investigate the LH response to Clomiphene by radioimmunoassay of the plasma LH level and estimate the timing of peak LH output. In patients with irregular cycles, this form of treatment is not so certain of success as the use of HCG, as there is a latent period before the LH levels begin to rise, and even if urinary estrogen levels are followed in each cycle, it is possible that the optimum point of follicular maturation may be missed.

In subjects in whom the 24-hr excretion of estrogen is below the normal range and not followed by a surge of LH release, it may be of value to give a course of estrogen therapy for 4 or 5 days from the 9th day of the cycle. This is initially followed by a fall in serum gonadotrophin levels, but when treatment is discontinued there is a rebound rise of LH.[15,86,128] This increase is greater at this phase of the cycle than if the dose is given earlier, and theoretically ovulation might occur. However, the rise in LH is not usually as great as a normal midcycle peak, and secondly, the fact that the rise in estrogen in the follicular phase was below the expected range suggests that follicular development was inadequate and ovulation would not be likely to occur. For these reasons,

Clomiphene is rather more likely to produce ovulation than estrogens alone.

3. Polycystic Ovary Syndrome

The object of therapy in the polycystic ovary syndrome is to suppress the abnormal steroid secretion of the ovarian tissue. A fall in the androgenic steroid levels results in a reduction of the effect of the hypothalamic centers, which allows a return toward normal in gonadotrophin secretion. Direct suppression of ovarian steroids is rarely successful, but suppressing the adrenal component of androgens is more readily achieved and results in improvement in most cases. Suppression with prednisone is successful at relatively low dosage, often as little as 5 or 10 mg daily, levels which are little more than twice the physiological daily production of cortisol. After therapy is commenced, measurements of 17-oxosteroid excretion show suppression when adequate doses are reached. In most patients a dose of 7.5 mg prednisone daily (usually given as 5 mg each morning and 2.5 mg each night) will produce a satisfactory reduction in oxosteroid output. It is necessary, in difficult cases, to monitor urinary excretion carefully to ensure adequate suppression, if necessary using higher doses of prednisone. It is interesting to note that in general there is no advantage to be gained in giving the suppressing dose at night, although this is almost essential in congenital adrenal hyperplasia. It is probable, therefore, that during treatment the suppressant effect on cortisol metabolism is kept to a minimum.

In many patients this dose of steroid will reduce the output of 17-oxosteroid to less than 5 mg daily. This will result in a return toward normal of menstrual habit, and in anovulatory patients will restore ovulation in a number. It may be possible in those individuals who still do not ovulate to induce ovulation by the addition of Clomiphene to the therapeutic regimen or by the use of HCG. However, it is necessary to document the hormonal response to steroid therapy in the first instance before introducing a luteinizing stimulus. Further, if Clomiphene is used it should be remembered that repeated use in high dosage may lead secondarily to polycystic ovarian changes. If HCG is used it must be given at the time shown to be optimum on measurement of urinary total estrogen excretion.

A high proportion of cases, particularly with milder forms of the disorder, will ovulate on steroid suppression only, and the addition of a luteotrophic stimulus is rarely needed. Some reports have suggested that steroid therapy for 6 months is as successful as wedge resection, but it has been agreed that severity of clinical symptoms and signs is no indication of the likelihood of success or failure of the various therapeutic managements.[52,63,95,110,113] In a small proportion of patients with this syndrome the periods may be so irregular that, whether anovulatory or ovulatory, it is impossible to achieve pregnancy. In these patients the most useful form of hormonal therapy is to use Human Menopausal or Human Pituitary Gonadotrophin to induce follicular maturation and HCG to stimulate ovulation.[24] Patients will respond to this form of treatment if serum levels of gonadotrophin are normal or low, but if gonadotrophin levels are persistently high, it is unlikely that HMG/HPG-HCG therapy will induce ovulation.

Failure of fertilization in the polycystic ovary patient may be due to failure of the release of the ovum through the sclerotic capsule surrounding the ovary. It is possible that the delay in release may mean that although the rise in pregnanediol and changes in estrogen may be seen, the ovum does not escape from the ruptured follicle until some hours or even days later. Thus, it may not survive until it reaches the fallopian tube where it should be fertilized. For this reason, if the regularity of the menstrual cycle is not improved by steroid therapy, or if, despite biochemical and basal body temperature evidence of ovulation, pregnancy does not result, then surgical wedge resection should be considered. Although it is not understood why such a procedure should succeed, various theories are advanced.[33,95] The most likely reason is that the surgical trauma to the gland itself is sufficient to produce a temporary interruption of the ovarian androgen and hormone production, and the effect on gonadotrophin secretion allowing restoration of the cyclical surges leads to the return of regular cycles and ovulation, which may persist for months or years before further sclerosis supervenes. It is obviously best to reserve this operation until pregnancy is sought, although other indications for surgery, such as menorrhagia, may occur. However, fertility is less often restored than is regular menstruation, possibly due to the capsular sclerosis.[52]

4. Obesity

The management of oligomenorrhea and infertility in the obese patient is by diet. Weight reduction is usually adequate to produce a return of normal gonadotrophin response and fertility. These patients usually have a greater emotional drive to abide by the restrictions of the dietary regime, in view of the hoped-for outcome of pregnancy. It is not uncommon to find that pregnancy begins before the expected ideal weight is achieved. It is rarely necessary to monitor the biochemical changes that follow the loss of weight, but if required, estimations of urinary excretion of estrogen or of plasma levels of gonadotrophin must be undertaken over periods of several weeks or months, with at least three estimations each week, in order that any cyclical surges of activity may be recognized and evidence of ovulation noted.

5. Systemic Disease

The treatment of infertility due to systemic disease is the treatment of the disease itself. Stabilization of the diabetic state is followed by a restoration of the usual pattern of menstruation for that individual, with the later return of fertility. The physician who has care of diabetic patients during pregnancy will recognize that fertility of the female diabetic population is not impaired by their condition if the degree of diabetic control is reasonable.

Remission of ulcerative colitis or of Crohn's disease may lead to restoration of menstruation, but such remissions are rarely adequate to lead to fertile ovulation in the severely wasted patient, although in the milder case pregnancy may occur in remission. If patients with ulcerative colitis or rheumatoid arthritis require steroids for their main condition, then fertile menstrual cycles are less likely to occur, as one may see in patients with abnormal endogenous production of steroids as in Cushing's syndrome.

Patients with malabsorption due to celiac disease will start to menstruate and will have a return of ovulation if they respond well to a gluten-free diet and have adequate replacement therapy with hematinic agents, calcium, and fat-soluble vitamins.

Bilateral adrenalectomy in patients with Cushing's syndrome will lead to remission of the signs and symptoms of cortisol excess, and when replacement therapy is established will be followed by the return of menstruation and ovulation. Hypophysectomy will, of course, lead to hypopituitarism and infertility, and in such patients treatment with HMG/HCG will be necessary. In the author's unit, it is usual practice in young females with bilateral adrenal hyperplasia to undertake the operation of bilateral adrenalectomy rather than hypophysectomy, provided there is no evidence of pituitary enlargement and if the patient is anxious to start or increase her family. In most patients on replacement therapy with cortisone acetate and fluorohydrocortisone, pregnancy proceeds uneventfully, although in a few cases difficulties may occur.[93]

6. Genetic

Although most of the genetic disorders of menstruation present with infantilism which can be treated by estrogen therapy, the resultant infertility is irremediable. In the case of congenital adrenal hyperplasia, however, the suppression of the abnormal adrenal secretion by the use of oral steroid therapy results in the development of cyclical ovarian activity, with the appearance of secondary sexual characteristics and physical maturation. Ovulation eventually occurs, and may be detected by the usual findings in basal temperature, cervical mucus, and urinary pregnanediol excretion.

The treatment of congenital adrenal hyperplasia is based on the exhibition of a glucocorticoid such as prednisone in doses sufficient to suppress the abnormal output of adrenal steroids. The commonest form of this syndrome is the absence of the C-21 hydroxylase enzyme, leading to the accumulation of 17-OHprogesterone which is excreted in the urine as pregnanetriol. The resultant increase in adrenal activity in an attempt to produce an adequate level of plasma cortisol leads to the production of large amounts of androgenic C-19 steroids which cause virilization in the patient. In normal individuals pregnanetriol excretion is less than 1.5 mg/24 hr, but in patients with congenital adrenal hyperplasia the rate of excretion is often greater than 20 mg/day. Untreated adults, however, are not often seen, for most cases present in childhood and are established on treatment at that time. In these subjects, if adequately treated, infertility is rarely a problem. In the untreated individual prednisone is commenced in the initial dose of 5 mg daily with increments of 1 mg every 2 or 3 days until suppression is achieved, with urinary excretion of pregnanetriol falling to less than 1.0 mg/24 hr. The required dose of prednisone is usually between 7 and 10 mg/day. It is best to give the whole as a single dose in the evening before retiring to bed, for then the greatest effect is obtained at a time when the normal diurnal pattern of adrenal activity is at its highest. Only occasionally is any benefit obtained by giving a small dose early in the day. Once the endogenous adrenal activity is suppressed by the oral prednisone, with the excretion of pregnanetriol at less than 1.0 mg and 17-oxo- and 17-oxogenic steroids within the normal range, normal hypothalamo-pituitary-ovarian activity shortly ensues. Secondary sexual characteristics develop and ovulation eventually begins. Management through pregnancy is that of the hypoadrenal patient, as these individuals are unable to respond to stress by increased adrenal activity.

7. Galactorrhea

Patients with galactorrhea due to pituitary tumor should be subjected to hypophysectomy, and then ovulation induced by the use of HMG/HPG-HCG. Patients with no evidence of pituitary prolactin-secreting tumor have, until recent years, presented rather more of a problem. Gonadotrophin secretion is suppressed by the abnormal production of prolactin; the therapeutic inhibition of the prolactin release was not achieved by use of steroids, and only some of the milder cases responded to therapy with Clomiphene with or without the addition of HCG. In many, ovulation may be induced by HMG/HPG-HCG therapy. In patients treated with Clomiphene cyclical courses of therapy should be given at intervals of 4 weeks, starting with doses of one tablet (50 mg) daily for 5 days, and with monitoring of urinary estrogen and pregnanediol excretion after the end of the tablet course. If possible, urinary gonadotrophin excretion should be monitored if radioimmunoassay services are available, particularly with a view to seeking the midcycle peak of LH secretion. If evidence of follicular development occurs but the expected peak of LH is not seen, despite increasing the dose of Clomiphene to 100 or even 150 mg daily for the 5-day course, then an injection of HCG in a dose of 5,000 to 10,000 units may be given when the total estrogen excretion rises to between 50 and 100 μg/day, and ferning of cervical mucus is marked. After the injection, parameters of ovulation should be sought by following the changes in basal body temperature, karyopyknotic index, ferning, and a rise in urinary pregnanediol excretion to a level of greater than 2.5 mg/day.

A small group of patients may respond to these measures, but a greater group will show no evidence of response. Until recently there was little to offer such cases, but it has been shown that L-dopa has the effect of increasing the release of hypothalamic releasing factors and hypothalamic prolactin release-inhibiting hormone. This management has been shown to induce a fall in plasma prolactin levels and in some patients has resulted in the return of cyclical gonadotrophin activity, with ovulation and menstruation.[1 2 2] Unfortunately, in many subjects the response is short lived.

For a number of cases that still fail to respond, hope has been raised by the introduction for research purposes of the compound 2-bromo-ergocryptine.[8,29] This appears to have greater promise as a therapeutic agent, but as its teratogenicity is unknown, it is not suitable at the present time for use in the management of infertility. However, its effect does appear to be more prolonged and more potent than L-dopa.

8. Postcontraceptive Pill Amenorrhea

Many patients with this condition will spontaneously start to menstruate again within 2 or 3 years, but there is a small group of patients who persist with amenorrhea beyond this time. Some of these subjects have evidence of early return of function in that some estrogen production is occurring, as shown by levels of estrogen excretion over 20 μg/day and by some estrogenization of the vagina. These patients may respond to one or two courses of Clomiphene, and in responding may reestablish regular cyclical activity. Those patients with no evidence of FSH production, as shown by lack of cyclical variation of plasma or urinary gonadotrophin levels or in the pattern of urinary estrogen excretion, will require treatment with human gonadotrophin as described below.

9. Hypopituitarism

The number of patients presenting with hypopituitarism due to pituitary adenoma following previous hypophysectomy or in idiopathic disease is quite small. Treatment with HPG/HMG-HCG will be necessary.

10. Specific Gonadotrophin Deficiency

Much larger is the group of patients with an apparent specific lack of gonadotrophin secretion without other evidence of disturbed pituitary function. It is possible that the abnormality of function lies within the hypothalamus, with failure of production of releasing hormones and possibly disturbance of the biological rhythm of cyclical surges of activity. These subjects are best treated by the use of HMG or HPG, both of which are mainly FSH. In adequate dosage, this will induce follicular development to a degree sufficient to allow ovulation when an LH stimulus is given in the form of HCG. Preparations of HMG and HPG contain a significant but lesser proportion of LH, but it has been recognized that response to FSH depends on the presence of a basal level of LH to ensure follicular response.[25,56] However, the content is insufficient to promote ovulation and this must be achieved by giving the much larger dose of HCG at the optimum time of follicular development.[32]

B. Clomiphene Therapy

The structure of Clomiphene resembles that of the nonsteroid estrogen chlorotrianisene (TACE), and is similar to stilbestrol. It has only weakly estrogenic activity, but there has been some evidence that there are direct effects on the ovary.[16] It has been shown that the uptake of radioactive estradiol by the uterus and pituitary in immature rats is diminished by Clomiphene, suggesting that it competes for the tissue-binding sites with endogenous estrogen. This displacement of estrogen would lead to an increase in gonadotrophin secretion by the pituitary.[84,104] As a result of the rise, there is a subsequent rise in estrogen excretion.[84,104,114] The effect of the hormonal secretion described is to lead to ovulation in patients with amenorrhea and anovulation.[55,92,99,124] The greatest responses have been reported in patients with proliferative endometrium or endometrial hyperplasia, and the least satisfactory results in those subjects with atrophic endometrium, suggesting that there was little stimulation by estrogen and therefore little response to gonadotrophin by the ovary.[75]

From the evidence of studies of the action of Clomiphene, it is possible to select the ideal patient for treatment. She should have evidence of anovulation, but with estrogen levels (or vaginal wall smears or endometrial biopsy) suggestive of follicular activity, with the ability to produce gonadotrophin upon stimulation and with evidence that other aspects of endocrine function are entirely normal. In such patients Clomiphene can be expected to promote a further rise in estrogen production, suggesting that the rise in FSH has stimulated further follicular activity, and a sharp rise in LH leading to ovulation. A high proportion of cases of Stein-Leventhal syndrome will respond satisfactorily to a course of Clomiphene therapy in that treatment is followed by a rise in basal body temperature and urinary pregnanediol excretion. The fact that luteinization of the follicle may occur without rupture must not be overlooked, however.

Treatment should be given using the lower dose of 50 mg each day, and each course should last 5 days. The subsequent response should be followed, and if the urinary steroid secretion suggests that the response is inadequate, the next course should then be increased to 100 mg daily for 5 days. The maximum dose for routine use is usually restricted to 150 mg daily for 5 days. On rare occasions the course may be prolonged up to 10 days,[75,76] but it is essential that the urinary excretion of estrogen and pregnanediol is carefully monitored in such cases so that hyperstimulation may be avoided.[116] This syndrome, associated with bilateral ovarian enlargement, ascites, and pleural effusion, has been seen after doses of 100 mg daily for 14 days. If careful monitoring of urinary estrogen is carried out this situation can be predicted as it is associated with high levels of estrogen production, suggesting that several follicles are stimulated to maturation and may be expected to lead to superovulation. Clomiphene hyperstimulation is unlike the hyperstimulation seen with gonadotrophin therapy, where, if very high levels of estrogen excretion are produced, the injection of HCG may be withheld, preventing ovulation and avoiding the hyperstimulation syndrome. In such cases the author has had experience of a case who had a total estrogen excretion of over 1,000 μg/24 hr without any symptoms, but with marked ovarian enlargement, which settled within 10 days. To have given HCG under such circumstances would have led to a disastrous situation. In the case of Clomiphene hyperstimulation, ovulation is likely to occur spontaneously, with inevitable production of hyperstimulation effects, in view of the release of LH occurring in response to the Clomiphene.

A more common side effect is the temporary enlargement of the ovaries which may occur following mild overdosage with Clomiphene. In such circumstances the dosage should be reduced in subsequent courses. Other side effects including abdominal symptoms, gastrointestinal upsets, hot flushes, depression, dizziness, etc. occur from time to time, but incidence is low. A contraindication to treatment is a previous history of liver disease, for Clomiphene is excreted by way of the biliary system. Some workers prefer to perform studies of liver function in patients before therapy begins.[76]

Although the ideal subject for therapy is the anovulatory patient with a relatively normal follicular phase shown by the evidence of estrogen excretion in the urine, some success has been obtained in patients with low estrogen production, provided that hypothalamic stimulation can lead to the release of gonadotrophins. Usually, low plasma FSH levels, low urinary total estrogen excretion, and a long history of amenorrhea are regarded as signs of poor prognosis with respect to Clomiphene therapy.[35,74,97] However, some work has shown that patients with "hypothalamic amenorrhea" may respond to 10- to 12-day courses of Clomiphene followed some 4 days later by 10,000 units HCG, and in this series one patient with Chiari-Frommel syndrome ovulated following this schedule of treatment, having failed to ovulate after Clomiphene alone.[75,76] It is probably wise, therefore, to use the Clomiphene stimulation test in all patients with anovulatory infertility, and measure the plasma levels of gonadotrophins for the 5 days during the Clomiphene test, if possible in parallel with plasma or urinary estrogen levels, and for 4 or 5 days following the cessation of treatment. A usual dose of 100 mg daily may, if necessary, be preceded by a similar study using 50 mg, or if no response is obtained to either course, then a longer course of 100 mg may be given up to 10 days. This will give guidance as to the most suitable patients and the most satisfactory dose schedule.

The correct timing of therapy in those patients with menstrual activity depends very much on individual response. Most commonly, treatment is begun on the 5th day of the cycle, although others prefer to commence therapy on the 8th to 10th day of the cycle,[76] which will lead to a rise of LH at about the 12th to the 15th day of the cycle. The latter seems to be the more physiological situation. In patients with amenorrhea, courses should be given at 28-day intervals. The success of the therapy depends on the selection of suitable patients, but in sensitive individuals there is a high incidence of ovulation. In a literature review,[76] ovulation occurred in 90% and pregnancy in 40%. Multiple births occurred once in every 16 deliveries. The spontaneous abortion rate of 20% compares with a figure of 10% in normally ovulating women.

Patients with anovulatory infertility who fail to respond to Clomiphene may be considered for HPG/HMG-HCG therapy. Patients with Stein-Leventhal syndrome should be considered for wedge resection, and if amenorrhea persists postoperatively then further Clomiphene or gonadotrophin therapy may be undertaken.

C. Gonadotrophin Therapy

Treatment with human gonadotrophin is expensive in material, laboratory, and clinical time, and if inadequately monitored is capable of producing considerable morbidity and even mortality among the patients under treatment. It is not a form of treatment to be undertaken lightly, and patient selection and preliminary investigation must be thorough. In properly controlled circumstances, however, it does offer the opportunity of childbearing to women in whom no other treatment will be of value, and does so in as nearly a physiological manner as possible.

Patients selected for treatment must be shown to have absent or reduced FSH and LH production, or such irregularity or infrequency of cyclical activity as to render it impossible to successfully time intercourse to allow fertilization. They must also have ovaries capable of response to the gonadotrophin stimulus, a uterus capable of response to the estrogens produced and free from endometrial, cervical, or vaginal infection, and have a husband who is fertile as demonstrated by seminal analysis. This means that all patients must be subjected to prolonged serial studies of gonadotrophin and estrogen production, assessment of other endocrine function with tests of adrenal, thyroid, and pituitary activity, laparoscopy or culdoscopy (or gynecography), tubal insufflation or dye instillation, curettage, repeated vaginal lateral wall smears, and, if necessary, endometrial and ovarian biopsy. During treatment, the patient must collect continuous 24-hr urine samples (or be subjected to daily venepuncture with resultant inconvenience and discomfort, especially if these are carried out over 14 to 21 days), have daily lateral vaginal wall smears (which may readily be taken by the patient herself), cervical mucus samples, and vaginal examination with bimanual palpation of the ovaries at each visit for injection, and keep a daily basal body temperature chart. On the day on which HCG is given, although some guide may be given by the presence of marked ferning and by increased cervical mucus with a spinnbarkeit of 10 cm or more,[26,62] with just palpable enlargement of one or both ovaries, it is far from safe to give the injection of HCG to induce ovulation without also having available the results of estrogen excretion assays on all 24-hr urine samples obtained from the 1st day of the treatment cycle up to the day prior to HCG injection (i.e., the 24-hr urine sample completed on the morning on which HCG is to be given). This requires the patient to deliver the sample to the laboratory early in the day, and the laboratory to be able to complete the assays within the time before the patient returns for the HCG injection. Without this information the risks of inducing the hyperstimulation syndrome are appreciable. Without such a service available it is probably unwise to undertake treatment with HMG/HPG.

Both HMG and HPG appear to be equally effective in inducing follicular maturation.[9]

Although it has been shown that ovarian response to HMG and HPG may be obtained by a single injection, most workers have found that repeated injections of the gonadotrophin produce a more repeatable and consistent response. There are many different regimes described which are reviewed in the literature,[33,49,119] but they fall mainly into two groups. First, injections may be given daily. The daily dose commences at the lowest level of 50 to 100 i.u. or one ampule per day (in the case of HMG 75 i.u. per ampule), increasing after each week by one ampule per day if there is no response of estrogen excretion, and continuing for 3 weeks followed by 7 days without treatment in each treatment cycle. This program is continued, gradually increasing the level of dose until a progressive rise in estrogen excretion is seen as evidence of follicular maturation. Although the threshold tends to be constant in an individual the variation between patients is great.[19] When the level of estrogen excretion reaches a point between 50 and 100 μg/day, 1 day without treatment is allowed to elapse, and HCG is given on the following day.

Second, injections of gonadotrophin may be given at intervals, and the most widely used example of this system is the injection of three doses given on alternate days, namely days 1, 3, and 5 of the treatment cycle. The usual starting dose is to give three ampules of HMG at each injection, increasing the dose in steps of approximately 50% in each subsequent course until a response is obtained (i.e., 5, 7, 10 ampules per injection). On days 6 and 7 of the treatment cycle the urinary estrogen excretion of between 50 and 100 μg total estrogens represents an adequate ovarian response. It should be noted, however, that rise of more than 50% per day in estrogen excretion probably represents evidence of overstimulation, even if the level of urinary estrogen has not reached 100 μg/day. These subjects are at risk from superovulation and multiple pregnancy on the one hand, and the hyperstimulation syndrome on the other. When an adequate and satisfactory estrogen response is obtained by days 6 and 7 with a suitably progressive rise over the previous days, HCG is given on day 8, usually the peak day of estrogen production.[13] As with the daily treatment system, the dose of HCG used has varied between 5,000 and 12,000 units in different centers, but many workers use 10,000 units as the standard dose to precipitate ovulation. It has been found best to give HCG as a single dose, as this is less likely to produce superovulation.[22,23] The dose of HCG required is least if estrogen response is optimal.[20]

It may be found on occasion that the rise in estrogen began at a late stage of the treatment cycle, say on the 4th day, and although a steady rise in estrogen excretion is occurring, the level on the 7th day is still under the therapeutic range. It is possible under such circumstances to give a further injection of HMG/HPG on day 8, with HCG being given on day 10.

After HCG has been given, the patient is advised to undertake intercourse on each of the next 3 succeeding days. Usually 12 to 24 hr after the HCG injection most patients experience discomfort in one or both iliac fossae, usually of sudden onset, and lasting an hour or so. This is the manifestation of ovulation corresponding to mittelschmerz, but is not adequate evidence that ovulation has taken place. It is necessary to seek biochemical evidence to confirm ovulation. The patient should continue to collect continuous 24-hr urine samples, and it will be seen that after ovulation the urinary estrogen excretion falls appreciably between the 10th and 12th days of the treatment cycle, before rising to the second peak on or after the 14th day, often equaling or even exceeding the preovulatory peak, and rather more prolonged. Urinary pregnanediol rises from the basal level of 1 mg/day or less in parallel with the second estrogen peak, often reaching levels in excess of 4 mg/day, although a level over 2.5 mg/24 hr in the author's laboratory is regarded as evidence of ovulation.

Following ovulation other parameters change. The basal body temperature rises by about 1°F (often over 24 hr but occasionally over 2 or 3 days), ferning is lost, spinnbarkeit decreases to 1 to 2 cm, and the changes of ovulation are seen in the karyopyknotic index. If fertilization and nidation do not occur, then menstruation follows some 10 to 14 days after the HCG injection. If the temperature and urinary pregnanediol levels remain raised for over 14 days, then pregnancy should be suspected. Although with careful patient selection ovulation may be induced in over 90% of cases, the frequency of the successful achievement of pregnancy is only 30 to 50% in the most carefully treated series.[49,119] The reasons for this are obscure. There are many possible barriers to conception which may occur and which are not routinely sought, such as sperm antibodies, failure of sperm capacitation, and failure of mucus penetration. Methods are available for these studies, but will need to be specially sought to investigate the unsuccessful case. One feature of patients treated with human gonadotrophin is the thick, viscid cervical mucus which is not uncommonly produced. One personal communication to the author describes two patients in whom repeated ovulation was achieved but without fertilization and conception. The introduction of a mucolytic agent in both cases resulted in pregnancy. The author has, however, been unable to repeat this success, and as far as he is aware no study of this treatment has been made elsewhere. Once a satisfactory dose has been achieved, it will be noticed that in most individuals the responses in succeeding courses diminish progressively, or may even cease abruptly. It is necessary to increase the dose in any further courses, usually with the return of response. Sometimes a return of previous sensitivity can be obtained by stopping treatment for one or two "cycles," when the reintroduction of the lowest dose to which a response was obtained will often result in ovulation.[27]

Once pregnancy is successfully achieved, it should not be forgotten that patients who become pregnant after treatment for infertility carry a greater risk of abortion than do normal subjects. The rate is said to be up to 25%,[49] well above the 10% quoted for normal conceptions. Similarly, all workers in this field find that despite the greatest care, multiple pregnancy is not uncommon. Single pregnancies account for about 50%, twin pregnancies for 30 to 40%, and triplets for most of the remainder.[47,48,96] This means that the risks of premature labor and the other complications of multiple pregnancy are present.

The two major complications of treatment are superovulation and hyperstimulation. Multiple ovulation or superovulation is most likely to occur when an excessive stimulating dose of FSH (HPG or HMG) has been given, or when repeated doses of LH (HCG) are given after the initial peak of estrogen secretion. In the former case the level of urinary total estrogen excretion will be in excess of 120 μg/24 hr on the 6th and 7th days of the treatment cycle, or will show a progressive rate of increase in which the daily estrogen excretion doubles or more than doubles each day. Under such circumstances multiple ovulation is almost certain to occur if HCG is given, and the injection should be withheld. The chances of such pregnancies proceeding to normal term are small, and the obstetric risks are high. In some centers, repeated injections of HCG have been suggested to induce ovulation, but in the author's own experience superovulation and severe ovarian enlargement will occur.

Hyperstimulation, with ascites and pleural effusions, severe ovarian enlargement, and often a degree of shock occur if HCG is given in the situation of excessive FSH (HPG or HMG) stimulation described above.[14,45,96] In particularly severe cases, thromboembolic states may occur and fatalities have taken place.[88] Only by most careful monitoring of hormone production by plasma levels or urinary excretion studies performed continuously through the initial phases of each cycle of treatment can hyperstimulation be avoided. Other parameters of estrogen production such as fern test, spinnbarkeit, and karyopyknotic index are not sufficiently sensitive to be used as alternatives to hormone assays.

Despite the difficulties of treatment and the

hazards to which the pregnancy is exposed, there can be no doubt that this management is at the moment the only treatment that can overcome the infertility of this group of patients (although the advent of the synthetic hypothalamic releasing hormones may eventually open a further and possibly more successful avenue),[69] and the dedication with which the patients apply themselves to the demanding treatment regime is such that the attending physician must show equal diligence if the greatest chance of conception with the least risk of iatrogenic ill effects is to be achieved.

D. Conclusion

The management of anovulatory infertility is far from a simple process. It is necessary to seek a precise diagnosis if time and resources are not to be wasted. Thorough investigation cannot be evaded, and only by careful delineation of all aspects of normal and abnormal function of the hypothalamo-pituitary-ovarian axis in the individual patient can the chance of designing the regime of management most likely to succeed in that particular subject be offered. If this is done the chances of success are good and the rewards for the patient great. Clinics for the investigation and management of this condition are best run jointly by an endocrinologist and an obstetrician, with the cooperation of a well-equipped and experienced hormone laboratory. Only in this way can the necessary expertise be made available, and only in this way can the experience from which new advances are made be easily obtained. There still remain a group of individuals in whom pregnancy is not achieved despite all measures and without any clear indication as to cause. It is in this group where the greatest work remains to be done.

REFERENCES

1. **Abraham, G. E., Odell, W. D., Swerdloff, R. S., and Hopper, K.,** Simultaneous radioimmunoassay of plasma FSH, LH, progesterone, 17-hydroxy-progesterone, and estradiol-17β during the menstrual cycle, *J. Clin. Endocrinol. Metab.,* 34, 312, 1972.
2. **Argonz, J. and Del Castillo, E. B.,** A syndrome characterised by estrogenic insufficiency, galactorrhoea and decreased urinary gonadotrophin, *J. Clin. Endocrinol. Metab.,* 13, 79, 1953.
3. **Axelrod, L. R. and Goldzieher, J. W.,** The polycystic ovary. III. Steroid biosynthesis in normal and polycystic ovarian tissue, *J. Clin. Endocrinol. Metab.,* 22, 431, 1962.
4. **Axelrod, L. R., Goldzieher, J. W., and Ross, S. D.,** Concurrent 3β-hydroxy-steroid dehydrogenase deficiency in adrenal and sclerocystic ovary, *Acta Endocrinol.* (Kbh.), 48, 392, 1965.
5. **Barraclough, C. A.,** Production of anovulatory sterile rats by single injections of testosterone propionate, *Endocrinology,* 68, 62, 1961.
6. **Bayliss, P. F. C. and Van't Hoff, W.,** Amenorrhoea and galactorrhoea associated with hypothyroidism, *Lancet,* 2, 1399, 1969.
7. **Besser, G. M., McNeilly, A. S., Anderson, D. C., Marshall, J. C., Harsoulis, P., Hall, R., Ormston, B. J., Alexander, L., and Collins, W. P.,** Hormonal responses to synthetic luteinizing hormone and follicle stimulating hormone – releasing hormone in man, *Br. Med. J.,* 3, 267, 1972.
8. **Besser, G. M., Parke, L., Edwards, C. R. W., Forsyth, I. A., and McNeilly, A. S.,** Galactorrhoea: successful treatment with reduction of plasma prolactin levels by bromo-ergocryptine, *Br. Med. J.,* 3, 669, 1972.
9. **Bettendorf, G.,** A comparison of human pituitary and menopausal gonadotrophins in the induction of ovulation, in *Developments in the Pharmacology and Clinical Uses of Human Gonadotrophins,* Butler, J. K., Ed., G. D. Searle, High Wycombe, Engl., 1970, 101.
10. **Boyar, R., Perlow, M., Hellman, L., Kapen, S., and Weitzman, E.,** Twenty-four hour pattern of LH secretion in normal men with sleep stage recording, *J. Clin. Endocrinol. Metab.,* 35, 73, 1972.
11. **Brooks, R. V., Jeffcoate, S. L., London, D. R., Prunty, F. T. G., and Smith, P. M.,** *Proc. 2nd Symp. Steroid Hormones,* Int. Congr. Ser., 101, 108, 1966.
12. **Brown, J. B.,** Urinary excretion of oestrogens during the menstrual cycle, *Lancet,* 1, 320, 1955.
13. **Butler, J. K.,** Collected results of the treatment of infertile women with Pergonal, in *The Clinical Uses of Gonadotrophins,* Butler, J. K., Ed., G. D. Searle and Co. Ltd., London, 1966, 63.

14. **Buxton, C. L. and Herrmann, W.,** Induction of ovulation in the human with human gonadotrophins, *Am. J. Obstet. Gynecol.,* 81, 584, 1961.
15. **Cargille, C. M., Vaitukaitis, J. L., Bermudez, J. A., and Ross, G. T.,** A differential effect of ethinyl estradiol upon plasma FSH and LH relating to time of administration in the menstrual cycle, *J. Clin. Endocrinol. Metab.,* 36, 87, 1973.
16. **Charles, D., Loraine, J. A., Bell, E. T., and Harkness, R. A.,** The mechanism of action of Clomiphene, in *Fertility and Sterility,* Proc. 5th World Congress on Fertility and Sterility, Stockholm, 1966, Excerpta Medica Foundation, Amsterdam, 1967, 92.
17. **Chiari, J., Brown, C., and Spaeth, J.,** in *Klin. Geburtsh. Gynaek.,* 1855, 371; Frommel R., Ueber puerperale atrophie des uterus, *Z. Geburtshilfe Gynaekol.,* 7, 305, 1882.
18. **Corral-Gallardo, J., Acevedo, H. A., de Salazar, J. L. P., Loria, M., and Goldzieher, J. W.,** The polycystic ovary. VI. A hilus cell tumour of the ovary associated with polycystic ovarian disease: in vivo and in vitro studies, *Acta Endocrinol.* (Kbh.), 52, 425, 1966.
19. **Crooke, A. C., Butt, W. R., and Bertrand, P. V.,** Clinical trial of human gonadotrophins. III. Variation in sensitivity between patients and standardisation of treatment, *Acta Endocrinol. [Suppl.]* (Kbh.), 111, 1966.
20. **Crooke, A. C., Butt, W. R., and Bertrand, P. V.,** Ovarian function and ovulation induction, *Lancet,* 2, 1030, 1966.
21. **Crooke, A. C., Butt, W. R., Palmer, R. F., Bertrand, P. V., Carrington, S. P., Edwards, R. L., and Anson, C. J.,** Clinical trial of human gonadotrophins. II. The effect of pituitary and urinary follicle stimulating hormone and chorionic gonadotrophin on patients with idiopathic secondary amenorrhoea, *J. Obstet. Gynaecol. Br. Commonw.,* 71, 571, 1964.
22. **Crooke, A. C., Butt, W. R., Morris, R., and Palmer, R. F.,** Pregnancy following treatment with human pituitary follicle-stimulating hormone and chorionic gonadotrophin, *Acta Endocrinol. [Suppl.]* (Kbh.), 67, 132, 1962.
23. **Crooke, A. C., Butt, W. R., Palmer, R. F., Morris, R., Edwards, R. L., and Anson, C. J.,** Clinical trial of human gonadotrophins. I. The effect of pituitary and urinary follicle-stimulating hormone and chorionic gonadotrophin on patients with idiopathic secondary amenorrhoea, *J. Obstet. Gynaecol. Br. Commonw.,* 70, 604, 1963.
24. **Crooke, A. C., Butt, W. R., Palmer, R., Morris, R., Edwards, R. L., Taylor, C. W., and Short, R. V.,** Effect of human pituitary follicle-stimulating hormone and chorionic gonadotrophin in Stein-Leventhal syndrome, *Br. Med. J.,* 1, 1119, 1963.
25. **Crooke, A. C., Butt, W. R., Palmer, R. F., Morris, R., and Morgan, D. B.,** The effect of human gonadotrophins on a patient with Simmonds's disease, *Acta Endocrinol.* (Kbh.), 46, 292, 1964.
26. **David, A., Insler, V., and Ber, R.,** The fern test following treatment with human gonadotrophins, *Proc. Tel Hashomer Hosp.,* 3, 193, 1964.
27. **Davies, D.,** personal observations, 1973.
28. **Deller, J. J., Wegienka, L. C., Conte, N. F., Rosner, J. M., and Forsham, P. H.,** Testosterone metabolism in idiopathic hirsutism, *Ann. Intern. Med.,* 63, 369, 1965.
29. **Del Pozo, E., Brun Del Re, R., Varga, L., and Friesen, H.,** The inhibition of prolactin secretion in man by CB-154 (2-Br-α-ergocryptine), *J. Clin. Endocrinol. Metab.,* 35, 768, 1972.
30. **Dignam, W. J., Pion, R. J., Lamb, E. J., and Simmer, H. H.,** Plasma androgens in women. II. Patients with polycystic ovaries and hirsutism, *Acta Endocrinol.* (Kbh.), 45, 254, 1964.
31. **Edwards, C. R. W., Forsyth, I. A., and Besser, G. M.,** Amenorrhoea, galactorrhoea and primary hypothyroidism with high circulating levels of prolactin, *Br. Med. J.,* 3, 462, 1971.
32. **Eshkol, A. and Lunenfeld, B.,** Purification and separation of follicle-stimulating hormone (FSH) and luteinizing hormone (LH) from human menopausal gonadotrophin (HMG), *Acta Endocrinol.* (Kbh.), 54, 91, 1967.
33. **Ferriman, D.,** *Anovulatory Infertility,* William Heinemann, London, 1969.
34. **Ferriman, D. and Purdie, A. W.,** Association of oligomenorrhoea, hirsuties and infertility, *Br. Med. J.,* 2, 69, 1965.
35. **Ferriman, D., Purdie, A. W., and Corns, M.,** Factors affecting the response to Clomiphene therapy, *Br. Med. J.,* 4, 444, 1967.
36. **Forbes, A. P., Henneman, P. H., Griswold, G. C., and Albright, F.,** Syndrome characterised by galactorrhoea, amenorrhoea, and low urinary FSH: comparison with acromegaly and normal lactation, *J. Clin. Endocrinol. Metab.,* 14, 265, 1954.
37. **Forsyth, I. A., Besser, G. M., Edwards, C. R. W., Francis, L., and Myres, R. P.,** Plasma prolactin and inappropriate lactation, *Br. Med. J.,* 3, 225, 1971.
38. **Foss, B. A., Horne, H. W., and Hertig, A. T.,** The endometrium and sterility, *Fertil. Steril.,* 9, 193, 1958.
39. **Franklin, R. R. and Dukes, C. D.,** Antispermatozoal antibody and unexplained infertility, *Am. J. Obstet. Gynecol.,* 89, 6, 1964.
40. **Franklin, R. R. and Dukes, C. D.,** Antibodies' role in infertility, *J.A.M.A.,* 189(3), 32, 1964.
41. **Friesen, H., Webster, B. R., Hwang, P., Guyda, H., Munro, R. E., and Read, L.,** Prolactin synthesis and secretion in a patient with Forbes-Albright syndrome, *J. Clin. Endocrinol. Metab.,* 34, 192, 1972.
42. **Gambrell, R. D., Greenblatt, R. B., and Mahesh, V. B.,** Post-pill and pill-related amenorrhoea-galactorrhoea, *Am. J. Obstet. Gynecol.,* 110, 838, 1971.
43. **Gates, R. B., Friesen, H., and Samaan, N. A.,** Inappropriate lactation and amenorrhoea; pathological and diagnostic considerations, *Acta Endocrinol.* (Kbh.), 72, 101, 1973.

44. **Geist, S. H., Gaines, J. A., and Salmon, U. J.,** The effect of gonadotrophins upon the human ovary, *Am. J. Obstet. Gynecol.,* 42, 619, 1941.
45. **Gemzell, C. A.,** Induction of ovulation with human gonadotrophins, *Proc. 2nd Int. Congr. Endocr.,* Int. Cong. Ser., 83, 805, 1964.
46. **Gemzell, C. A.,** Induction of ovulation with human gonadotrophin, *Recent Progr. Horm. Res.,* 21, 179, 1965.
47. **Gemzell, C. A.,** Multiple births following treatment with human gonadotrophins, in *Fertility and Sterility,* Proc. 5th World Congress on Fertility and Sterility, Stockholm, 1966, Excerpta Medica Foundation, Amsterdam, 1967, 43.
48. **Gemzell, C. and Roos, P.,** Pregnancies following treatment with human gonadotrophins with special reference to the problem of multiple births, *Am. J. Obstet. Gynecol.,* 94, 490, 1966.
49. **Gemzell, C. A., Roos, P., and Loeffler, F. E.,** Follicle stimulating hormone extracted from human pituitary, in *Progress in Infertility,* Behrman, S. J. and Kistner, R. W., Eds., Little, Brown, Boston, 1968.
50. **Gillam, J. S.,** Study of inadequate secretion-phase endometrium, *Fertil. Steril.,* 6, 18, 1955.
51. **Goering, R. W. and Herrmann, W. L.,** Effect of ovarian stimulation on estrogen production rates, *J. Clin. Endocrinol. Metab.,* 26, 65, 1966.
52. **Goldzieher, J. W.,** Polycystic ovarian disease, in *Progress in Infertility,* Behrman, S. J. and Kistner, R. W., Eds., Little, Brown, Boston, 1968.
53. **Goldzieher, J. W. and Green, J. A.,** The polycystic ovary. I. Clinical and histological features, *J. Clin. Endocrinol. Metab.,* 22, 325, 1962.
54. **Grant, A., McBride, W. G., and Moyes, J. M.,** Luteal phase defect in infertility, *Int. J. Fertil.,* 4, 315, 1959.
55. **Greenblatt, R. B., Barrfield, W. E., Jungck, E. C., and Ray, A. W.,** Induction of ovulation with MRL-41, *J.A.M.A.,* 178, 101, 1961.
56. **Greep, R. O.,** Physiology of the anterior hypophysis in relation to reproduction, in *Sex and Internal Secretions,* Young, W. C., Ed., Williams and Wilkins, Baltimore, 1961, 240.
57. **L'Hermite, M.,** The present status of prolactin assays in clinical practice, *Clinics Endocrinol. Metab.,* 2, 423, 1973.
58. **Horth, C. E. and Palmer, R. F.,** Measurement of progesterone in human plasma, *Clin. Endocrinol.* (Lond.), 1, 199, 1972.
59. **Horton, R.,** Estimation of androstenedione in human peripheral blood with ^{35}S-thiosemicarbazide, *J. Clin. Endocrinol. Metab.,* 25, 1237, 1965.
60. **Horton, R. and Neisler, J.,** Plasma androgens in patients with polycystic ovary syndrome, *J. Clin. Endocrinol. Metab.,* 28, 479, 1968.
61. **Iannaccone, A., Gabrilove, J. L., Sohval, A. R., and Soffer, L. J.,** The ovaries in Cushing's syndrome, *N. Engl. J. Med.,* 261, 775, 1959.
62. **Igarashi, M. and Matsumoto, S.,** Induction of human ovulation by individualized gonadotrophin therapy in two phases, *Am. J. Obstet. Gynecol.,* 73, 1294, 1957.
63. **Jeffcoate, T. N. A.,** The androgenic ovary, with special reference to the Stein-Leventhal syndrome, *Am. J. Obstet. Gynecol.,* 88, 143, 1964.
64. **Jones, G.,** Luteal phase defects, in *Progress in Infertility,* Behrman, S. J. and Kistner, R. W., Eds., Little, Brown, Boston, 1968, 299.
65. **Jones, H. W. and Jones, G. E. S.,** The gynaecological aspects of adrenal hyperplasia and allied disorders, *Am. J. Obstet. Gynecol.,* 68, 1330, 1954.
66. **Judd, H. L. and Yen, S. S. C.,** Serum androstenedione and testosterone levels during the menstrual cycle, *J. Clin. Endocrinol. Metab.,* 36, 475, 1973.
67. **Kapen, S., Boyar, R., Hellman, L., and Weitzman, E. D.,** Episodic release of luteinizing hormone at mid-menstrual cycle in normal adult women, *J. Clin. Endocrinol. Metab.,* 36, 724, 1973.
68. **Kastin, A. J., Gual, C., and Schally, A. V.,** Clinical experience with hypothalamic releasing hormones. Part 2. Luteinizing hormone-releasing hormone and other hypophysiotropic releasing hormones, *Recent Progr. Horm. Res.,* 28, 201, 1972.
69. **Kastin, A. J., Zarate, A., Midgley, A. R., Canales, E. S., and Schally, A. V.,** Ovulation confirmed by pregnancy after infusion of porcine LH-RH, *J. Clin. Endocrinol. Metab.,* 33, 980, 1971.
70. **Keye, W. R., Yuen, B. H., and Jaffe, R. B.,** New concepts in the physiology of the menstrual cycle, *Clinics Endocrinol. Metab.,* 2, 451, 1973.
71. **Kinch, R. A. H., Plunkett, E. R., and Devlin, M. C.,** Post-partum amenorrhoea-galactorrhoea of hypothyroidism, *Am. J. Obstet. Gynecol.,* 105, 766, 1969.
72. **Kirschner, M. A. and Jacobs, J. B.,** Combined ovarian and adrenal vein catheterization to determine the site(s) of androgen overproduction in hirsute women, *J. Clin. Endocrinol. Metab.,* 33, 199, 1971.
73. **Kirschner, M. A., Sinhamahapatra, S., Zucker, I. R., Loriaux, L., and Nieschlag, E.,** The production, origin, and role of dehydroepiandrosterone and Δ^5-androstenediol as androgen prehormones in hirsute women, *J. Clin. Endocrinol. Metab.,* 37, 183, 1973.
74. **Kistner, R. W.,** Induction of ovulation with clomiphene citrate (Clomid), *Obstet. Gynecol. Surv.,* 20, 873, 1965.
75. **Kistner, R. W.,** Further observations on the effects of Clomiphene citrate in anovulatory females, *Am. J. Obstet. Gynecol.,* 92, 380, 1965.

76. **Kistner, R. W.,** Induction of ovulation with clomiphene citrate, in *Progress in Infertility*, Behrman, S. J. and Kistner, R. W., Eds., Little, Brown, Boston, 1968.
77. **Krieger, D. T., Ossawski, R., Fogel, M., and Allen, W.,** Lack of circadian periodicity of human serum FSH and LH levels, *J. Clin. Endocrinol. Metab.*, 35, 619, 1972.
78. **Laitinen, O. and Pesonen, S.,** Some aspects of the function of ovaries in secondary amenorrhoea, *Acta Endocrinol.* (Kbh.), 50, 254, 1965.
79. **Lloyd, C. W., Lobotsky, J., Baird, D. T., McCracken, J. A., Weisz, J., Pupkin, M., Zanartu, J., and Puga, J.,** Concentration of unconjugated estrogens, androgens and gestagens in ovarian and peripheral venous plasma of women: the normal menstrual cycle, *J. Clin. Endocrinol. Metab.*, 32, 155, 1971.
80. **Lloyd, C. W., Lobotsky, J., Segre, E. J., Kobayashi, T., Taymor, M. L., and Batt, R. E.,** Plasma testosterone and urinary 17-ketosteroids in women with hirsutism and polycystic ovaries, *J. Clin. Endocrinol. Metab.*, 26, 314, 1966.
81. **Loraine, J. A. and Bell, E. T.,** Hormone excretion during the normal menstrual cycle, *Lancet*, 1, 1340, 1963.
82. **Mahesh, V. B. and Greenblatt, R. B.,** Steroid secretions of the normal and polycystic ovary, *Recent Progr. Horm. Res.*, 20, 341, 1964.
83. **Marcus, C. C. and Marcus, S. L.,** The cervical factor, in Infertility, McLane, C. M., Ed., *Clinics Obstet. Gynecol.*, 8(1), 15, 1965.
84. **Mellinger, R. C., Thompson, R. J., and Manson, J. A.,** Pituitary-gonadal stimulating effect of clomiphene, *Clin. Res.*, 11, 224, 1963.
85. **Moghissi, K. S., Dabich, D., Levine, J., and Neuhaus, O. W.,** Mechanism of sperm migration, *Fertil. Steril.*, 15, 15, 1964.
86. **Monroe, S. E., Jaffe, R. B., and Midgley, A. R., Jr.,** Regulation of human gonadotropins. XII. Increase in serum gonadotropins in response to estradiol, *J. Clin. Endocrinol. Metab.*, 34, 342, 1972.
87. **Montgomery, D. A. D., Welbourn, R. B., McCaughey, W. T. E., and Gleadhill, C. A.,** Pituitary tumours manifested after adrenalectomy for Cushing's syndrome, *Lancet*, 2, 707, 1959.
88. **Mozes, M., Bogokowsky, H., Antebi, E., Lunenfeld, B., Rabau, E., Serr, D. M., David, A., and Salomy, M.,** Thromboembolic phenomena after ovarian stimulation with human gonadotrophins, *Lancet*, 2, 1213, 1965.
89. **Naftolin, F., Yen, S. S. C., Perlman, D., Tsai, C. C., Parker, D. C., and Vargo, T.,** Nocturnal patterns of serum gonadotrophins during the menstrual cycle, *J. Clin. Endocrinol. Metab.*, 37, 6, 1973.
90. **Nankin, H. R. and Troen, P.,** Repetitive luteinising hormone elevations in serum of normal men, *J. Clin. Endocrinol. Metab.*, 33, 558, 1971.
91. **Nasr, H., Mozaffarian, G., Pensky, J., and Pearson, O. H.,** Prolactin-secreting pituitary tumours in women, *J. Clin. Endocrinol. Metab.*, 35, 505, 1972.
92. **Naville, A. H., Kistner, R. W., Wheatley, R. E., and Rock, J.,** Induction of ovulation with clomiphene citrate, *Fertil. Steril.*, 15, 290, 1964.
93. **Normington, E. A. M. and Davies, D.,** Hypertension and oedema complicating pregnancy in Addison's disease, *Br. Med. J.*, 1, 148, 1972.
94. **Pfeiffer, C. A.,** Sexual differences of the hypophyses and their determination by the gonads, *Am. J. Anat.*, 58, 195, 1936.
95. **Prunty, F. T. G.,** Hirsutism, virilism and apparent virilism and their gonadal relationship, part II, *J. Endocrinol.*, 38, 203, 1967.
96. **Rabau, E., David, A., Serr, D. M., Maschiach, S., and Lunenfeld, B.,** Human menopausal gonadotrophins for anovulation and sterility, *Am. J. Obstet. Gynecol.*, 98, 92, 1967.
97. **Rabau, E., Serr, D. M., Maschiach, S., Insler, V., Salomy, M., and Lunenfeld, B.,** Current concepts in the treatment of anovulation, *Br. Med. J.*, 4, 446, 1967.
98. **Rice, B. F. and Savard, K.,** Steroid hormone formation in the human ovary. IV. Ovarian stromal compartment; formation of radioactive steroids from Acetate-1-^{14}C and action of gonadotrophins, *J. Clin. Endocrinol. Metab.*, 26, 593, 1966.
99. **Riley, G. M. and Evans, T. N.,** Effects of clomiphene citrate on anovulatory ovarian function, *Am. J. Obstet. Gynecol.*, 89, 97, 1964.
100. **Rosenfield, R. L., Ehrlich, E. N., and Cleary, R. E.,** Adrenal and ovarian contributions to the elevated free plasma androgen levels in hirsute women, *J. Clin. Endocrinol. Metab.*, 34, 92, 1972.
101. **Ross, F. and Nusynowitz, M. L.,** A syndrome of primary hypothyroidism, amenorrhoea, and galactorrhoea, *J. Clin. Endocrinol. Metab.*, 28, 591, 1968.
102. **Ross, G. T., Cargille, C. M., Lipsett, M. B., Rayford, P. L., Marshall, J. R., Strott, C. A., and Rodbard, D.,** Pituitary and gonadal hormones in women during spontaneous and induced ovulatory cycles, *Recent Progr. Horm. Res.*, 26, 1, 1970.
103. **Roth, J. C., Kelch, R. P., Kaplan, S. L., and Grumbach, M. M.,** FSH and LH response to LHRF in prepubertal and pubertal children, adult males, and patients with hypogonadotropic and hypergonadotropic hypogonadism, *J. Clin. Endocrinol. Metab.*, 35, 926, 1972.
104. **Roy, S., Greenblatt, R. B., Mahesh, V. B., and Jungck, E. C.,** Clomiphene citrate: further observations on its use in induction of ovulation in the human and on its mode of action, *Fertil. Steril.*, 14, 575, 1963.

105. **Ryan, K. J. and Petro, Z.**, Steroid biosynthesis by human ovarian granulosa and thecal cells, *J. Clin. Endocrinol. Metab.*, 26, 46, 1966.
106. **Ryan, K. J. and Smith, O. W.**, Biosynthesis of steroid hormones in the human ovary, *Recent Progr. Horm. Res.*, 21, 367, 1965.
107. **Savard, K., Marsh, J. M., and Rice, B. F.**, Gonadotrophins and ovarian steroidogenesis, *Recent Progr. Horm. Res.*, 21, 285, 1965.
108. **Shearman, R. P.**, Prolonged secondary amenorrhoea after oral contraceptive therapy, *Lancet*, 2, 64, 1971.
109. **Shearman, R. P. and Cox, R. I.**, Clinical and chemical correlations in the Stein-Leventhal syndrome, *Am. J. Obstet. Gynecol.*, 92, 747, 1965.
110. **Shearman, R. P. and Cox, R. I.**, The enigmatic polycystic ovary, *Obstet. Gynecol. Surv.*, 21, 1, 1966.
111. **Shearman, R. P. and Turtle, J. R.**, Secondary amenorrhoea with inappropriate lactation, *Am. J. Obstet. Gynecol.*, 106, 818, 1970.
112. **Siler, T. M. and Yen, S. S. C.**, Augmented gonadotrophin response to synthetic LRF in hypogonadal state, *J. Clin. Endocrinol. Metab.*, 37, 491, 1973.
113. **Smith, K. D., Steinberger, E., and Perloff, W. H.**, Polycystic ovarian disease, *Am. J. Obstet. Gynecol.*, 93, 994, 1965.
114. **Smith, O. W., Smith, G. V., and Kistner, R. W.**, Action of MER-25 and clomiphene on the human ovary, *J.A.M.A.*, 184, 878, 1963.
115. **Solish, G. I., Gershowitz, H., and Behrman, S. J.**, Occurrence and titer of isohaemagglutinins in secretions of the human uterine cervix, *Proc. Soc. Exp. Biol. Med.*, 108, 645, 1961.
116. **Southam, A. L. and Janovski, N. A.**, Massive ovarian hyperstimulation with clomiphene citrate, *J.A.M.A.*, 181, 443, 1962.
117. **Stein, I. F. and Leventhal, M. L.**, Amenorrhea associated with bilateral polycystic ovaries, *Am. J. Obstet. Gynecol.*, 29, 181, 1935.
118. **Strott, C. A., Cargille, C. M., Ross, G. T., and Lipsett, M. B.**, The short luteal phase, *J. Clin. Endocrinol. Metab.*, 30, 246, 1970.
119. **Taymor, M. L.**, Human menopausal gonadotrophin, in *Progress in Infertility*, Behrman, S. J. and Kistner, R. W., Eds., Little, Brown, Boston, 1968.
120. **Trybuchowski, H.**, Effect of ampicillin on the urinary output of steroidal hormones in pregnant and non-pregnant women, *Clin. Chim. Acta*, 45, 9, 1973.
121. **Turkington, R. W.**, Secretion of prolactin by patients with pituitary and hypothalamic tumours, *J. Clin. Endocrinol. Metab.*, 34, 159, 1972.
122. **Turkington, R. W.**, Inhibition of prolactin secretion and successful therapy of Forbes-Albright syndrome with L-dopa, *J. Clin. Endocrinol. Metab.*, 34, 306, 1972.
123. **Vaitukaitis, J. L., Bermudez, J. A., Cargille, C. M., Lipsett, M. B., and Ross, G. T.**, New evidence for an anti-estrogenic action of clomiphene citrate in women, *J. Clin. Endocrinol. Metab.*, 32, 503, 1971.
124. **Vorys, N., Gantt, C. L., Hamwi, G. J., Copeland, W. E., and Ullery, J. C.**, Clinical utility of chemical induction of ovulation, *Am. J. Obstet. Gynecol.*, 88, 425, 1964.
125. **Woolever, C. A.**, Daily plasma progesterone levels during the menstrual cycle, *Am. J. Obstet. Gynecol.*, 85, 981, 1963.
126. **Yen, S. S. C., Rebar, R., Vandenberg, G., and Judd, H.**, Hypothalamic amenorrhea and hypogonadotropinism: responses to synthetic LRF, *J. Clin. Endocrinol. Metab.*, 36, 811, 1973.
127. **Yen, S. S. C., Rebar, R., Vandenberg, G., Naftolin, F., Ehara, Y., Engblom, S., Ryan, K. J., and Benirschke, K.**, Synthetic luteinizing hormone-releasing factor: a potent stimulator of gonadotropin release in man, *J. Clin. Endocrinol. Metab.*, 34, 1108, 1972.
128. **Yen, S. S. C. and Tsai, C. C.**, Acute gonadotropin release induced by exogenous estradiol during the mid-follicular phase of the menstrual cycle, *J. Clin. Endocrinol. Metab.*, 34, 298, 1972.
129. **Yen, S. S. C., Tsai, C. C., Naftolin, F., Vandenberg, G., and Ajabor, L.**, Pulsatile patterns of gonadotropin release in subjects with and without ovarian function, *J. Clin. Endocrinol. Metab.*, 34, 671, 1972.
130. **Yen, S. S. C., Vandenberg, G., Rebar, R., and Ehara, Y.**, Variation of pituitary responsiveness to synthetic LRF during different phases of the menstrual cycle, *J. Clin. Endocrinol. Metab.*, 35, 931, 1972.
131. **Yen, S. S. C., Vela, P., and Rankin, J.**, Inappropriate secretion of follicle-stimulating hormone and luteinising hormone in polycystic ovarian disease, *J. Clin. Endocrinol. Metab.*, 30, 435, 1970.

THE BIOCHEMISTRY OF CONTRACEPTION

G. W. Pennington and S. Naik

TABLE OF CONTENTS

I. ORAL CONTRACEPTION

More than 20 million women throughout the world are currently taking oral contraceptives. Conflicting reports in both medical and lay press regarding physical, psychological, and pathological alterations in body function have produced confusion in the minds of both medically qualified and lay people.

II. STRUCTURE OF THE STEROID CONTRACEPTIVES

The most commonly used pill is a combination of an estrogen and a progestogen. The two estrogens widely used are ethinyl estradiol or its 3-methyl ether, mestranol (Figure 1). The synthetic progestogens are related to either 19-nortestosterone or 17α-hydroxyprogesterone, although modifications have been made to each of these compounds. The 19-norsteroids are partially metabolized to estrogenic compounds,[1,2] although only norethynodrel and ethynodiol diacetate demonstrate estrogenic activity of clinical importance. The 17-hydroxy progestogens are devoid of estrogenic or androgenic effects, but like the nortestosterone group they oppose the effect of estrogen at a cellular level. The two types of derivative are shown in Figure 1. The combined contraceptive is usually taken as a daily tablet for 21 consecutive days, starting where possible on the 5th day of the menstrual cycle. On completion of the 21-day course of tablets, medication is stopped for 7 days, and whether or not menstruation has

A. Estrogens

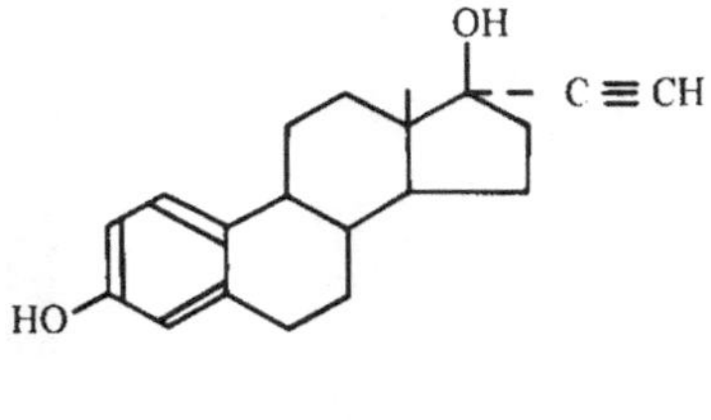

Ethinyl estradiol

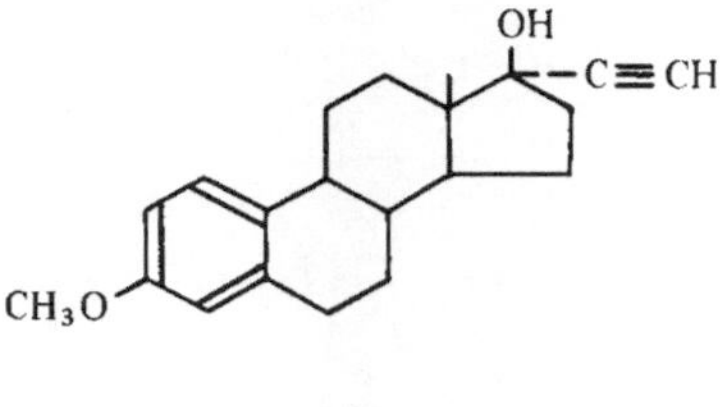

Mestranol

B. Progestogens

i. 19-Nortestosterone derivatives

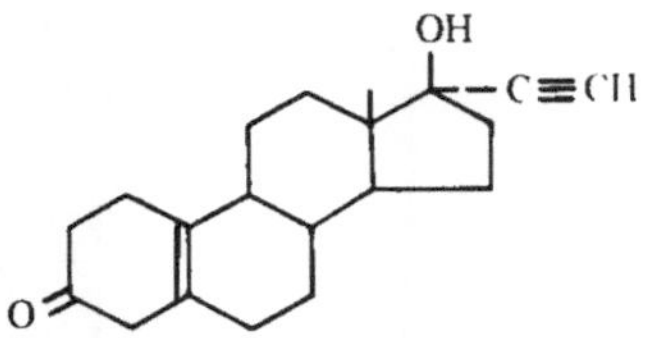

Norethynodrel

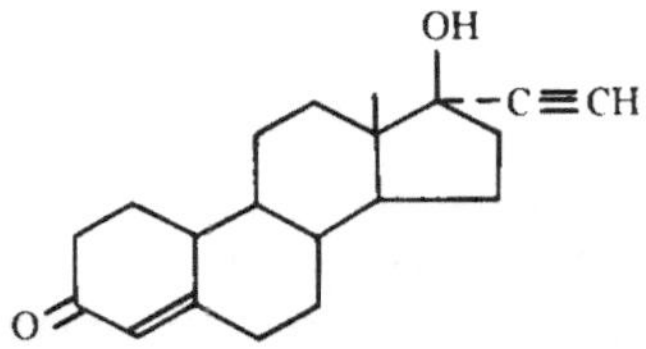

Norethisterone

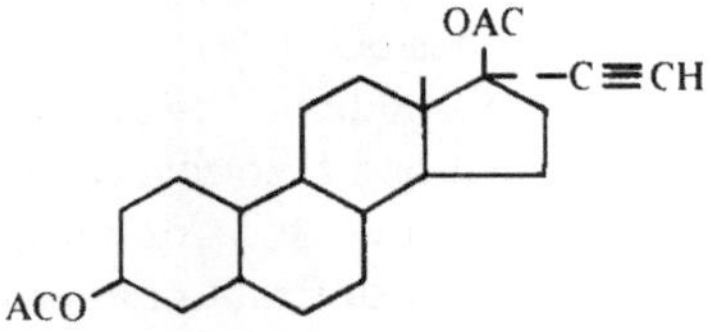

Ethynodiol acetate

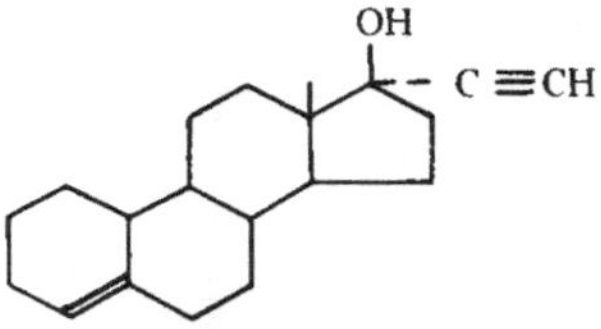

Lynestrenol

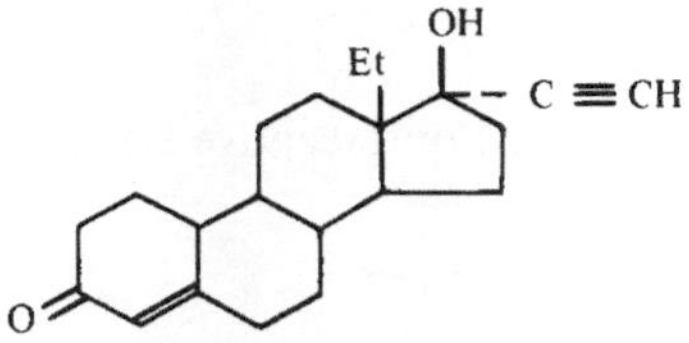

Norgestrol

ii. 17-Hydroxyprogesterone derivatives

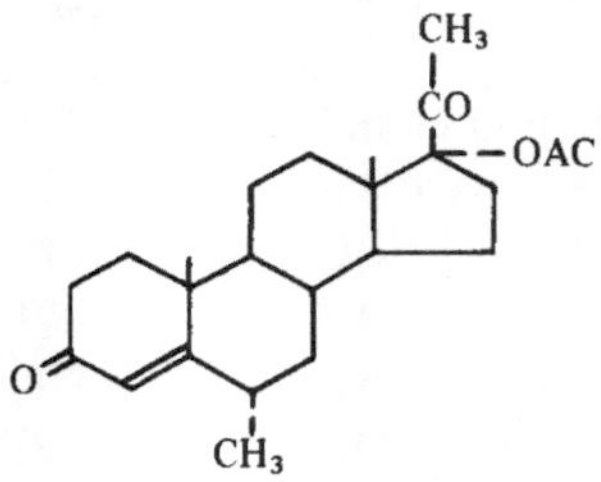

Medroxyprogesterone acetate

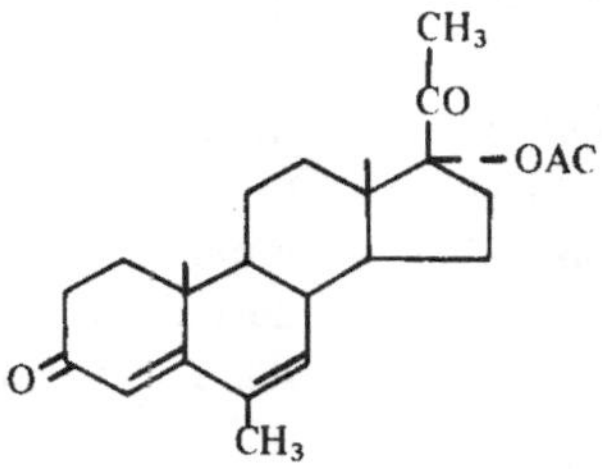

Megestrol acetate

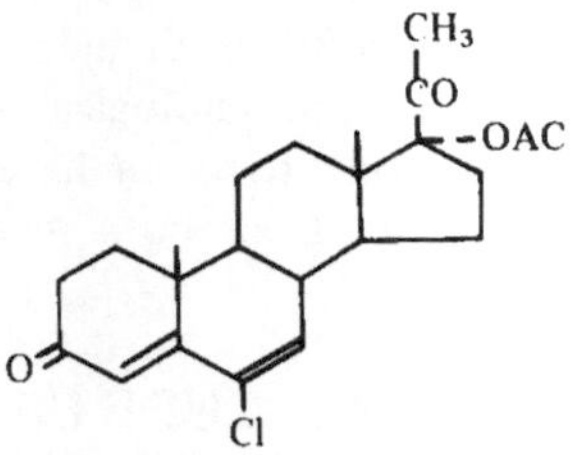

Chlormadinone acetate

FIGURE 1. Estrogen and progestogen components of the oral contraceptive pill.

occurred, is started again after this interval. Normally, menstruation occurs 2 to 3 days after the last pill. Other types of oral contraceptive include the sequential pill, where an effort is made to mimic the normal physiological sequence of hormonal effects when an estrogen alone is administered for 21 days with an estrogen/progestogen combined pill for the next 7 days. Again, 7 clear days are allowed for bleeding to occur prior to commencement of further medication. This method has a relatively high failure rate and, as a higher estrogen dosage is required, it is not widely used. A continuous progestogen-only pill, "the minipill," has also been tried and has been successful in some areas. Its failure rate is, however, higher than the combined pill and there is a tendency toward erratic menstrual bleeding in some women.

III. MODES OF ACTION

The main mode of action of the combined pill is in the suppression of ovulation by inhibition of the hypothalamus and the consequent reduction in FSH/LH releasing factors. It is, however, evident that there are also other effects at the level of the endometrium, ovary, and cervical mucus. All progestogens affect the endometrium to produce a pseudodecidual reaction which prevents implantation. They also act directly on the cervical mucus,[3,4] making it impenetrable to sperm. The effect on the ovary appears to be produced via the corpus luteum by a direct interference with its function.[5,6]

As might be expected, the hormonal changes observed during the contraceptive cycle are predictable, namely, an absence of the normal midcycle peaks of luteinizing hormone and follicle-stimulating hormone and the loss of the biphasic estrogen excretion pattern together with an absent progesterone elevation in the luteal phase of the cycle.

IV. MAJOR SIDE EFFECTS

Major side effects that have been reported include hypertension, jaundice, and thromboembolic disease.

A. Hypertension

The substantial incidence of hypertension in the general population makes it difficult to identify with certainty the women in whom the administration of oral contraceptives has led to an increase in the preexisting hypertension or has produced hypertension in previously normal women. In some reported cases,[7-9] blood pressure has returned to normal following cessation of treatment. A prospective study reported by Kunin et al.[10] indicated that provided weight difference was taken into account, the blood pressure was slightly higher in those women on the pill and that this complication affected a small proportion of the total. The precise cause of the hypertension is unknown, although it is of interest that Laragh et al.[11] demonstrated an increased aldosterone excretion and found renin substrate elevated in nine of ten hypertensive women on oral contraceptives. More recently, Beckerhoff, et al.[12] have conclusively demonstrated a marked increase in plasma aldosterone levels in the first week of medication, which rises still further during the third month. These workers were able to demonstrate an increase in plasma renin activity and in renin substrate coupled with a diminution in plasma renin concentration. This striking alteration of the renin/angiotension/aldosterone systems is usually well compensated, but may obviously accentuate or precipitate the development of hypertension in susceptible women.

B. Thromboembolic Disease

Isolated case reports of the occurrence of thromboembolic disease first appeared in 1961.[13] Despite variable earlier reports, a definite relationship has been established between the contraceptive pill and thromboembolic disease.[14,15] Inman and Vessey[14] estimated an annual death rate of 1.5 per 100,000 healthy nonpregnant women aged 20 to 34 years and 3.9 per 100,000 aged 35 to 44 among contraceptive pill users, as compared with 0.2 and 0.5 per 100,000, respectively, in nonpill users. Vessey and Doll[15] demonstrated that with respect to nonfatal idiopathic deep-vein thrombosis or pulmonary embolism, 45% of the study group had been taking oral contraceptives, compared with only 9% of the controls. Dugdale and Mason[16] have summarized the reports of many investigators and have demonstrated a substantial increase in platelets in a majority of oral contraceptive users. A possible link between thrombosis and altered serum lipid levels during oral contraceptive therapy has been provided by Bolton et al.[17] who found altered platelet behavior caused by an abnormality of the low-density lipoprotein, lecithin, similar to that seen in arteriosclerotic patients. Reports referring to the various clotting tests are confused, although there may well be an increase in various coagulation factors (VII, IX, and X). It is not known, however, if such increases facilitate intravascular clotting or thrombosis.[18] Zuck and Bergin[19] have observed shifts in the balance between the generation of thrombin and its activation by antithrombin III, both in patients on oral contraceptives and also in patients with clinical thrombosis not on oral contraceptive therapy. They suggest that a small group of women may have a thrombotic disposition which is accentuated by oral contraceptives. A recent study by the Boston Collaboration Drug Surveillance Programme (1973) has indicated that the risk of thromboembolic disease was 11 times greater in the women taking oral contraceptives as compared with the noncontraceptive user.[20]

Perusal of the very extensive literature, therefore, leaves little doubt that the use of oral contraceptives does incur an increased risk of thromboembolism, particularly with those pills having a high estrogen content. A slightly greater risk appears to be present with the use of the sequential regime, but the risks taken do not appear to be greater than the risk of death through thromboembolic disease throughout a normal pregnancy. Nevertheless, the Committee on Safety of Drugs (Great Britian) has recommended[21] a reduction to 50 μg in the estrogenic component of the contraceptive pill. The confirmation of a reduction in risk following implementation of this recommendation will be known when current studies are completed.[22]

C. Jaundice and Disturbances in Liver Function

The liver plays a central role in the metabolism of both estrogens and progestogens and function may be impaired if administration of oral contraceptives is continued over a long period of time. Despite previous conflicting evidence, it is now clear that there is a relationship between oral contraceptive therapy and liver damage. Although it is difficult to arrive at a true incidence, Schaffner[23] suggested that the incidence in the U.S. is 1 case of jaundice per 10,000 women taking oral contraceptives. The spectrum of hepatic abnormalities that may appear with the use of these compounds is wide, although they may be classified[24] into abnormalities in nonjaundiced women and abnormalities associated with jaundice.

1. Hepatic Abnormalities without Jaundice Occurring in Women on Oral Contraceptives

Alterations in some liver function tests may occur in normal women on oral contraceptives, although in some instances these alterations may be only transient. The abnormalities so far recognized concern bromsulphthalein retention, plasma enzyme activity, and the plasma proteins.

a. Bromsulphthalein Retention

Approximately 15% of women on oral contraceptives initially demonstrate retention, although in the majority it disappears.[25,26] This increase in retention is associated with a decrease in the maximal transport of the dye into the bile; there does not appear to be a change in storage capacity.[27]

b. Aminotransferases

Various authors have stated that between nil and 18% of women have raised aminotransferase levels while on oral contraceptive therapy. It has been shown that increased levels often disappear within 1 month of therapy; thus, recognition will demand early investigation.[23] That this reaction occurs at all after oral contraceptive administration suggests that cellular organelles may have disorganized to some extent. However, electron microscopy has not demonstrated this.

c. Alkaline Phosphatase and Other Enzymes

The occurrence of an alkaline phosphatase during pregnancy that is heat-stable has been well recognized and is used as a test of placental function. The raised alkaline phosphatase level that occurs during estrogen-progestogen administration may be of hepatic origin and may be related to functional alterations involving the hepatic excretory mechanisms.[28,29] Further support for this is demonstrated by the increase in serum 5-nucleotidase activity, which also occurs. This enzyme, like alkaline phosphatase, is located in the excretory surface of the liver cells and is considered highly specific for hepatic disorders. Various oral contraceptives appear to depress serum cholinesterase activity,[30] while the activity of ornithine transcarbamylase is raised during the first month of treatment.

d. Plasma Proteins

A decrease in total serum protein has been demonstrated in women on oral contraceptives.[31] Starch gel electrophoresis shows a zone which migrates in the α_2 globulin region similar to that which occurs in the pregnant woman. Fibrinogen levels are also increased.[32] An increased serum concentration of thyroxine and cortisol-binding globulins[31,33] has been demonstrated, as has an increase in testosterone-binding protein;[34] this gives rise to the finding of increased plasma levels of these three hormones during oral contraceptive administration. Recent work has served to indicate that the effects of oral contraceptives on plasma proteins are variable and are dependent on the actual constituents of the contraceptive pill used.[35]

2. Jaundice Associated with the Use of Oral Contraceptives

Jaundice has now been recorded on more than 100 occasions in women on the contraceptive pill. The case histories in many instances are remarkably similar. The jaundice appears early, often in the first month; the woman suffers from anorexia, malaise, nausea, and pruritis, after which the jaundice occurs accompanied by a dark urine. In most instances, the laboratory investigations are not helpful and demonstrate a serum bilirubin between 3 and 10 mg /100 ml with markedly raised aspartate and alanine aminotransferases to levels between 1,100 and 1,600 I.U. Liver biopsy demonstrates a minimal or absent inflammatory reaction, but a variable degree of hepatocellular degeneration and necrosis. There is also canalicular and hepatocellular bile stasis. It has been suggested that features of hypersensitivity may be present because of the occurrence of eosinophilia and tissue reactions such as erythema nodosum. Clinical remission normally occurs when drug therapy is stopped, although the abnormal liver function tests may persist for a longer period.[36] It is of interest that in a number of reported cases, the jaundice recurred following repeated administration of the oral contraceptive[37,38] and that the disturbed liver function is similar to that observed in some women during the later stages of pregnancy. Evidence that the estrogenic component is primarily responsible for the disturbance includes observations that in a woman with a history of cholestatic jaundice of pregnancy, the jaundice could be induced by the combined pill (estrogen plus progestogen), but not by progestogen alone.[37] Other workers demonstrated that hepatic dysfunction was caused by the estrogenic component.[39-41] There is, however, some evidence that it is possible for the progestogenic component to produce hepatic disturbances. Norethynodrel has been shown to cause bromsulphthalein retention[42] and lynestrenol has been associated with raised aspartate aminotransferase activity.[43] Progestational agents, in general, have been shown to have an effect on glucuronyl transferase, the enzyme conjugating bilirubin with uridine 5-pyrophosphate glucuronic acid. Estrone, estradiol, and estriol do not appear to have this effect, as judged by the in vitro experiments of Lathe and Walker.[44]

In summary, therefore, it appears that the estrogens can certainly impair hepatic excretory function, although the evidence about progestogens is less definite. Since liver function may be adversely affected in some people taking oral contraceptives, certain limitations in the use of these agents are advisable. It is suggested that where there is a history of cholestatic jaundice of pregnancy or where acute or chronic disturbances of liver function are known, oral contraception is contraindicated.

D. The Effect of Oral Contraceptives on Cortisol Metabolism

The effects produced on cortisol metabolism by the administration of oral contraceptives relate largely to the estrogen component, for the administration of progestogens alone has not been shown to have any consistent effect. The estrogen component appears to increase the total plasma cortisol levels, but does not increase the urinary excretion of cortisol metabolites.[45,46] The explanation for this is the increase in cortisol binding to plasma proteins which occurs following estrogen administration.[47,48] A concomitant increase in plasma concentration of corticosteroid-binding globulin (CBG) also occurs,[49-51] which is apparently due to an increased hepatic synthesis of CBG in parallel with the increase in other specific carrier proteins induced by estrogen.[31] It seems likely that the nonprotein bound as well as the protein-bound and total cortisol levels are all increased.[52] These raised levels usually fall to normal within 2 or 3 weeks of cessation of therapy. It has also been shown that estrogens modify the metabolic pathways of cortisol.[53]

E. Effect of Oral Contraceptives on Carbohydrate Metabolism

Despite a number of conflicting reports, the balance of evidence suggests that estrogen-progestogen contraceptives impair glucose tolerance in many women taking these drugs.[54-61] Mean plasma insulin levels, however, do not appear to be significantly different from normal following the administration of either oral or intravenous glucose. It is likely that the impaired glucose tolerance is of the "steroid diabetes" type caused by a raised plasma cortisol level[62] secondary to the increase in the hepatic synthesis of corticosteroid-binding globulin (CBG). It remains unclear whether these changes may ultimately lead to a clinically manifest disease state such as diabetes mellitus or if, for example, the continued impairment of glucose tolerance over a long period will accelerate the rate of development of atherosclerosis.[63-65] Considerable time must elapse before answers to these questions are available.

F. The Effect of Oral Contraceptives on the Serum Lipid Levels

There have been varying reports of the effects of estrogens and progestogens separately and in combination on the blood lipids. Raised fasting plasma triglycerides, cholesterol, low-density, and very low-density lipoprotein levels have been demonstrated in certain women receiving oral contraceptive therapy.[66-73] However, a few investigators have found little change in plasma triglyceride or cholesterol levels.[74,75]

There is evidence to suggest that the increase in serum triglyceride during therapy occurs rapidly. Hazzard et al.[71] found a 47% increase in serum triglyceride after 2 weeks' administration. After the initial rapid rise the level remains at an elevated plateau,[67] while following discontinuation of the drug the blood concentration returns to normal within 6 months.[66] Doar and Wynn[76] have not found a relation between the change in serum lipid levels during therapy and the subject's age or degree of obesity, nor could they find a correlation between changes in serum triglyceride level and duration of therapy. They did, however, note that the lipid response could alter in subjects with abnormally raised serum lipids compared with the change in normal subjects.

In general, estrogens alone decrease total serum cholesterol and serum low-density lipoprotein levels and raise serum high density lipoprotein and triglyceride levels.[77] Mestranol and ethinyl estradiol have identical effects at the same dosage, while the pill with the greatest progestational effect induces the highest cholesterol values, the levels being enhanced or decreased by the estrogen present.[76]

Many reports show no significant effect on serum lipid levels in man during progestogen administration. Chlormadinone does not appear to affect the triglyceride level,[78] while a fall in the level of cholesterol and lipoprotein triglyceride in women taking norethisterone has been reported.[79,80] Androgens, some of which bear a chemical similarity to the progestogens used in oral contraceptive preparations, usually increase serum low-density and decrease serum high-density lipoprotein levels; serum cholesterol levels are little affected.[81,82]

As steriod combinations appear to act in a synergistic rather than in an additive manner on serum lipid levels,[83] it is often difficult to predict the effect of any given estrogen-progestogen combination.

While not all subjects receiving oral contraceptive therapy develop serum lipid levels outside the normal range, the general trend is for the lipoprotein patterns of premenstrual women on oral contraceptive therapy to change to a pattern resembling that of men or of postmenopausal women. The precise mechanism by which serum lipid levels become raised during oral contraceptive therapy is unknown. It is not clear whether the raised serum lipid levels during therapy are due to a primary effect of estrogen and/or progestogen on lipoprotein metabolism or secondary to increased circulating levels of other hormones such as cortisol, thyroxine, or growth hormone. It has been suggested that the rate of hepatic apoprotein synthesis might be increased by estrogen therapy, resulting in the increased release of lipoprotein into the circulation.[76] The rapid rise in serum triglyceride levels during oral contraceptive therapy[71] resembles that of other carrier proteins during estrogen therapy.[84]

G. The Effect of Oral Contraceptives on Tryptophan Metabolism

It has been found that following treatment with estrogens or with combined estrogen-progestogen therapy changes in tryptophan metabolism occur via the nicotinic acid ribonucleotide pathway. This is due to induction of hepatic tryptophan oxygenase by the estrogen.[85]

After an oral load of tryptophan, women taking oral contraceptive preparations excrete increased amounts of the tryptophan metabolites, xanthurenic acid,[86] kynurenine, 3-hydroxykynurenine, 3-hydroxyanthranilic acid, kynurenic acid, and acetyl kynurenine in the urine.[87] No such elevation is observed when progesrogen alone is administered.[88]

Pyridoxal 5-phosphate, the coenzyme derived from various forms of vitamin B_6, is a required cofactor for several of the enzymatic reactions in the tryptophan metabolic pathway. The altered excretion of metabolites in oral contraceptive-treated women is similar to that found in nutritional vitamin B_6 deficiency.[89] Rose et al.[90] have shown that about 80% of women taking oral contraceptives have abnormal tryptophan metabolism, indicating a relative vitamin B_6 deficiency and that about 20% have evidence of absolute deficiency of this vitamin.

H. Occurrence of Cancer

A time interval of at least two decades is necessary before carcinogenic potential can be excluded. At the present time, there is no valid evidence that oral contraceptives are carcinogenic. Indeed, progestogens produce dramatic remission of endometrial cancer and these agents may exert a protective influence in patients with a typical endometrial hyperplasia[91,92] and with breast cancer.[93,94]

I. Minor Side Effects

Usually the minor side effects encountered by some women taking the combined pill improve with the duration of the treatment or when another estrogen-progestogen combination is substituted. The following side effects have been attributed to the estrogen component of the pill: nausea and vomiting, cervical erosion, leukorrhea, and fluid retention. Acne, greasiness of the skin, hirsutism, and depression can be caused by the 19-nortestosterone compounds.

IV. PROSTAGLANDINS

Prostaglandins (PG) are a group of modified hydroxy fatty acids with a wide distribution in mammalian tissue. The basic structure common to all prostaglandins is a prostanoic acid skeleton which is a 20 carbon atom molecule containing a cyclopentane ring (Figure 2). All prostaglandins have a 15-hydroxy substituent and a trans double bond in the 13 position; the degree of unsaturation (i.e., the number of double bonds) of the side chain is indicated by the subscript numeral after the letter. Differences occurring in the structure of the 5 carbon ring are used to subdivide the prostaglandins into four naturally occurring groups, E, F, A, and B (Figure 3). "Unnatural" but active prostaglandins containing an uneven number of carbon atoms have also been described. Prostaglandins influence many human and animal systems.

A major role of prostaglandins may be to influence the contractile response of smooth muscle in the reproductive tract. The way in which prostaglandins act at the muscle cell level is not clear, but it is known that unimpaired oxidative metabolism is required to obtain a contractile response of smooth muscle to prostaglandin E (PGE).[95] It is thought that the action of prostaglandins is dependent on the chelation of Ca^{++} with transport of these ions across the cell membrane, resulting in depolarization and consequent contraction.[96,97]

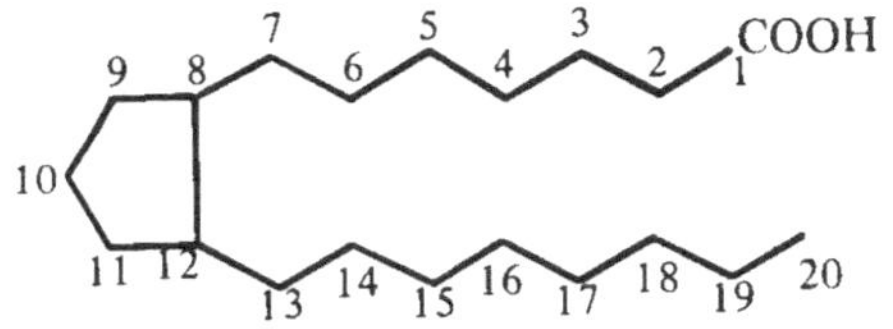

FIGURE 2. Prostanoic acid. Basic structure common to all prostaglandins.

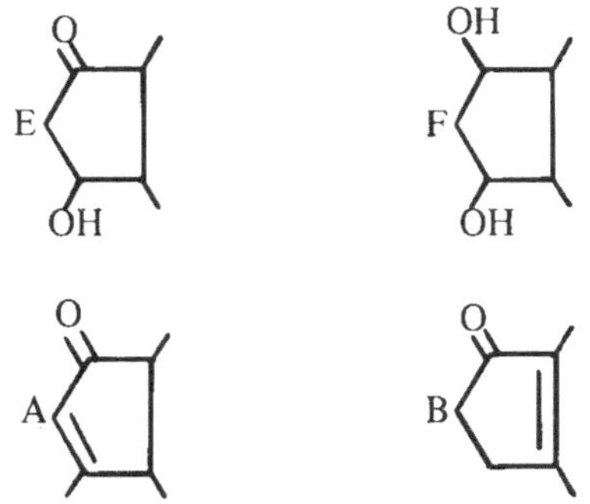

FIGURE 3. Essential structural differences between prostaglandins of E, F, A, and B series.

Both the PGE and the PGF series act on the gastrointestinal tract by increasing intrinsic intestinal activity.

Vasodilation occurs in response to infusion of PGA or PGE.[98] However, the isolated umbilical artery at term contracted when exposed to PGE_2 and $PGF_{1\alpha}$.[99] Direct renal artery infusion in dogs increased urine flow, sodium excretion, and free water clearance.[100]

Prostaglandins have been shown to relax isolated human bronchial muscle.[101] However, Brenner et al.[102] have shown that during intravenous infusion of $PGF_{2\alpha}$ for therapeutic abortion, many patients exhibit changes compatible with bronchoconstriction and reduction in vital capacity occurs.

The naturally occurring prostaglandins are local hormones and it is possible that the tissue level achieved at the site of action may be critical. Where high systemic levels are required to produce adequate tissue concentrations, side effects can be expected.

The use of prostaglandins for fertility control has aroused interest in the last few years. The advantages of a once-a-month contraceptive with abortifacient, luteolytic, or menses-inducing properties would include convenience of administration and freedom from side effects because only periodic short-term administration is required.

Prostaglandins appear to terminate pregnancy by stimulating rhythmic contractions of uterine muscle. The effectiveness of the prostaglandins in early pregnancy of 7 to 8 weeks duration has been established.[103] In contrast to oxytocin, prostaglandins produce effective contractions even early in gestation and in the nonpregnant uterus. Roth-Brandel and co-workers[104] studied the effects of prostaglandins E_1 and $F_{2\alpha}$ given by intravenous injection; doses above the threshold level always resulted in stimulation of the uterus. They also found that both substances induced elevation of uterine tone and increased uterine motility.[105] In a further study, they concluded that the degree of response was, in general, dose dependent.[106] The strong and frequent uterine contractions that result could dislodge the fertilized implanted ovum and initiate bleeding.[107] Bygdeman et al.[108] have noted cases where the pregnancy continued despite menstrual-like bleeding being induced and they assumed that this was due to damage of the vascular bed surrounding the early ovum. The significance of prostaglandins in terms of teratogenicity is not known should the pregnancy continue.

Experimental evidence in guinea pigs, rats, and rabbits indicates that prostaglandins possess luteolytic activity, indicating the possibility that they could be used to prevent nidation by terminating the necessary progestational environment.[109] Changes have not been found in peripheral plasma progesterone levels or in luteal phase length after intravenous infusion of $PGF_{2\alpha}$.[110-112] Hillier et al.[113] found that administration of high doses of $PGF_{2\alpha}$ depressed plasma progesterone levels in the late luteal phase of the menstrual cycle and caused concomitant side effects. However, the luteal phase length was not reduced.

Prostaglandins are effective when given intravenously, subcutaneously, or intramuscularly.[114] Intravenous infusion has to be maintained for some hours and the dose increased to a level where side effects, usually in terms of gastrointestinal symptoms, become apparent. The method, therefore, seems useful mainly for clinical research.

Intravaginal administration has similar effects to those seen after intravenous administration. Effective uterine stimulation has been associated with frequent generalized side effects, indicating absorption into the systemic circulation. Vaginal suppositories are available containing $PGF_{2\alpha}$ and PGE_2. The results are consistent and encouraging, but the incidence of side effects in terms of vomiting and diarrhea or fever remain high. Karim[115] used intravaginal prostaglandin E_2 and $F_{2\alpha}$ in 12 women who had passed their expected date of menstruation. Menstrual-like bleeding and eventual abortion was induced in 11 women. Bygdeman et al.[108] treated eight pregnant women with vaginal administration of $PGF_{2\alpha}$. Despite the fact that menstrual-like bleeding was induced, only three women aborted. All patients suffered side effects during treatment.

The potential for the use of prostaglandins in postconceptional fertility control is obvious, but at the present time the side effects minimize the usefulness of the method.

V. INTRAUTERINE CONTRACEPTIVE DEVICES

Intrauterine devices (I.U.D.'s) are playing an increasingly important role in modern contraceptive practice. The intrauterine device should be the almost perfect contraceptive; the method is cheap, insertion is simple, and the device can be used for as long as the patient desires fertility control. Patient error cannot occur as it can with oral contraceptives, which makes the clinical efficacy of these compounds less than the theoretical.

Numerous hypotheses have been proposed regarding the contraceptive action of intrauterine devices. The concept that the I.U.D. functions by simple distortion of the uterine cavity[116] fails to explain the usefulness of small devices or even thread-like I.U.D.'s which are quite incapable of distending the uterine cavity significantly. It is also evident, from both human and animal experiments, that so long as the intrauterine device is functioning as a contraceptive, the natural ovulatory cycle continues undisturbed.[117] Hall et al.[118] disproved the abortifacient effect of the I.U.D. since endometrial biopsy and hysterectomy specimens failed to reveal any products of gestation. They further found that the structural characteristics of uterine mucosa, typical of each day of the normal cycle, were maintained in the presence of the intrauterine device. The discrepancy between chronological and histological dates was not more marked in the study group than in the control group and the glycogen production and enzymatic processes were unaffected by the presence of the device. Similarly, the accelerated tubal transport hypothesis advanced by Mastroianni and Russeau[119] has been abandoned.[120]

The fact that links both human and animal observations is the induction of a mild chemotactic response at the interface between the device and the endometrium. Sagiroglu and Sagiroglu[121] showed that in human beings the I.U.D. attracts a coating of histiocytes. The concept that a local surface interaction between the device and the endometrium occurs regularly in human beings and other mammals is well established. The device becomes coated with a thin layer of white cells and examination of the tissue adjacent to the device shows leukocytes migrating between the epithelial cells.[122] Leukocytes are also found in the endometrial fluid and the presence of a low intrauterine pH indicates that an inflammatory process accompanies the presence of the device.[123]

Moyer and Mishell[124] found significant increases in the numbers of neutrophils and mononuclear cells (macrophages and lymphocytes) in the uterine flushings and endometrial biopsy specimens in the early weeks after insertion. High concentrations of plasma cells were present in the endometrium from the 2nd to the 15th week after insertion and thereafter their numbers were reduced, disappearing in 80 to 90% of patients 6 months following insertion of the device.

A marked increase in the total protein and nonprotein-nitrogen levels in the uterine fluid of women using the intrauterine device has been noted.[124,125] This may be due to cellular degradation of neutrophils and macrophages. The way in which the leukocytes, or their products, exert their antifertility effect is not clear. Sagiroglu found that the many millions of leukocytes present destroyed the spermatozoa and that phagocytosis was complete within 16 hr of sexual intercourse.[126]

The reason for the marked chemotactic effect of the intrauterine device is unknown; it is theorized, however, that the action depends on its ability to stimulate the production of antibodies with the subsequent formation of immune complexes. Holub et al.[127] found that the serum immunoglobulins G and M (IgG) and (IgM) were increased in women using I.U.D.'s; these levels increased with the duration of use. The findings were confirmed by Gump et al.,[128] who found that the immunoglobulin G rise occurred earlier than that of immunoglobulin M and was noticeable even in those women with the device in place for less than 2 months. The IgM rise was more gradual and was not marked until the I.U.D. had been in use for over 1 year. No changes in serum immunoglobulin A levels were seen. Limited studies to determine if micoplasma or viral infection could be implicated in this immunoglobulin rise were negative.

This surface interaction hypothesis has been tested in the human by examining the relationship between the surface area of the device in contact with the endometrium and its efficiency as a contraceptive. When standardized material is used for the manufacture of I.U.D.'s, the relationship between the surface area of plastic available to elicit an endometrial response and contraceptive action is demonstrable.[129]

The type of material in contact with the endometrial surface is also very important. Zipper et al.[130] reported that a small copper or zinc loop placed in the lower portion of one of the uterine horns of rabbits produced a decrease in the number of implantation sites without producing a significant alteration in the control uterus. In 1970,[131] they utilized this observation to design a copper-containing intrauterine device and reported that clinical effectiveness was related to the amount of copper present. The precise mechanism of the copper as an antifertility agent is still unknown. The effect of copper has been investigated in rodents,[132] when it was demonstrated that ovarian function was not inhibited by the presence of the metal in the uterus. There is, however, a change in the intrauterine environment which may interfere with the development of the morula into blastocysts. Zincl and Rudel[132] have speculated that the antifertility effect of metallic copper in rodents and possibly in humans may be due to an interference with the metal-containing enzymes, such as the carbonic anhydrase, which is directly associated with the progestational response in many species.

From blood analysis it has been shown that there is no rise of copper levels in the circulating blood as a consequence of the copper T device in the uterus.[133]

Rosado et al.[134] found a significant decrease of ribonucleic acid in the endometrium of women wearing an I.U.D. in both the proliferative and luteal phases of the menstrual cycle; they suggested that this indicates that the metabolic and kinetic activities of the tissue are lower than in the control subjects. This decrease in the ribonucleic acid/deoxyribonucleic acid ratio may suggest an interference with hormonal action upon the endometrium. They also found an increased concentration of calcium in the endometrium of I.U.D. wearers. This agrees with the findings of calcium salt deposits on the I.U.D. after prolonged use.[135,136] It has been shown that relatively small concentrations of calcium ions interfere with some important metabolic processes in human spermatozoa, such as oxygen uptake and pyruvate and adenylate kinase activities. These calcium changes may be a factor in the uterine bleeding which is often associated with the use of the I.U.D.

Intrauterine devices are playing an increasingly important role in modern contraception. The lack of systemic complication associated with the I.U.D. method combined with the good results of improved devices makes this method of contraception one of the most acceptable today.

REFERENCES

1. **Brown, J. B. and Blair, H. A. F.,** Urinary oestrogen metabolites of 19-norethisterone and its esters, *Proc. R. Soc. Med.*, 53, 433, 1960.
2. **Paulsen, C. A.,** Progestin metabolism: special reference to estrogenic pathways, *Metabolism*, 14, 313, 1965.
3. **Elstein, M.,** in *Chlormadinone Acetate*, Christie, G. A. and Moore-Robinson, M., Eds., Excerpta Medica, Amsterdam, 1969, 46.
4. **McDonald, R. R.,** in *Chlormadinone Acetate*, Christie, G. A. and Moore-Robinson, M., Eds., Excerpta Medica, Amsterdam, 1969, 33.
5. **Hammerstein, J.,** The influence of continuously administered chlormadinone acetate at low dosage on hormonal excretion during the menstrual cycle in three healthy women, *Acta Endocrinol.*, 56(Suppl. 119), 145, 1967.
6. **Foss, G. L., Svenson, E. K., Fotherby, K., and Richards, D. J.,** Contraceptive action of continuous low doses of norgestrol, *Br. Med. J.*, 4, 489, 1968.
7. **Bret, A. J., Blanchier, H., and Sanchez, R. J.,** Oral progestogens and hypertension, *Rev. Franc. Gynecol. Obstet.*, 63, 703, 1968.
8. **Owen, G.,** Hypertension associated with oral contraceptives, *Canad. Med. Assoc. J.*, 95, 167, 1966.
9. **Tyson, J. E. A.,** Oral contraception and elevated blood pressure, *Am. J. Obstet. Gynecol.*, 100, 875, 1968.
10. **Kunin, C. M., McCormack, R. C., and Abernathy, J. R.,** Oral contraceptives and blood pressure, *Arch. Intern. Med.*, 123, 362, 1969.

11. **Laragh, J. H., Sealey, J. E., Ledingham, J. G. G., and Newton, M. A.,** Oral contraceptives. Renin, aldosterone, and high blood pressure, *J.A.M.A.*, 201, 918, 1967.
12. **Beckerhoff, R., Vetter, W., Armbruster, H., Leutscher, J. A., and Siegenthaler, W.,** Plasma aldosterone during oral contraceptive therapy, *Lancet*, 1, 1218, 1973.
13. **Jordan, W. M.,** Pulmonary embolism, *Lancet*, 2, 1146, 1961.
14. **Inman, W. H. W. and Vessey, M. P.,** Investigation of deaths from pulmonary, coronary and cerebral thrombosis and embolism in women of child-bearing age, *Br. Med. J.*, 2, 193, 1968.
15. **Vessey, M. P. and Doll, R.,** Investigation of relation between use of oral contraceptives and thrombo-embolic disease, *Br. Med. J.*, 2, 199, 1968.
16. **Dugdale, M. and Masi, A. T.,** Effects of the oral contraceptives on blood clotting. Second report on the oral contraceptive, Advisory Committee on Obstetrics and Gynecology, Food and Drug Administration, Washington, D.C., 1969, 43.
17. **Bolton, L. H., Hampton, J. R., and Mitchell, S. R. A.,** Effect of oral contraceptive agents on platelets and plasma phospholipids, *Lancet*, 1, 1336, 1968.
18. **Drill, V. A.,** *Oral Contraceptives*, McGraw-Hill, New York, 1966, 73.
19. **Zuck, T. F. and Bergin, J. T.,** Thrombotic predisposition associated with oral contraceptives, *Obstet. Gynecol.*, 41, 427, 1973.
20. Boston Collaborative Drug Surveillance Programme, Gall bladder disease and breast tumours, *Lancet*, 2, 1399, 1973.
21. Committee on Safety of Drugs, Combined Oral Contraceptives, A statement by the Committee, *Br. Med. J.*, 2, 231, 1970.
22. **Vessey, M. P. and Inman, W. H. W.,** Speculation about mortality trends from venous thromboembolic disease in England and Wales and their relation to the pattern of oral contraceptive use, *J. Obstet. Gynaecol. Br. Commonw.*, 80, 562, 1973.
23. **Schaffner, F.,** The effect of oral contraceptives on the liver, *J.A.M.A.*, 198, 1019, 1966.
24. **Hargreaves, T.,** Oral contraceptives and liver function, *J. Clin. Pathol.*, Suppl. 23, 3, 1, 1970.
25. **Goldzieher, J. W.,** Newer drugs in oral contraception, *Med. Clin. North Am.*, 48, 529, 1964.
26. **Rice-Wray, E.,** Oral contraceptives and liver damage, *Br. Med. J.*, 2, 1011, 1964.
27. **Kleiner, G. J., Kresch, L., and Arias, I. M.,** Studies of hepatic excretory function. II. The effect of norethynodrel and mestranol on bromsulfalein sodium metabolism in women of childbearing age, *N. Engl. J. Med.*, 273, 420, 1965.
28. **Mueller, M. N. and Kappas, A.,** Oestrogen pharmacology. I. The influence of oestradiol and oestriol on hepatic disposal of sulphobromophthalein (B.S.P.) in man, *J. Clin. Invest.*, 43, 1905, 1964.
29. **Eisalo, A., Järvinen, P. A., and Luukkainen, T.,** Liver function tests during intake of contraceptive tablets in pre-menopausal women, *Br. Med. J.*, 1, 1416, 1965.
30. **Robertson, G. S.,** Serum protein and cholinesterase changes in association with contraceptive pills, *Lancet*, 1, 232, 1967.
31. **Musa, B. U., Doe, R. P., and Seal, U. S.,** Serum protein alterations produced in women by synthetic oestrogens, *J. Clin. Endocrinol.*, 27, 1463, 1967.
32. **Phillips, L. L. and Skrodelis, V.,** The fibrinolytic enzyme system in normal, haemorrhagic and disease states, *J. Clin. Invest.*, 37, 965, 1958.
33. **Musa, B. U., Seal, U. S., and Doe, R. P.,** Elevation of certain plasma proteins in man following estrogen administration: a dose-response relationship, *J. Clin. Endocrinol.*, 25, 1163, 1965.
34. **Pearlman, W. H., Crépy, O., and Murphy, M.,** Testosterone binding levels in the serum of women during the normal menstrual cycle, pregnancy and the post-partum period, *J. Clin. Endocrinol.*, 27, 1012, 1967.
35. **Barbosa, J., Seal, U. S., and Doe, R. P.,** Antioestrogens and plasma proteins II, Contraceptive drugs and oestrogens, *J. Clin. Endocrinol. Metab.*, 36, 706, 1973.
36. **Boake, W. C., Schade, S. G., Morrissey, J. F., and Schaffner, F.,** Intrahepatic cholestatic jaundice of pregnancy followed by Enovid-induced cholestatic jaundice – report of a case, *Ann. Intern. Med.*, 63, 302, 1965.
37. **Adlercreutz, H. and Ikonen, E.,** Oral contraceptives and liver damage, *Br. Med. J.*, 2, 1133, 1964.
38. **Thulin, K. E. and Nermark, J.,** Seven cases of jaundice in women taking an oral contraceptive, Anovlar, *Br. Med. J.*, 1, 584, 1966.
39. **Eisalo, A., Järvinen, P. A., and Luukkainen, T.,** Hepatic impairment during the intake of contraceptive pills: clinical trial with postmenopausal women, *Br. Med. J.*, 2, 426, 1964.
40. **Kreek, M. J., Weser, E., Sleisenger, M. H., and Jeffries, G. H.,** Idiopathic cholestasis of pregnancy. The response to challenge with the synthetic estrogen, ethinylestradiol, *N. Engl. J. Med.*, 277, 1391, 1967.
41. **Urban, E., Frank, B. W., and Kern, F., Jr.,** Liver dysfunction with mestranol but not with norethynodrel in a patient with Enovid-induced jaundice, *Ann. Intern. Med.*, 68, 598, 1968.
42. **Marquardt, G. H., Fisher, C. I., Levy, P., and Dowben, R. M.,** Effect of anabolic steroids on liver function tests and creatine excretion, *J.A.M.A.*, 175, 851, 1961.
43. **Stoll, B. A., Andrews, J. T., and Motteram, R.,** Liver damage from oral contraceptives, *Br. Med. J.*, 1, 960, 1966.
44. **Lathe, G. H. and Walker, M.,** Inhibition of bilirubin conjugation in rat liver slices by human pregnancy and neonatal serum and steroids, *Quart. J. Exp. Physiol.*, 43, 257, 1958.

45. **Robertson, M. E., Stiefel, M., and Laidlaw, J. C.,** The influence of oestrogen on the secretion, disposition and biologic activity of cortisol, *J. Clin. Endocrinol.*, 19, 1381, 1959.
46. **Peterson, R. E., Nokes, E., Chen, P. S., Jr., and Black, R. L.,** Oestrogens and adrenocortical function in man, *J. Clin. Endocrinol.*, 20, 495, 1960.
47. **Sandberg, A. A. and Slaunwhite, W. R., Jr.,** Transcortin: a corticosteroid-binding protein of plasma. II. Levels in various conditions and the effects of oestrogens, *J. Clin. Invest.*, 38, 1290, 1959.
48. **Sandberg, A. A., Slaunwhite, W. R., Jr., and Carter, A. C.,** Transcortin: a corticosteroid-binding protein of plasma. III. The effects of various steroids, *J. Clin. Invest.*, 39, 1914, 1960.
49. **Daughaday, W. H., Adler, R. E., Mariz, I. K., and Rasinski, D. C.,** Measurement of the binding capacity of corticosteroid-binding globulin in human plasma, *J. Clin. Endocrinol.*, 22, 704, 1962.
50. **Doe, R. P., Fernandez, R., and Seal, U. S.,** Measurement of corticosteroid-binding globulin in man, *J. Clin. Endocrinol.*, 24, 1029, 1964.
51. **De Moor, P. Steeno, O., Brosens, I., and Hendrikx, A.,** Data on transcortin activity in human plasma as studied by gel filtration, *J. Clin. Endocrinol.*, 26, 71, 1966.
52. **Burke, C. W.,** The effect of oral contraceptives on cortisol metabolism, *J. Clin. Pathol.*, Suppl. 23, 3, 11, 1970.
53. **Katz, F. H., Lipman, M. M., Frantz, A. G., and Jailer, J. W.,** The physiologic significance of 6β hydroxycortisol in human corticoid metabolism, *J. Clin. Endocrinol. Metab.*, 22, 71, 1962.
54. **Gershberg, H., Javier, Z., and Hulse, M.,** Glucose tolerance in women receiving an ovulatory suppressant, *Diabetes*, 13, 378, 1964.
55. **Wynn, V. and Doar, J. W. H.,** Some effects of oral contraceptives on carbohydrate metabolism, *Lancet*, 2, 715, 1966.
56. **Wynn, V. and Doar, J. W. H.,** Longitudinal studies of the effects of oral contraceptive therapy on plasma glucose, non-esterified fatty acid, insulin and blood pyruvate levels during oral and intravenous glucose tolerance tests, in *Metabolic Effects of Gonadal Hormones and Contraceptive Steroids*, Salhanick, H. A., Kipnis, D. M., and Vande Wiele, R. L., Eds. Plenum Press, New York, 1969, 155.
57. **Wynn, V. and Doar, J. W. H.,** Some effects of oral contraceptives on carbohydrate metabolism, *Lancet*, 2, 761, 1969.
58. **Spellacy, W. N.,** A review of carbohydrate metabolism and the oral contraceptives, *Am. J. Obstet. Gynecol.*, 104, 448, 1969.
59. **Spellacy, W. N.,** The effect of ovarian steroids on glucose, insulin and growth hormone, in *Metabolic Effects of Gonadal Hormones and Contraceptive Steroids*, Salhanick, H. A., Kipnis, D. M., and Vande Wiele, R. L., Eds., Plenum Press, New York, 1969, 126.
60. **Beck, P.,** Effects of gonadal hormones and contraceptive steroids on glucose and insulin metabolism, in *Metabolic Effects of Gonadal Hormones and Contraceptive Steroids*, Salhanick, H. A., Kipnis, D. M., and Vande Wiele, R. L., Eds., Plenum Press, New York, 1969, 97.
61. **Beck, P. and Wells, S. A.,** Comparison of the mechanisms underlying carbohydrate intolerance in subclinical diabetic women during pregnancy and during post-partum oral contraceptive steroid treatment, *J. Clin. Endocrinol.*, 29, 807, 1969.
62. **Wynn V. and Doar, J. W. H.,** Effects of oral contraceptives on carbohydrate metabolism, *J. Clin. Pathol.*, Suppl. 23, 3, 19, 1970.
63. **Keen, H., Rose, G., Pyke, D. A., Boyns, D., Chlouverakis, C., and Mistry, S.,** Blood sugar and arterial disease, *Lancet*, 2, 505, 1965.
64. **Epstein, F. D.,** Hyperglycemia. A risk factor in coronary heart disease, *Circulation*, 36, 609, 1967.
65. **Stout, R. W. and Vallance-Owen, J.,** Insulin and atheroma, *Lancet*, 1, 1078, 1969.
66. **Aurell, M., Cramer, K., and Rybo, G.,** Serum lipids and lipoproteins during long term administration of an oral contraceptive, *Lancet*, 1, 291, 1966.
67. **Wynn, V., Doar, J. W. H., and Mills, G. L.,** Some effects of oral contraceptives on serum lipid and lipoprotein levels, *Lancet*, 2, 720, 1966.
68. **Wynn, V., Doar, J. W. H., Mills, G. L., and Stokes, T.,** Fasting serum triglyceride, cholesterol and lipoprotein levels during oral contraceptive therapy, *Lancet*, 2, 756, 1969.
69. **Stokes, T. and Wynn, V.,** Serum lipids in women on oral contraceptives, *Lancet*, 2, 677, 1971.
70. **Osman, M., Toppozada, H., Ghanem M., and Guergis, F.,** The effect of an oral contraceptive on serum lipids, *Contraception*, 5, 105, 1972.
71. **Hazzard, W. R., Spiger, M. J., Bagdade, J. D., and Bierman, E. L.,** Studies of the mechanism of increased plasma triglyceride levels induced by oral contraception, *N. Engl. J. Med.*, 280, 471, 1969.
72. **Sachs, B. A., Wolfman, L. and Herzig, N.,** Plasma lipid and lipoprotein alteration during oral contraception administration, *Obstet. Gynecol.*, 34, 530, 1969.
73. **Spellacy, W. N., Buhi, W. C., Birk, S. A., and Cabal, R.,** The effects of oestrogen, progestogen, oral contraceptives and intra-uterine devices on fasting triglyceride and insulin levels, *Fertil. Steril.*, 24, 178, 1973.
74. **Brody, S., Hogdahl, A. M., Nilsson, L., Svanborg, A., and Vikrot, O.,** The effect of some ovulation inhibitors on the lipid metabolism, preliminary report, *Acta Med. Scand.*, 179, 501, 1966.
75. **Pincus, G.,** Ed., in *The Control of Fertility*, Academic Press, New York, 1965, 277.

76. **Doar, J. W. H. and Wynn, V.**, Serum lipid levels during oral contraception and glucocorticoid administration, *J. Clin. Pathol.*, Suppl. 23, 3, 55, 1970.
77. **Wynn, N. and Doar, J. W. H.**, Fasting serum triglyceride and cholesterol levels during oral contraceptive therapy, in *Metabolic Effects of Gonadal Hormones and Contraceptive Steroids*, Salhanick, H. A., Kipnis, D. M., and Vande Wiele, R. L., Eds., Plenum Press, New York, 1969, 219.
78. **Barton, O. M. G., Freeman, P. R., and Lawson, J. P.**, Oral contraceptives and serum lipids, *J. Obstet. Gynaecol. Br. Commonw.*, 77, 551, 1970.
79. **Gowland, E. and Jones, R.**, Effect of oral norethisterone on serum lipid levels in pre-menopausal women, *J. Obstet. Gynaecol. Br. Commonw.*, 80, 357, 1973.
80. **Brody, S., Kerstell, J., Nilsson, L., and Svanborg, A.**, The effect of some ovulation inhibitors on the different plasma lipid fractions, *Acta Med. Scand.*, 183, 1, 1968.
81. **Russ, E. M., Elder, H. A., and Barr, D. P.**, Protein lipid relationships in human plasma. I. In normal individuals, *Am. J. Med.*, 11, 468, 1951.
82. **Furman, R. H., Howard, R. P., Norcia, L. N. and Keaty, E. C.**, The influence of androgens, oestrogens and related steroids on serum lipids and lipoproteins, *Am. J. Med.*, 24, 80, 1958.
83. **Hood, B. and Cramer, K.**, Effects on serum lipoprotein and cholesterol of oestrogen in combination with Δ^4 androstenedione, testosterone and methyl testosterone, *Acta Med. Scand.*, 165, 459, 1959.
84. **Doe, R. P., Mellinger, G. T., Swain, W. R., and Seal, U. S.**, Oestrogen dosage effects on serum proteins; a longitudinal study, *J. Clin. Endocrinol.*, 27, 1081, 1967.
85. **Braidman, I. P. and Rose, D. P.**, Effect of sex hormones on three glucocorticoid-inducible enzymes concerned with amino acid metabolism in rat liver, *Endocrinology*, 89, 1250, 1971.
86. **Rose, D. P.**, Excretion of xanthurenic acid in the urine of women taking progestogen-oestrogen preparations, *Nature*, 210, 196, 1966.
87. **Price, J. M., Thornton, M. J., and Mueller, L. M.**, Tryptophan metabolism in women using steroid hormones for ovulation control, *Am. J. Clin. Nutr.*, 20, 452, 1967.
88. **Brown, R. R., Rose, D. P., Price, J. M., and Wolf, H.**, Tryptophan metabolism as affected by anovulatory agents, *Ann. N.Y. Acad. Sci.*, 166, 44, 1969.
89. **Yess, N., Price, J. M., Brown, R. R., Swan, P. B., and Linkswiler, H.**, Vitamin B6 depletion in man: urinary excretion of tryptophan metabolites, *J. Nutr.*, 84, 229, 1964.
90. **Rose, D. P., Strong, R., Adams, P. W., and Harding, P. E.**, Experimental vitamin B_6 deficiency and the effect of oestrogen containing oral contraceptives on tryptophan metabolism and vitamin B_6 requirements, *Clin. Sci.*, 42, 465, 1972.
91. **Kistner, R. W.**, Histological effects of progestins on hyperplasia and carcinoma in situ of the endometrium, *Cancer*, 12, 1106, 1959.
92. **Kistner, R. W., Baginsky, S., Craig, J. M., and Bigler, P.**, Effect of progestins on induced endometrial cancer in the rabbit, *Surg. Forum*, 13, 410, 1962.
93. **Hertz, R.**, The problem of possible effect of oral contraceptives on cancer of the breast, *Cancer*, 24, 1140. 1969.
94. **Allen, W.**, Possible hazards in oestrogen administration, *Cancer*, 24, 1137, 1969.
95. **Coceani, F. and Wolfe, F.**, On the action of prostaglandin E_1 and prostaglandins from brain on the isolated rat stomach, *Canad. J. Physiol. Pharmacol.*, 44, 933, 1966.
96. **Vogt, W.**, Pharmacologically active acidic phospholipids and glycolipids, *Biochem. Pharmacol.*, 12, 415, 1963.
97. **Strong, C. G. and Bohr, D. F.**, Effects of prostaglandins E_1 E_2 A_1 F on isolated vascular smooth muscle, *Am. J. Physiol.*, 213, 725, 1967.
98. **Bergstrom, S., Duner, H., Euler, U. S., Von Pernow, B., and Sjovall, J.**, Observations on the effects of infusion of prostaglandins E in man, *Acta Physiol. Scand.*, 45, 145, 159.
99. **Karim, S. M. M.**, The identification of prostaglandins in human umbilical cord, *Br. J. Pharmacol. Chemother.*, 29, 230, 1967.
100. **Johnston, H. H., Herzog, J. P., and Lauler, D. P.**, Effect of prostaglandin E_1 on renal hemodynamics, sodium and water excretion, *Am. J. Physiol.*, 213, 939, 1967.
101. **Sweatman, W. J. F. and Collier, H. O. J.**, Effect of prostaglandins on human bronchial muscle, *Nature*, 217, 69, 1968.
102. **Brenner, W. E., Hendricks, C. H., Fishburne, J. J., and Braaksma, J. T.**, Intravenous Prostaglandin $F_{2\alpha}$ for therapeutic abortion, in Prostaglandins in Fertility Control, Reports from conference on analytical procedures, metabolism and clinical evaluation, Stockholm, January 18-20, 1972, 156.
103. **Wiqvist, N. and Bygdeman, M.**, Induction of therapeutic abortion with intravenous prostaglandin $F_{2\alpha}$, *Lancet*, 1, 889, 1970.
104. **Roth-Brandel, V., Bygdeman, M., and Wiqvist, N.**, Effect of intravenous administration of prostaglandin E_1 and $F_{2\alpha}$ on the contractility of the non-pregnant human uterus in vivo, *Acta Obstet. Gynecol. Scand.*, 49, 19, 1970.
105. **Roth-Brandel, V., Bygdeman, M., and Wiqvist, N.**, Effect of prostaglandins on the contractility of the non-pregnant human uterus in vivo, *Acta Obstet. Gynecol. Scand.*, 50, 35, 1971.
106. **Wiqvist, N., Roth-Brandel, V., and Bygdeman, M.**, The effect of prostaglandin E_1 and $F_{2\alpha}$ on the non-pregnant human uterus in vivo, *Int. J. Gynecol. Obstet.*, 8, 165, 1970.

107. **Karim, S. M. M.**, Prostaglandins, *Br. J. Hosp. Med.*, 5, 555, 1971.
108. **Bygdeman, M., Beguin, F., Toppozada, M., Wide, L., and Wiqvist, N.**, Postconceptional fertility control by prostaglandin $F_{2\alpha}$, in Prostaglandins in Fertility Control, Reports from conference on analytical procedures, metabolism and clinical evaluation, Stockholm, January 18-20, 1972, 175.
109. **Blatchley, F. R. and Donavan, B. T.**, Luteolytic effect of prostaglandin in the guinea pig, *Nature*, 221, 1065, 1969.
110. **Lemaire, W. J. and Shapiro, A. G.**, Prostaglandin $F_{2\alpha}$: its effect on the corpus luteum of the menstrual cycle, *Prostaglandins*, 1, 259, 1972.
111. **Jewelewicz, R., Cantor, B., Dyrenfurth, I., Warren, M. P., and Vande Wiele, R. L.**, Intravenous infusion of prostaglandin $F_{2\alpha}$ in the midluteal phase of the normal human menstrual cycle, *Prostaglandins*, 1, 443, 1972.
112. **Jones, G. S. and Wentz, A. C.**, The effect of prostaglandin $F_{2\alpha}$ infusion on corpus luteum function, *Am. J. Obstet. Gynecol.*, 114, 393, 1972.
113. **Hillier, K., Dutton, A., Corker, C. S., Singer, A., and Embrey, M. P.**, Plasma steroid and luteinizing hormone levels during prostaglandin $F_{2\alpha}$ administration in luteal phase of menstrual cycle, *Br. Med. J.*, 4, 333, 1972.
114. **Karim, S. M. M., Hillier, K., Somers, K., and Trussell, R. R.**, The effects of prostaglandins E_2 and $F_{2\alpha}$ administered by different routes on uterine activity and the cardio-vascular system in pregnant and non-pregnant women, *J. Obstet. Gynaecol. Br. Commonw.*, 78, 172, 1971.
115. **Karim, S. M. M.**, Once-a-month vaginal administration of prostaglandins E_2 and $F_{2\alpha}$ for fertility control, *Contraception*, 3, 173, 1971.
116. **Rozin, S., Schwartz, A., and Schenker, J. G.**, The mode of action of an intrauterine contraceptive device, *Int. J. Fertil.*, 14, 174, 1969.
117. **Eckstein, P.**, Mechanisms of action of intrauterine contraceptive devices in women and other mammals, *Br. Med. Bull.*, 26, 52, 1970.
118. **Hall, H. H., Sedlis, A., Chabon, I., and Stone, M. L.**, Effect of intra-uterine stainless steel ring on endometrial structure and function, *Am. J. Obstet. Gynecol.*, 93, 1031, 1965.
119. **Mastroianni, L. and Russeau, C. H.**, Influence of the uterine coil on ovum transport and sperm distribution in the monkey, *Am. J. Obstet. Gynecol.*, 93, 416, 1965.
120. **Kelly, W. A., Marston, J. H., and Eckstein, P.**, Effect of an intrauterine device on endometrial morphology and deciduomal reaction in the Rhesus monkey, *J. Reprod. Fertil.*, 19, 331, 1969.
121. **Sagiroglu, N. and Sagiroglu, E.**, Biological mode of action of the Lippes loop in intrauterine contraception, *Am. J. Obstet. Gynecol.*, 106, 506, 1970.
122. **Potts, M. and Pearson, R. M.**, A light and electron microscope study of cells in contact with intrauterine contraceptive devices, *J. Obstet. Gynaecol. Br. Commonw.*, 74, 129, 1967.
123. **Sedlis, A. and Reyniak, J. V.**, Endometrial leucocytes in patients using intrauterine contraceptive devices, *Am J. Obstet. Gynecol.*, 108, 1209, 1970.
124. **Moyer, D. I. and Mishell, D. R., Jr.**, The intrauterine device; a bacteriologic study of the endometrial cavity, *Am. J. Obstet. Gynecol.*, 110, 633, 1971.
125. **Kar, A. B., Engineer, A. D., Goel, R., Kamboj, V., Dasgupta, P. R., and Choudhary, S. R.**, Marked increase in total-protein and non-protein nitrogen levels in the uterine fluid of women with Lippes loop, *Am. J. Obstet. Gynecol.*, 101, 966, 1968.
126. **Sagiroglu, N.**, Phagocytosis of spermatozoa in the uterine cavity of women using intrauterine devices, *Int. J. Fertil.*, 16, 1, 1971.
127. **Holub, W. R., Reyner, F. C., and Forman, G. H.**, Increased levels of serum immunoglobulins, G. and M. in women using intrauterine contraceptive devices, *Am. J. Obstet. Gynecol.*, 110, 362, 1971.
128. **Gump, D. W., Mead, P. B., Horton, E. I., Lamborn, K. R., and Forsyth, B. R.**, Intrauterine contraceptive device and increased serum immunoglobulin levels, *Obstet. Gynecol.*, 41, 259, 1973.
129. **Davis, H. J. and Lesinski, J.**, Mechanism of action of intrauterine contraceptives in women, *Obstet. Gynecol.*, 36, 350, 1970.
130. **Zipper, J., Medel, M., and Prage, R.**, Suppression of fertility by intrauterine copper and zinc in rabbits. A new approach to intrauterine contraception, *Am. J. Obstet. Gynecol.*, 105, 529, 1969.
131. **Zipper, J., Tatum, H. J., Medel, M., Pastene, L., and Rivera, M.**, Presented at the 15th ›bel Symposium on Control of Human Fertility, Stockholm, May 27-29, 1970.
132. **Zincl, F. A. and Rudel, H.**, Possible mechanism of the anti-fertility effect of copper, in *Current Problems in Fertility*, Ingelman-Sundberg A. and Lunell, N. O., Eds., Plenum Press, New York, 1971, 187.
133. **Fortier, L., Lefebure, Y., Larose, M., and Lanctot, R.**, Canadian experience with a copper covered intrauterine contraceptive device, *Am. J. Obstet. Gynecol.*, 115, 291, 1973.
134. **Rosado, A., Hicks, J., Aznar, R., and Martinez-Manautou, J.**, Effect of the intrauterine contraceptive device upon the biochemical composition of human endometrium, *Am. J. Obstet. Gynecol.*, 114, 88, 1972.
135. **Yang, T. and Yang, W.**, Calcium carbonate deposition on intrauterine contraceptive devices, *Am. J. Obstet. Gynecol.*, 109, 664, 1971.
136. **Ingemanson, C. A.**, Intrauterine devices – side effects, in *Current Problems in Fertility*, Ingelman-Sundberg, A. and Lunell, N. O., Eds., Plenum Press, New York, 1971, 191.

HORMONAL CHANGES IN PREGNANCY AS INDICATORS OF FETOPLACENTAL FUNCTION

D. Tulchinsky

TABLE OF CONTENTS

I. INTRODUCTION

Recent developments in methodology have brought about an influx of new information on the fetoplacental unit. It is beyond the scope of this chapter to deal with the wide subject of hormonal changes in pregnancy, but the reader is referred to several reviews.[6,41,46] Growing interest, however, in adapting new hormonal methods for monitoring intrauterine fetal well-being has resulted in an array of new information. This chapter will review the subject of blood hormones as they apply to monitoring fetal well-being.

II. ADVANCES IN METHODOLOGY

A major technological advance in hormone measurement occurred in 1960 when Yalow and Berson[72] introduced the principle of competitive binding assay for measurement of plasma insulin. The method that utilizes the principle of competition between labeled and unlabeled hormones on a limited number of binding sites allows the measurement of minute amounts of hormone in small volumes of blood. The adoption of this principle for the measurement of various plasma hormones has resulted in an array of new methods. It is important to analyze such methods in regard to their specificity, precision, and simplicity.

It is difficult to prove that a method specifically measures one hormone only and the evidence in most instances is indirect. Specific antiserum, if available, would diminish the chances that other hormones would interfere with the assay. The specificity of binder is determined, however, by studies of cross-reactivity in which the ability of various hormones to compete on a limited number of binding sites is tested. If only one hormone binds to an antiserum, then it can be assumed that all the other tested hormones would not interfere with measurement of this hormone. Since not every single hormone is tested, it is always possible that there are untested hormones that could interfere in the assay procedure. Likewise, a hormone that cross-reacts only minimally with a given hormone but whose plasma concentrations exceed that of the measured hormone by manyfold may cause falsely high hormone determinations. Moreover, since concentrations of the tested hormone may be negligible in normal state but enormously high in a disease, one may have to reevaluate the specificity of the method for the abnormal state.

The problem may be even more complicated if nonspecific binders are used because removal of all competing hormones then becomes necessary. Such separations whether done by chromatography, solvent partition, or extraction are time-consuming and tedious. Moreover, there is always the possibility that such a separation has not removed untested hormones with physical and chemical properties similar to that of the measured hormone.

Radioimmunoassay (RIA) methods do not distinguish between biologically active and inactive hormones. Such RIA methods may detect hormone metabolites or split products of hormones which have only immunologic but no biologic activities. Results of radioimmunoassay do not, therefore, always correlate with bioassays.[3 5]

Despite all these potential problems, it should be emphasized that the sensitivity, precision, and reproducibility of radioligand and radioimmunoassay methods usually exceed those of biological and most other chemical methods. Large numbers of samples can be measured simultaneously in minute amounts of plasma and results are available in a short period of time.

III. BLOOD VS. URINARY DETERMINATIONS

The desire to replace urinary methods of hormone determinations by simple methods of plasma determination is based on several premises:

1. It is easier to collect blood than urine.
2. Errors of collecting blood samples are minimal.
3. Blood determinations can be done and reported promptly, whereas urinary measurements must be delayed until urinary collection has been completed.
4. The biologic potency of hormones may be reflected more accurately by blood measurements since such measurements are made closer to the site of hormone production and action.
5. Because of the short half-life of many blood hormones, any acute changes in production or secretion of a hormone are likely to be readily detected by frequent blood determinations. Urinary studies are unlikely to detect pulsatile patterns of hormone secretion and more pronounced and persistent changes of hormone secretion may not be reflected by urinary measurement for many hours or days.
6. Pathologic process affecting an organ that participates in clearing blood hormones (e.g., kidneys) may result in an erroneously high blood value and erroneously low urinary value of the hormone. Determinations of hormone concentration in blood would be advantageous over urinary measurement of the hormone only if the hormone in blood is cleared first by an organ (such as the liver) which has not been affected by the disease.

Determination of hormones in blood may not be always ideal. A single hormone blood determination may occasionally be misleading if there are marked minute-to-minute or hour-to-hour fluctuations in its plasma concentrations. Although inconvenient, serial blood determinations may overcome these difficulties. Results of urinary measurements may be more meaningful, however, in that they would indicate the average hormone secretion for the entire period.

IV. HORMONES OF EARLY PREGNANCY

A. Normal Pregnancy

1. HCG and Steroid Secretion by the Corpus Luteum*

HCG is believed to be responsible for continuous function of the corpus luteum into the

*Human chorionic gonadotropin.

early weeks of pregnancy. HCG is first detectable by presently utilized methods at day 8 to 10 after ovulation or immediately after implantation of the ovum. Its plasma levels continue to rise rapidly until the 10th to 11th week of pregnancy, after which they decline (Figure 1). The maximal stimulatory effect of HCG on the corpus luteum is limited, however, to a period of 3 to 4 weeks only. HCG administration to pregnant patients was shown capable of extending the life span of the corpus luteum for an additional period of only 3 weeks.

In early pregnancy the peak steroid plasma levels of 17-OHP,* which is mainly an ovarian product, are reached at the 5th to 6th week of pregnancy (3 to 4 weeks after ovulation), which is much sooner than the peak HCG level (Figure 1). Therefore, it appears that the life span of the corpus luteum is limited and that it starts regressing after the 6th week of pregnancy, despite the continuous increase of HCG levels. It must be pointed out, however, that the corpus luteum in many instances may continue functioning throughout pregnancy, although its relative importance in providing steroids is lost.

2. Corpus Luteum to Placental Shifts of Steroid Secretion

After the 6th week of pregnancy, when the corpus luteum starts regressing, the plasma concentrations of hormones of primarily ovarian origin, such as 17-OHP, start falling rapidly. Plasma levels of other steroid hormones are maintained, however, mainly through their increased secretion by the placenta. The shift from corpus luteum to placental secretion of E_2** has probably been completed by the 7th week of pregnancy when E_2 plasma levels start rising rapidly. By the 10th week of pregnancy the plasma levels of E_2 are already sevenfold higher than those observed during the peak activity of the corpus luteum (Figure 1). Changes in the plasma concentration of progesterone (P) are less dramatic. From the 6th to the 10th week of pregnancy, progesterone plasma levels remain essentially unchanged. Only after the 10th week of pregnancy do P*** plasma levels start rising again. By the end of the first trimester, P plasma levels are only 50% higher than those found at the 5th week of pregnancy (Figure 1). Considering that some of this increase in plasma P concentration is due to increased plasma levels of corticosteroid binding globulin (CBG)[68] during the first trimester and assuming that the clearance of P remains unchanged,[24] it can be extrapolated that by the end of the first trimester, the placenta secretes progesterone in amounts which only slightly exceed those secreted by the corpus luteum during its peak activity.

The shift from corpus luteum to placental secretion of E_2 and P plasma is rather smooth in most normal pregnancies. Individual patients, however, may have marked minute-to-minute and day-to-day fluctuations in their plasma progesterone levels.[59] A consistent fall in P plasma concentrations between the 6th and the 9th week of pregnancy is not observed, however, in most normal patients.[62] Reports claiming the occurrence of such a fall are based primarily on studies of drug-induced pregnancies[73] in whom the fall of P plasma levels represents a shift from a state of overstimulated ovarian function to a state of normal placental secretion of these steroids. Other studies which included patients scheduled to undergo therapeutic abortion may have included many abnormal pregnancies that may have otherwise terminated by spontaneous abortion. The lower levels of progesterone at the 9th week of pregnancy in these study groups may, therefore, reflect the inevitability of many such pregnancies. It may also reflect the nonspecificity of the method for P determination, and it is possible that the fall of P may have been due to a fall of 17-OHP plasma levels.

3. Steroid Secretion by the Trophoblast

It is presently unknown how soon after implantation the trophoblast starts secreting estriol.

Prior to the 9th week of pregnancy, unconjugated estriol (E_3) is circulating in plasma in very low concentrations (less than 50 pg/ml).[62] It is doubtful whether it is being secreted in any appreciable amount by the corpus luteum of the menstrual cycle and urinary E_3 found during the menstrual cycle has been considered to be derived primarily from plasma E_1 and E_2. Its appearance

*17α-Hydroxyprogesterone
**Estradiol-17β
***Progesterone

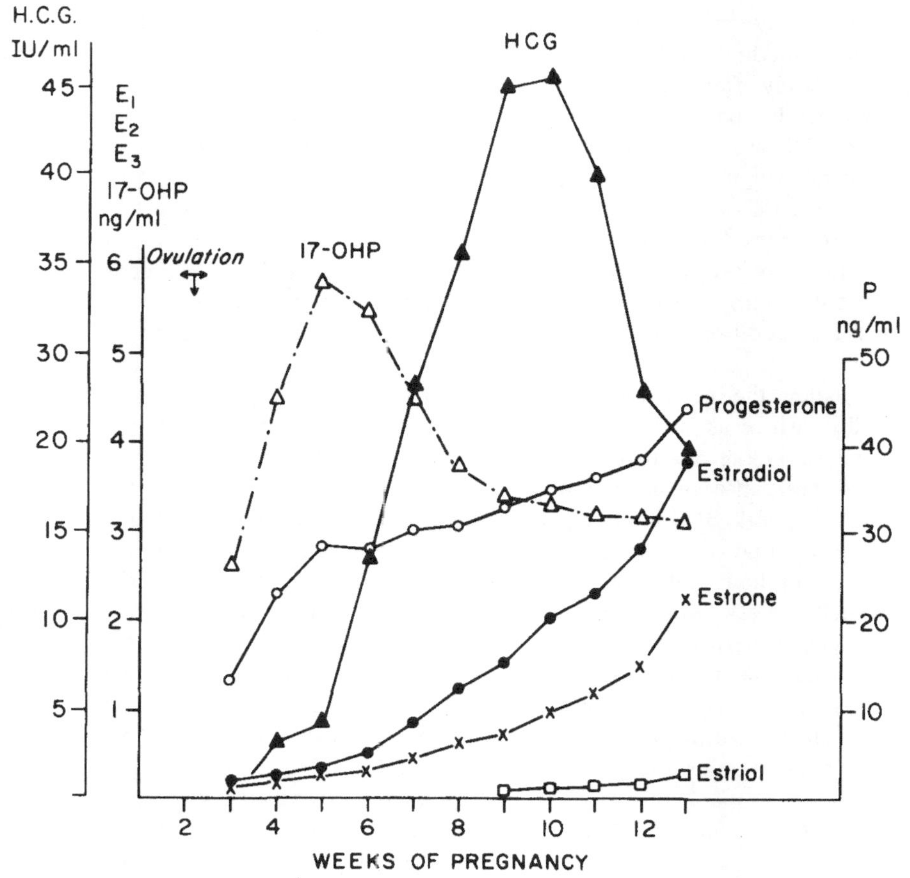

FIGURE 1. The mean plasma concentrations of human chorionic gonadotropin (H.C.G.) progesterone (P), 17-α-hydroxyprogesterone (17-OHP) and unconjugated estrone (E_1), estradiol (E_2), and estriol (E_3) during the first trimester of pregnancy. (Modified from Tulchinsky, D. and Hobel, C. J., *Am. J. Obstet. Gynecol.*, 117, 884, 1973. With permission.)

in measurable quantities after the 9th week of pregnancy may represent the beginning of its synthesis by the maternal fetoplacental unit or it may be simply a reflection of the rising plasma levels of E_1 and E_2 which are partially converted into plasma E_3. By the end of the first trimester, the mean plasma unconjugated E_3 level is rather low and is only 7% of that of E_2 (Figure 1).

It is also unknown how soon after implantation the trophoblast starts secreting progestogens and estrogens, or what is the relative contribution of the placenta to E_2 and P plasma levels at each week of gestation. Attempts to assess this contribution by looking at the magnitude of fall of plasma level of hormone following luteectomy or removal of the placenta are subject to various methodologic errors. First, such surgical procedures may affect organs other than the one removed. Second, the plasma levels that are maintained following such procedures may misrepresent the contribution of the remaining organ because the plasma concentrations of sex hormone binding globulin (SHBG) and corticosteroid binding globulin (CBG) remain elevated and can account for the increase in the plasma levels of E_2 and P. The ability, however, to maintain pregnancy after removal of the corpus luteum with essentially unchanged levels of progesterone in blood suggests that corpus luteum function is not critical after the 7th week of pregnancy when placental function is well established.

4. Steroids and the Diagnosis of Pregnancy

During the first 6 weeks of pregnancy, most plasma E_1 and E_2 values fall within ranges reported during the menstrual cycle.[62] Levels of P and 17-OHP plasma concentrations of many patients also fall within the range found during the menstrual cycle.[30,62] The diagnosis of pregnancy cannot be consistently made on the basis of plasma steroid determination, although higher-than-normal plasma levels of P, 17-OHP, or E_2 would certainly suggest such a possibility. Confirmation of pregnancy should, therefore, be done best by HCG determination.

B. Abnormal Pregnancy

1. Spontaneous Abortions

a. Hormonal Pattern

Most spontaneous abortions during the first trimester are due to abnormalities of fetal and placental development. The question whether any of the very early abortions may be due to inadequate corpus luteum function has not yet been resolved. Patients who abort spontaneously during the first trimester are usually found to have normal P, 17-OHP, and E_2 plasma levels during the first 6 weeks of pregnancy, suggesting normal corpus luteum function[22] (Figure 2). A gradual fall of their concentrations to subnormal levels usually occurs only after the 8th week of pregnancy and is usually associated with low or subnormal HCG and HPL levels,[22,34] suggesting inappropriate placental function. Subnormal plasma P and 17-OHP, if found during the first 6 weeks of pregnancy, would suggest inadequate corpus luteum function. This may be due to either an abnormal response of the corpus luteum to HCG stimulation or production of HCG in amounts which are insufficient to bring about adequate corpus luteum response.

b. Therapy for Threatened Abortions

There is no reason to treat patients who threaten to abort with hormones, since most of these patients would either continue their pregnancy to term without any therapy or would abort spontaneously despite such therapy.

Previous attempts to select patients for therapy on the basis of low pregnanediol excretion have not been found to be of any value. Such studies may have included many patients with inevitable abortions. Most spontaneous abortions are due to either fetal or placental abnormalities and should not, theretore, be treated. Hormonal therapy may become justified only in patients found to have inadequate corpus luteum function.

At present all modes of hormonal therapy

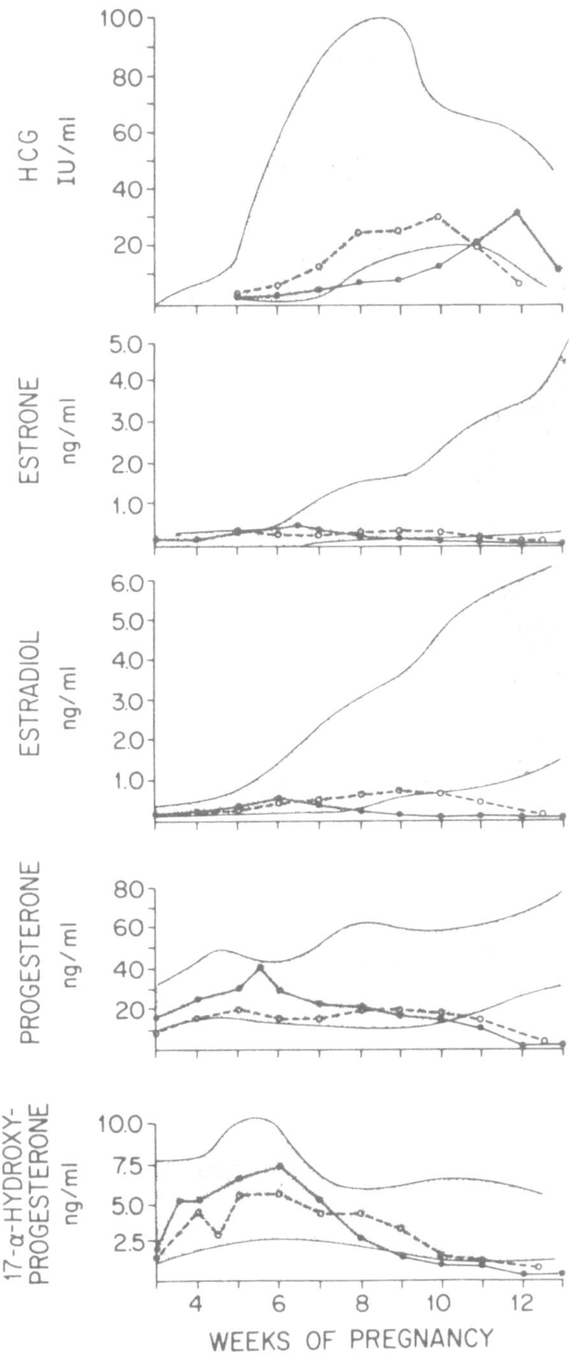

FIGURE 2. Plasma human chorionic gonadotropin (HCG), estrone, estradiol, progesterone, and 17-α-hydroxyprogesterone of two patients who subsequently aborted. The shaded areas represent the 95% confidence limits for normal values.[59]

should be withheld until methods for the identification of such entities become more refined. It should also be noted that indiscriminate use of hormones in early pregnancy may have adverse effects on the offspring.

2. Hydatidiform Mole

Certain endocrine patterns are common in molar pregnancies.

1. HCG levels are usually abnormally high and continue to increase even beyond the 11th week of pregnancy.
2. HPL plasma levels are substantially below the normals for the given stage of pregnancy. As a result, the ratio between HCG and HPL may be up to 100-fold higher than that found in normal pregnancy.
3. Estriol levels are usually subnormal, but at times may approach normal levels. Under such circumstances, E_3 is most likely to be derived from two sources: either from placental aromatization of 16-α-dehydroepiandrosterone sulfate derived from maternal sources or from conversion of maternal circulating E_1 and E_2 into E_3.
4. Estetrol urinary levels have been reported to be substantially subnormal or undetectable.[21]

The follow-up of such patients after evacuation of the uterus or following therapy can best be assessed by serial quantitative measurements of HCG in plasma utilizing specific radioimmunoassay of the B subunit of HCG, which would differentiate this hormone from LH.*[67] A rapid disappearance of this hormone from plasma would indicate good prognosis, whereas the persistence or further increase in the HCG levels would suggest persistent disease.[15]

V. HORMONES OF LATE PREGNANCY

During the second half of human pregnancy, the placenta secretes various steroids and polypeptide hormones into the maternal circulation, having utilized either fetal or maternal precursors for their production. Some circulating hormones have been studied extensively in the hope that they may reflect the functional status of either the placenta or the fetus. Many other placental hormones have not yet been thoroughly studied and many others probably have not yet even been identified. This discussion would deal, therefore, only with several steroids and polypeptide hormones.

*Luteinizing hormone

A. Placental Hormones Dependent on Maternal Precursors

1. Human Placental Lactogen (HPL) Human Chorionic Somatomammotropin (HCS)

a. Physiologic Considerations

This polypeptide hormone is known to have lactogenic, mammotrophic, luteotrophic, diabetogenic, and growth hormone-like activities in a variety of animal species.[2,16,41] It is responsible for some of the metabolic alterations of pregnancy. It is produced by the synctiotrophoblast and is secreted primarily into the maternal circulation. It is detected in serum beginning at the 5th week of pregnancy[30,47] and its levels rise progressively to a plateau at the 34th week of pregnancy.[42,44,51]

Several features of this hormone suggest its usefulness as an indicator of the functional status of the placenta.

1. A good correlation between plasma HPL levels and placental weight.
2. The first half-life of this hormone is in the order of 11 to 30 min, indicating rapid clearance of this hormone from plasma.
3. There are no diurnal variations in its plasma levels.
4. There is a narrow range of normal values at term.
5. HPL is present in plasma in high concentrations.

b. Methods

Various methods, such as complement fixation, hemagglutination inhibition, radioimmunoassay, and radiological receptor assays, are available for measurement of HPL in plasma.[11,16,20,42,44,52,56] RIA is the most commonly used method for HPL determination in plasma. Because of the high plasma concentrations of HPL, only minute amounts of plasma are required for the assay and a large number of samples can be assayed within a 1-day period. Simple commercial immunodiffusion kits, which do not require the use of radioactive materials or any sophisticated equipment, are also available.

c. Clinical Considerations

HPL estimations may be useful in assessing placental function. Since fetal well-being depends largely upon normal placental function, this test may also reflect fetal status. Several retrospective studies have shown patients with chronic hypertension, pre-eclampsia, and intrauterine growth retardation to have falling or subnormal HPL levels prior to intrauterine fetal death.[11,42,44,47,50,51,56] It is conceivable that serial HPL determinations in patients with these disorders, which are commonly associated with compromised placenta, could be used as a test of placental function and a preliminary report of a prospective study suggests that the use of HPL plasma determinations may indeed bring about a decrease in the perinatal mortality rate.[50] It should be emphasized, however, that in disorders such as diabetes mellitus not complicated by hypertensive disease and in erythroblastosis fetalis, placental function is not usually compromised and fetal distress under such circumstances is likely to be due to causes unrelated to placental function. Therefore, *normal HPL levels* when found in these disorders *should not be interpreted as indicating good fetal prognosis.*

2. Human Chorionic Gonadotrophin (HCG)

a. Physiologic Considerations

The placenta synthesizes HCG throughout pregnancy. The pattern of increased HCG in plasma is different, however, from the growth pattern of the placenta and the peak plasma levels of HCG are reached at approximately the 10th to 11th week of gestation, after which they fall rapidly, remaining at low levels until delivery. HCG has a half-life of 5 to 11 hr. The nature and factors that control the production or release of HCG are not known. Its role may be not only in maintaining corpus luteum function, but also in suppressing the immunologic response of the mother.[1] Whether it may have a role in fetal and placental metabolism of hormones is presently unknown.

b. Clinical Applications

HCG assays are most useful in the diagnosis of pregnancy and in the management of trophoblastic disease. They have generally not been shown to be valuable in managing complicated pregnancies.

3. Progesterone

a. Source

During the last trimester of pregnancy the

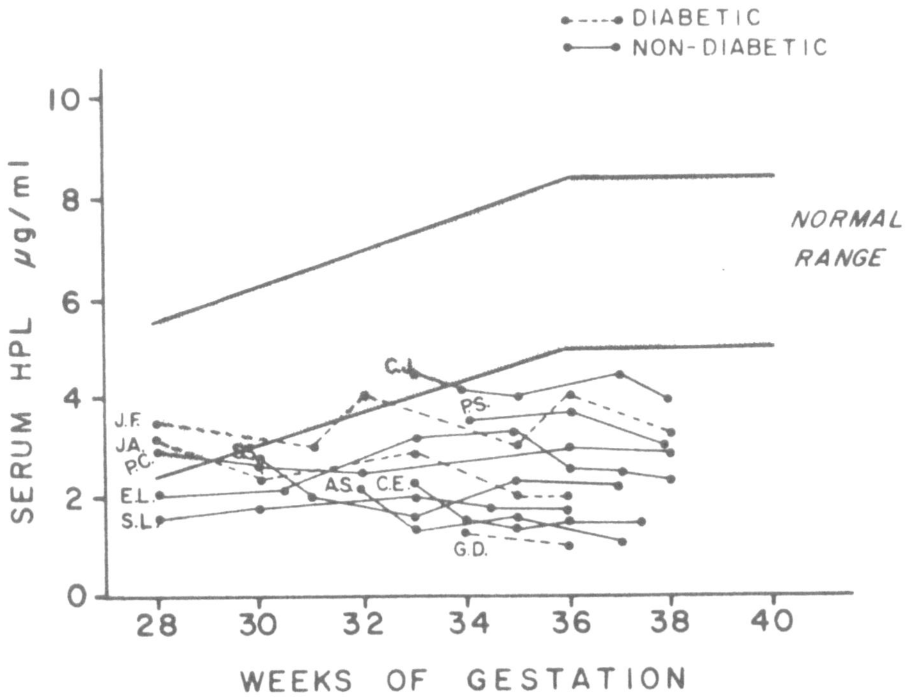

FIGURE 3. Serum HPL levels in chronic placental insufficiency. (From Saxena, B. N., Emerson, K., Jr., and Selenkow, H. A., *N. Engl. J. Med.*, 281, 225, 1969. With permission.)

placenta produces large amounts of progesterone which are secreted largely into the maternal circulation. For the formation of progesterone the placenta primarily utilizes maternal plasma cholesterol which is then hydroxylated at carbons 20 and 22. Cleavage of the side chain between these two carbons produces pregnenolone which is then converted into progesterone. The fetus also participates in progesterone turnover by providing the placenta with precursors, such as pregnenolone and pregnenolone sulfate, and by metabolizing a portion of the progesterone which is secreted by the placenta into its circulation. Since, however, fetal death does not have any appreciable effect on maternal progesterone plasma levels[59] and since mothers of anencephalic fetuses are found to have normal progesterone plasma concentrations, it appears that the overall effect of the fetus on placental production of progesterone is negligible.

b. Physiologic and Clinical Considerations

During the second half of human pregnancy, progesterone plasma levels continue to increase. There are studies reporting that progesterone plasma concentrations continue to increase until term (Figure 4).[64] Others suggest that the increase slows down plateaus or even decreases toward the end of pregnancy.[19] The rate of increase of progesterone plasma levels during the second half of human pregnancy seems to correlate well with the rate of placental growth. Good correlation is also found between individual placental weights and maternal plasma progesterone levels. All of this as well as the short half-life of progesterone in plasma suggested that plasma progesterone could be a good indicator of placental function. However, the wide range of normal plasma levels of progesterone in the last trimester as well as wide fluctuations in its level throughout the day makes follow-up

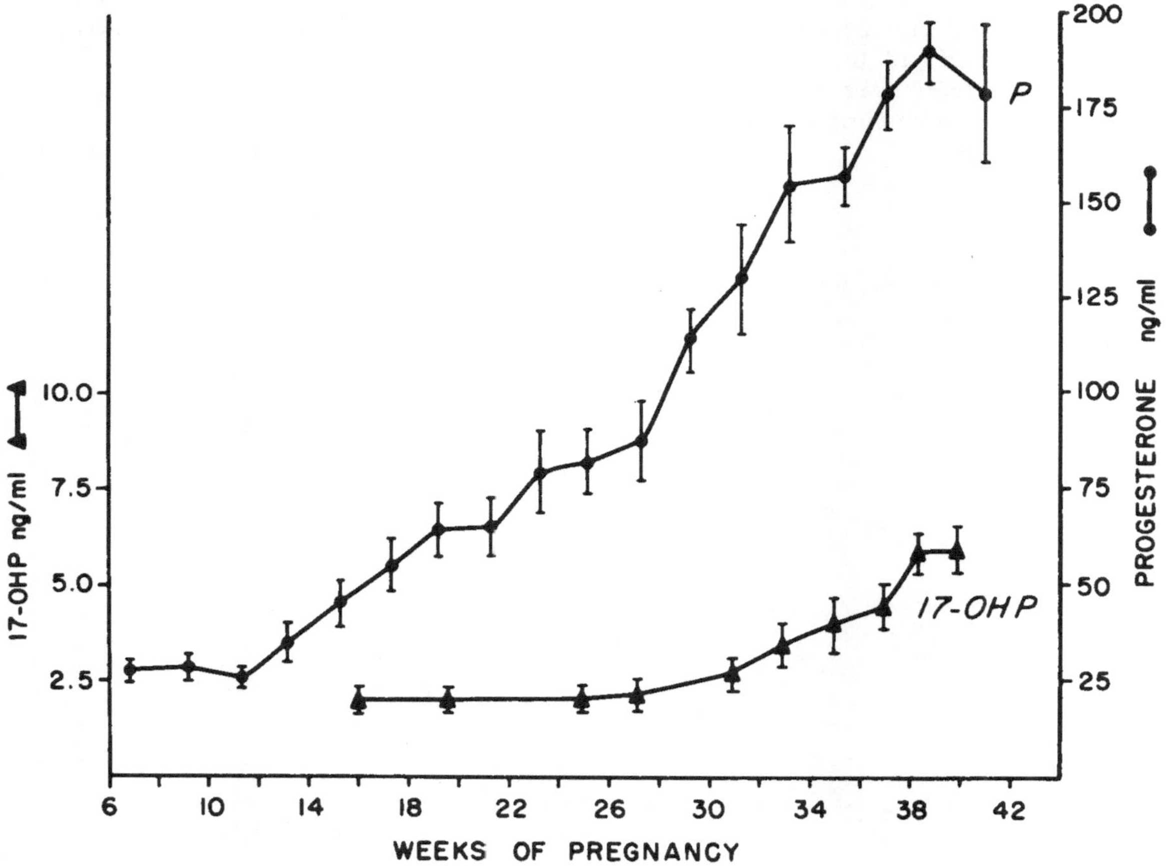

FIGURE 4. Mean plasma concentration of progesterone and 17α-hydroxyprogesterone (17-OHP) in human pregnancy. (Modified from Tulchinsky, D., Hobel, C. J., Yeager, E., and Marshall, J. R., *Am. J. Obstet. Gynecol.*, 112, 1095, 1972. With permission.)

studies of such patients difficult to interpret. In many instances, normal progesterone levels are found despite apparent severe placental insufficiency, leading to fetal death (Figure 5). In other instances, it is difficult to interpret whether the fall of progesterone plasma levels is of significant enough magnitude to indicate placental function deterioration.

B. Placental Hormones Dependent on Fetal Precursors

1. Estriol (E_3)

a. Physiologic Considerations

At term pregnancy over 90% of maternal estriol (E_3) is produced by the placenta from fetal adrenal precursors, primarily dehydroepiandrosterone-sulfate (DHEAS) (Figure 6). DHEAS undergoes 16-α-hydroxylation in fetal liver and the 16-α-OH-DHEAS reaches the placenta where it is hydrolyzed; the liberated 16-α-OH-DHEAS then undergoes aromatization forming E_3. E_3 is then secreted into both the maternal and the fetal circulation primarily as unconjugated E_3. In the maternal circulation, E_3 undergoes conjugation in the liver and the conjugated E_3 is then excreted in the urine primarily as E_3 glucuronide. A major portion of conjugated E_3 appears, however, to enter the enterohepatic circulation where it is excreted into the intestines where it probably undergoes hydrolysis by the intestinal flora only to reenter the maternal circulation again as unconjugated E_3.[36]

In maternal plasma only 10% of total circulating E_3 is unconjugated. The half-life of both conjugated and unconjugated E_3 is approximately 20 min, indicating rapid clearance of E_3 from plasma. Fluctuations in the plasma levels of both conjugated and unconjugated E_3 throughout the day usually fall within a range of 20% of the mean daily value.[57,63]

b. Conjugated vs. Unconjugated E_3

Measurements of plasma unconjugated E_3 may offer certain advantages over measurements of total plasma estriol. First, the placenta primarily secretes unconjugated E_3. Measurement of unconjugated E_3 may, therefore, reflect placental production of this hormone more closely. Second, unconjugated E_3 is cleared by the liver, which is

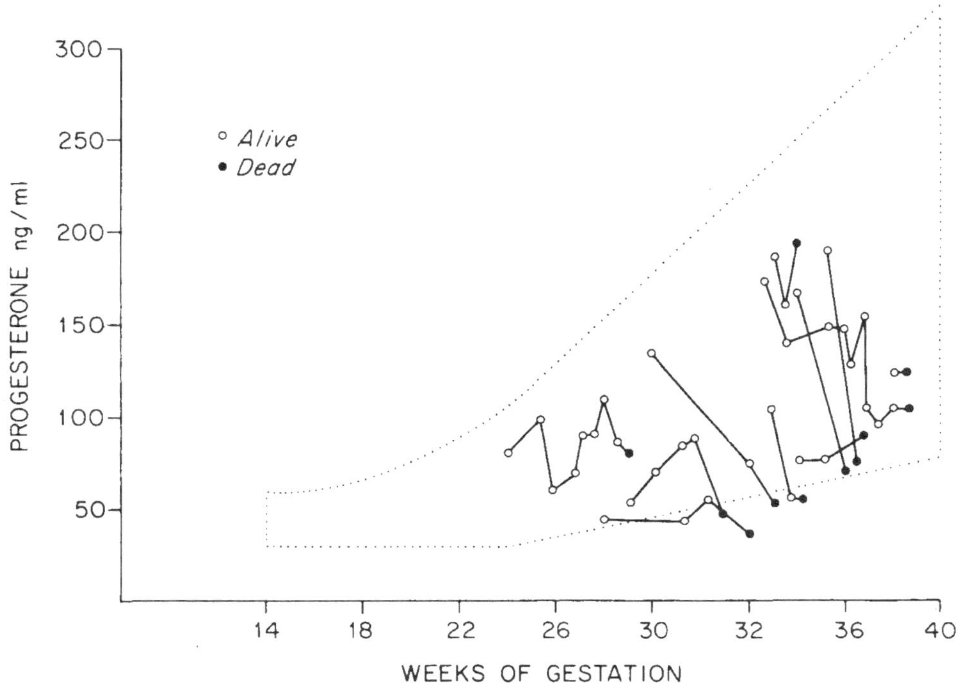

FIGURE 5. Serial plasma progesterone determinations in patients with apparent placental insufficiency leading to fetal death. The dotted line represents the 95% confidence limit of normal values.[59]

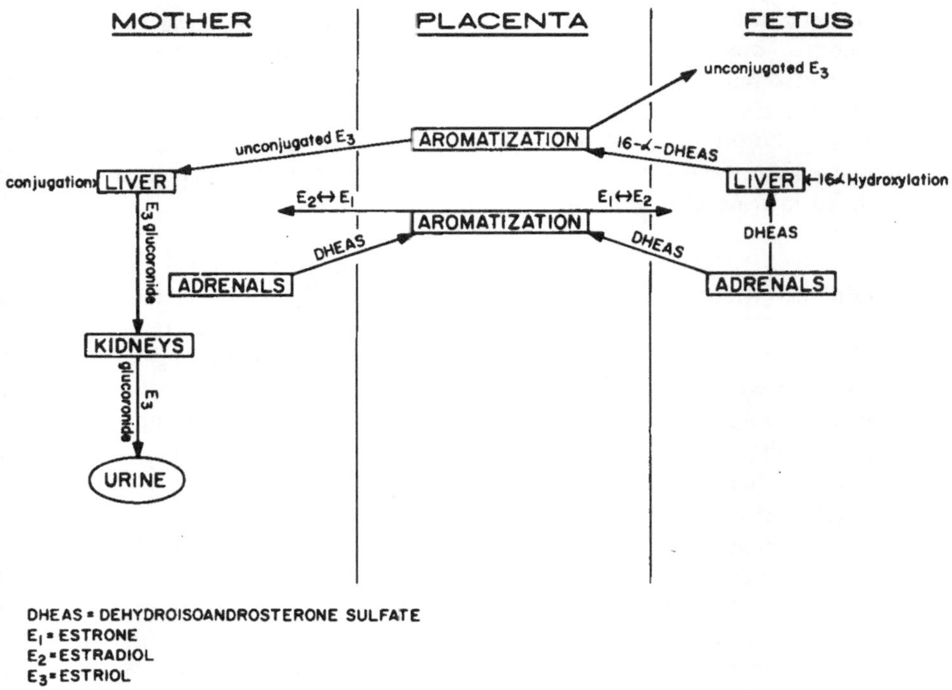

FIGURE 6. Pathway for estrogen synthesis from DHEAS.

only rarely affected by a pathologic process during pregnancy, whereas conjugated E_3 is cleared by the kidneys which can be affected by pathologic process. The chances of having erroneously high plasma unconjugated E_3 as a result of maternal liver disease are small, whereas the chances of having erroneously high plasma conjugated (or total) E_3 levels and low urinary E_3 values as a result of renal disease are real[32] and may mislead the obstetrician to believe that estriol production is good (or bad).

c. Methods for E_3 Determinations

Methods for urinary E_3 determinations have been available since the early 1930's. Practical methods for E_3 blood determinations that require much more sensitive methods became available only in the last decade. They belong to two main groups.

1. Methods for Determining Total Plasma E_3

Since unconjugated E_3 in plasma consists of only 10% of total E_3, these methods primarily measure conjugated E_3. Fluorimetric, colorimetric, gas chromatography, mass spectrometry, and radioimmunoassay methods are available for its measurement.[4,5,7,8,14,26–29,32,33,37,39,40,49,55,70] Many of these methods require prior hydrolysis of the conjugated E_3 and the liberated E_3 is then measured. Results can be available within a day.

2. Methods for Measurements of Unconjugated E_3

These include primarily radioligand and radioimmunoassay.[28,61,63,69,71] E_3 cannot yet be measured in unextracted plasma, but the availability of specific E_3 antiserum[25] allows measurement of E_3 following simple ether extraction.[60] A large number of samples can be assayed simultaneously and results can be available in 4 hr.

d. Low Plasma Estriol: Theoretical and Clinical Considerations

The formation of estriol depends on the integrity of several organs and multiple enzymatic systems. Low estriol values can, therefore, be obtained under various conditions.

1. When fetal adrenals are putting out less than normal amounts of E_3 precursors. This may be a result of (a) primary congenital adrenal hypoplasia, (b) hypoplasia secondary to ACTH deficiency (anencephalic), or (c) suppression of

fetal ACTH by corticosteroid administration to the mother.

2. In cases of fetal liver dysfunction (cirrhosis).

3. In cases of placental abnormalities that may be either (a) a congenital absence of an enzymatic system such as the sulfatase or (b) an acquired pathologic process which is capable of either compromising the blood supply of the placenta or decreasing the functional mass of the placenta (placental infarcts).

4. Impaired maternal liver function (liver cirrhosis) may result in impaired conjugation. In such cases, high plasma unconjugated E_3 and low plasma conjugated E_3 may be expected.

5. Interruption of the hepato-intestinal circulation by antibiotic administration may eliminate normal intestinal flora and bring about a decrease in urinary and plasma estriol. This is most likely due to losses of the sulfated conjugates of E_3 in stool.

In cases of fetal distress, low estriol levels may be explained by either decreased output of E_3 precursors by the fetal adrenals,[3,38] decreased conjugation by the fetal liver, or impaired placental function.[10] It may well be that impairment of placental function is the most important rate-limiting step in the production of E_3. The finding of diminished secretion of E_3 precursors by infants born to such mothers[3,38] does not preclude this possibility since the decreased availability of E_3 precursors may also be secondary to impairment of placental function. Indeed E_3 determinations are helpful indicators of fetal condition in diseases which are frequently associated with impaired placental function (such as pre-eclampsia, chronic hypertension, intrauterine growth retardation, and postmaturity syndrome) (Figure 7). They have questionable value, however, in predicting fetal distress in diseases such as diabetes and Rh isoimmunization[37,65] which are not usually associated with placental dysfunction. It may be argued that in diabetes mellitus fetal deterioration may be so sudden that even daily plasma E_3 determinations may not be frequent enough to detect it. However, this thesis is not accepted by all investigators. Another possibility is

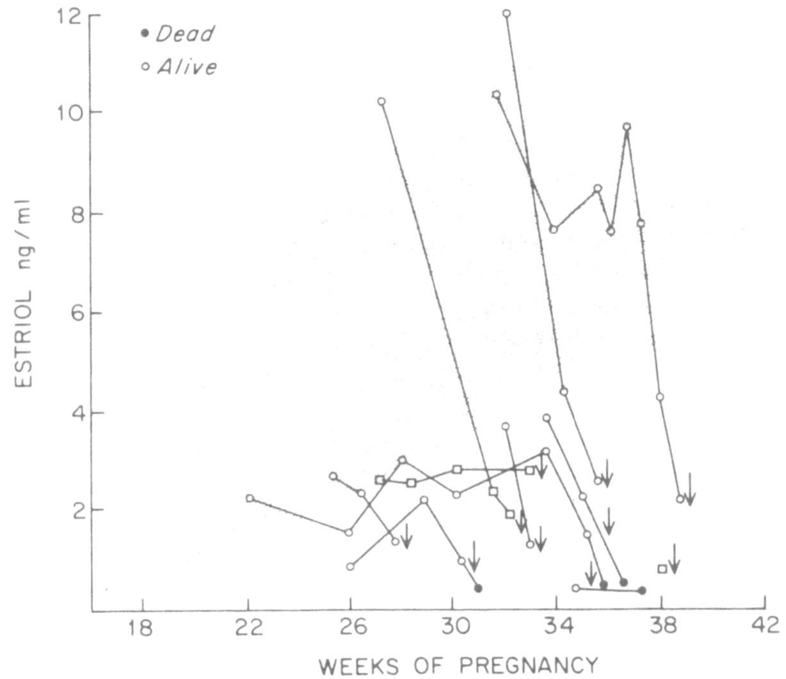

FIGURE 7. Serial plasma estriol determinations in patients with pre-eclampsia and chronic hypertension who had intrauterine fetal demise while under observation. The shaded area represents the 95% confidence limits for normal values. The arrow indicates the time of fetal death.[59]

that in the distressed diabetic fetus the system responsible for the production of E_3 remains intact. In erythroblastosis fetalis there is evidence that the very large and hypertrophic placenta, even if provided with diminished amounts of precursors,[3] would be capable of converting them into E_3 more efficiently than normal placenta,[4,5] resulting in normal plasma E_3 levels.

In clinical practice low E_3 values should always be considered as suggesting fetal distress unless proven otherwise. This should ideally be confirmed, however, by serial E_3 determinations and not by a single E_3 determination. Other causes of low plasma E_3 should be ruled out. The ingestion of corticosteroids or antibiotics should be ruled out after obtaining a history. At times it is more difficult to rule out other more rare causes of low E_3, such as placental sulfatase deficiency, except that the obstetrical history of such patients is usually unremarkable. Good conversion rate into E_2 of labeled or unlabeled DHEA infused into the maternal circulation with poor conversion of DHEAS may confirm the diagnosis. Therefore, the decision to terminate pregnancy should not be based on results of E_3 values only. The entire clinical picture should be taken into consideration, including results of other tests such as lung maturity tests that could help assess chances of fetal survival outside the uterus.

2. Plasma Estradiol (E_2)

a. Physiologic Consideration

E_2 appears to be the main unconjugated estrogen in human plasma. At term pregnancy its plasma levels are 60% higher than that of unconjugated E_3 and twofold higher than that of unconjugated E_1 (Figure 8). Approximately half of maternal circulating E_2 at term is derived from fetal adrenal precursors, primarily DHEAS. Upon reaching the placenta, DHEAS is hydrolyzed and

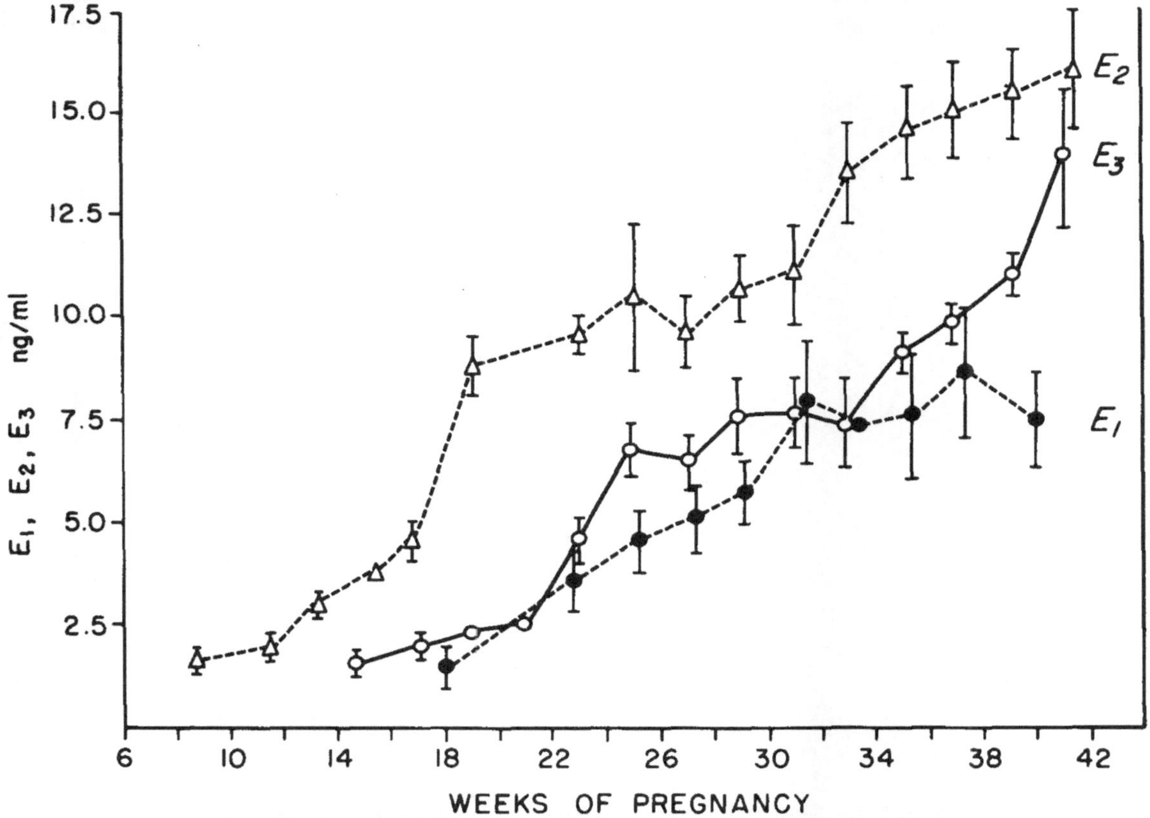

FIGURE 8. Mean plasma concentration of unconjugated estrone (E_1), estradiol (E_2), and estriol (E_3) in human pregnancy. (Modified from Tulchinsky, D., Hobel, C. J., Yeager, E., and Marshall, J. R., *Am. J. Obstet. Gynecol.*, 112, 1095, 1972. With permission.)

the liberated DHEA is then aromatized, producing E_2. Unconjugated E_2 is secreted by the placenta into the maternal and fetal circulation and it remains in maternal plasma mostly unconjugated. After being metabolized, it is excreted in urine primarily as E_3.

b. Correlations Between E_2 and E_3

1. The main precursor of both E_2 and E_3 is DHEAS. Maternal adrenals provide approximately one half of E_2 precursors, whereas E_3 is derived almost totally (over 90%) from fetal precursors.
2. Fetal liver participates in the production of E_3 but not in that of E_2.
3. Unlike E_3, E_2 remains in plasma primarily as unconjugated hormone.
4. Both E_2 and E_3 have a short half-life in plasma of approximately 20 min.
5. The plasma level of E_2 throughout the day appears to be more stable than that of E_3.

c. Methods

Gas chromatography, double isotope derivative methods, and other methods[31,39,48,49,53] have generally given way to radioligand[54,66] and radioimmunoassay techniques.[58,60,71] Measurement of E_2 directly in unextracted pregnancy plasma became possible with the use of uterine cytosol, which binds E_2 preferentially.[66] The use of specific E_2 antiserum[25] allows the measurement of E_2 following simple ether extraction.[60] A large number of samples can be processed and reported in 3 to 4 hr and only minute amounts of plasma are required (less than 0.02 μl).

d. Clinical Considerations

Interest in testing E_2 as an index of fetoplacental function arose when simple methods of E_2 determination first became available preceding, in that regard, methods for E_3 determinations. Many patients with pre-eclampsia and chronic hypertension were found to have subnormal or serially falling E_2 values prior to intrauterine fetal death.[54,58,66] Difficulties in interpretation of the results are often encountered because normal E_2 levels are found in some instances even when the fetus is dead[66] and because it is difficult to decide whether a given fall of E_2 is significant enough to indicate fetal distress. Because of these difficulties and with present availability of methods for rapid E_3 determinations, it is unlikely that E_2 plasma determinations would have a real place in monitoring fetal well-being.

3. 15-α-Hydroxyestriol (Estetrol) (E_4)

a. Physiologic Consideration

In late pregnancy, estetrol is excreted in urine in amounts which exceed those of all the other estrogens, except estriol and 16-α-hydroxyestrone.[18] In plasma, however, unconjugated E_4 is found in late pregnancy in concentrations of 1 to 4 ng/ml[9,12] which are two to tenfold lower than that of unconjugated E_1, E_2, or E_3. Estetrol seems to originate at term pregnancy from fetal plasma E_2 and, to a less extent, from fetal plasma E_3. These hormones are secreted by the placenta into the fetal circulation, then reach the fetal liver where they undergo either 15-α- and 16-α-hydroxylation (in case of E_2) or 15-α-hydroxylation (in case of E_3) forming E_4, which is then transported, unchanged, into the maternal circulation.[17,43]

b. Comparing E_3 and E_4

The placenta and fetal liver both participate in the formation of E_3 and E_4. Both are produced by the fetoplacental unit. The biogenesis of E_4 is different, however, from that of E_3 in that approximately one half the precursors of estradiol, which is the main E_4 precursor, are derived from maternal sources, whereas E_3 precursors are derived almost solely from fetal sources. In fetal liver 15 α hydroxylation is an obligatory step for the formation of E_4. Another more important difference between the two is that the fetal liver is only an intermediate organ in the production of E_3 whereas it is the rate-limiting organ in the production of E_4. At first glance these differences may not seem to be important since levels of E_3 and E_4 could both be expected to be low in cases of severe placental insufficiency. However, in cases where fetal liver disorders are associated with placental hypertrophy (such as erythroblastosis fetalis), E_4 may reflect fetal condition more accurately than E_3. Under such circumstances, diminished amounts of 16-α-hydroxylated E_3 precursors (16-α-DHEAS) may be converted more efficiently into E_3 by the hypertrophic placenta, resulting in normal E_3 production and normal E_3 levels despite the severity of the fetal condition. The production of E_4, however, should theoretically remain subnormal since the diseased liver is likely to be limited in its capacity to produce E_4 even if supplied with an excess of E_2 or E_3. The good correlation reported between urinary E_4 excretion and fetal hemoglobin levels in erythroblastosis fetalis[18] suggests that in this disease E_4 determinations may be helpful in assessing fetal well--being. It is not known whether E_4 determination could be helpful in predicting fetal distress in cases not associated with placental dysfunction or fetal liver disorders. It has been suggested to be useful in the diagnosis of molar pregnancies.

VI. THE SELECTION OF HORMONES FOR THE ASSESSMENT OF FETAL WELL-BEING

Hormones are presently used for the assessment of fetal well-being only because of our present inability to get access into the fetal compartment and to determine other biochemical and physiologic indices of fetal well-being. These hormones, besides being produced by an interplay of several organs and various enzymatic systems, are also subjects of metabolism by maternal organs. Malfunction of any of these systems may lead to subnormal hormone levels. At our present state of knowledge we can usually only guess which system has been affected in any particular disease. Difficulties in interpreting results of such studies are, therefore, numerous. Despite all these problems, the clinician has to depend on these tests because they are one of few objective criteria which the clinician presently has at hand for aiding in the management of high-risk pregnancies.

To make the best use of such hormone determinations, first the clinician must ascertain that he has selected the proper test for the proper patient. Thus, patients with diabetes mellitus would best be followed by serial plasma estriol determinations, whereas patients with pre-eclampsia may be followed by either plasma estriol or HPL determinations. Second, the reliability of each laboratory should be established, preferably by comparing results of repeated analysis of the same plasma samples. Third, the range of normal values should always be obtained for each laboratory because there may be wide variations in values reported from one laboratory to another.

Daily plasma determinations are much preferable to less frequent studies. For economic reasons, however, many commercial laboratories prefer to carry out such studies only two or three times a week. Under such circumstances, daily blood samples should still be obtained and run simultaneously, because they could establish whether the trend is upward or downward. Termination of pregnancy should not be initiated, however, on the basis of these laboratory data only; the entire clinical picture should be taken into consideration. Chances of the fetus to survive outside the uterus should be assessed with the aid of lung maturity tests and weighed against chances of intrauterine death. Once a decision to terminate the pregnancy has been made, the clinician should be aware that he may be dealing with a compromised fetus who may not withstand another insult. Care must, therefore, be taken to monitor such fetuses very closely during labor and to be ready to resort promptly to abdominal delivery should signs of acute fetal distress develop.

The search for better methods of assessing fetal well-being is likely to continue. Hormones of fetal origin that are not subject to metabolism by the maternal organism may provide more accurate information on fetal well-being. Dynamic studies that would test the efficiency of enzymatic systems of the placenta in converting either labeled or unlabeled precursors into end hormones should be further evaluated. It should also be realized that improving our ability to detect intrauterine fetal distress and to determine the cause of this insult may not only decrease prenatal mortality by early termination of pregnancy, but may perhaps allow in the future to continue such pregnancies via institution of proper therapy.

There is a need to carry out *prospective* studies that would indicate the value of such tests not only for improving perinatal mortality but also for improving the mental quality of the newborns. Only then could the real value of following high-risk patients or perhaps all pregnant patients with such endocrine tests could be established.

REFERENCES

1. **Adcock, E. W., Teasdale, F., August, C. S., Cox, S., Meschia, G., Battaglia, F. C., and Naughton, M. A.,** Human chorionic gonadotropin: its possible role in maternal lymphocyte suppression, *Science*, 181, 845, 1973.
2. **Brody, S.,** Protein hormones and hormonal peptides from the placenta, in *Foetus and Placenta*, Klopper, A. and Diczfalusky, E., Eds., Blackwell, Oxford, 1969, 299.
3. **Cleary, R. E., Depp, R., and Pion, R.,** Relation of C_{19} steroid sulfates in cord plasma to maternal urinary estriol, *Am. J. Obstet. Gynecol.*, 106, 534, 1970.
4. **Cooper, W., Coyle, M. G., and Mills, J. A.,** Determination of plasma oestriol in normal pregnancy and labor using gas-liquid chromatography, *J. Endocrinol.*, 51, 447, 1971.
5. **Corker, C. S. and Naftolin, F.,** A rapid method for the measurement of oestriol in pregnancy plasma by competitive protein binding analysis, *J. Obstet. Gynaecol. Br. Commonw.*, 78, 330, 1971.
6. **Diczfalusy, E.,** *The Foeto-Placental Unit*, Pecile, A. and Finzi, C., Eds., Excerpta Medica Foundation, Amsterdam, 1969, 65.
7. **Dubin, N. H., Crystle, C. D., Grannis, G. F., and Townsley, J. D.,** Comparison of maternal serum estriol and urinary estrogen determinations as indices of fetal health, *Am. J. Obstet. Gynecol.*, 115, 835, 1973.
8. **Fischer-Rasmussen, W.,** Plasma oestrogens and the fetal outcome, *Acta Obstet. Gynecol. Scand.*, 50, 301, 1971.
9. **Fishman, J. and Guzik, H.,** Radioimmunoassay of 15α-hydroxyestriol in pregnancy plasma, *J. Clin. Endocrinol. Metab.*, 35, 892, 1972.
10. **Gant, N. F., Hutchinson, H. T., Siiteri, P. K., and MacDonald, P. C.,** Study of the metabolic clearance rate of dehydroisoandrosterone sulfate in pregnancy, *Am. J. Obstet. Gynecol.*, 111, 555, 1971.
11. **Genazzani, A. R., Cocola, F., Neri, P., and Fioretti, P.,** Human chorionic somatomammotropin (HCS) plasma levels in normal and pathological pregnancies and their correlation with the placental function, *Acta Endocrinol.*, Suppl. 167, 5, 1972.
12. **Giebenhaim, M. E., Tagatz, G. E., and Gurpide, E.,** Serum levels of unconjugated estetrol during human pregnancy, *J. Steroid Biochem.*, 3, 707, 1972.
13. **Goebelsmann, U., Freeman, R. K., Mestman, J. H., Nakamura, R. M., and Woodling, B. A.,** Estriol in pregnancy. II. Daily urinary estriol assays in the management of the pregnant diabetic woman, *Am. J. Obstet. Gynecol.*, 115, 795, 1973.
14. **Goebelsmann, U. and Gorwill, R. H.,** The use of silica gel impregnated glass fiber sheets in plasma estriol assays, *Steroids*, 16, 635, 1970.
15. **Goldstein, D. P.,** The management of gestational trophoblastic disease, in *Progress in Gynecology*, Vol. V, Meigs, J. V. and Sturgis, S. H., Eds., Grune & Statton, New York, 1970, 397.
16. **Grumbach, M. M., Kaplan, S. L., Sciarra, J. J., and Burr, I. M.,** Chorionic growth hormone-prolactin (CGP): secretion, disposition, biologic activity in man and postulated function as the "growth hormone" of the second half of pregnancy, *Ann. N.Y. Acad. Sci.*, 148, 501, 1968.
17. **Hagen, A. A.,** Formation of 15α-hydroxyestriol from 4-^{14}C-17β-estradiol and 6,7-^{3}H estriol by an anencephalic, *J. Clin. Endocrinol. Metab.*, 30, 763, 1970.
18. **Heikkila, J. and Luukkainen, T.,** Urinary excretion of estriol and 15α-hydroxyestriol in complicated pregnancies, *Am. J. Obstet. Gynecol.*, 110, 509, 1971.
19. **Johansson, E. D. B.,** Plasma levels of progesterone in pregnancy measured by a rapid competitive protein binding technique, *Acta Endocrinol.*, 61, 607, 1969.
20. **Josimovich, J. B., Kosor, B., Boccella, L., Mintz, D. H., and Hutchinson, D. L.,** Placental lactogen in maternal serum as an index of fetal health, *Obstet. Gynecol.*, 36, 244, 1970.
21. **Kaplan, H. G. and Hreschyshyn, M. M.,** Estetrol in molar pregnancies quantitated by a rapid gas chromatographic method in conjunction with estrone, estradiol-17β, estriol and pregnanediol, *Am. J. Obstet. Gynecol.*, 115, 803, 1973.
22. **Karow, W. G., Gentry, W. C., and Tulchinsky, D.,** Corpus luteum function in abnormal and induced pregnancies, Pacific Coast Fertility Society, Palm Springs, October 1973.
23. **Levitz, M. and Selinger, M.,** Plasma estriol levels in class D diabetes of pregnancy, *Am. J. Obstet. Gynecol.*, 108, 82, 1970.
24. **Lin, T. J., Lin, S. C., Erlenmeyer, F., Kline, I. T., Underwood, R., Billiar, R. B., and Little, B.,** Progesterone production rates during the third trimester of pregnancy in normal women, diabetic women, and women with abnormal glucose tolerance, *J. Clin. Endocrinol. Metab.*, 34, 287, 1972.
25. **Lindner, H. R., Perel, E., Friedlander, A., and Zeitlin, A.,** Specificity of antibodies to ovarian hormones in relation to the site of attachment of the steroid hapten to the peptide carrier, *Steroids*, 19, 357, 1972.
26. **Macourt, D., Corker, C. S., and Naftolin, F.,** Plasma oestriol in pregnancy, *J. Obstet. Gynaecol. Br. Commonw.*, 78, 335, 1971.
27. **Masson, G. M.,** Plasma oestriol in pre-eclampsia, *J. Obstet. Gynaecol. Br. Commonw.*, 80, 206, 1973.
28. **Masson, G. M. and Klopper, A.,** Changes in plasma oestriol concentration associated with the onset of labor, *J. Obstet. Gynaecol. Br. Commonw.*, 79, 970, 1972.

29. **Mathur, R. S., Chestnut, S. K., Leaming, A. B., and Williamson, H. O.,** Application of plasma estriol estimation in the management of high-risk pregnancies, *Am. J. Obstet. Gynecol.*, 117, 210, 1973.
30. **Mishell, D. R., Thorneycroft, I. H., Nagata, Y., Murata, T., and Nakamura, R. M.,** Serum gonadotropin and steroid patterns in early human gestation, *Am. J. Obstet. Gynecol.*, 117, 631, 1973.
31. **Munson, A. K., Mueller, J. R., and Yannone, M. E.,** Quantification of free 17β-estradiol in pregnancy plasma by gas-liquid chromatography using electron-capture detection, *Biochem. Med.*, 3, 187, 1969.
32. **Nachtigall, L., Bassett, M., Hogsander, U., and Levitz, M.,** Plasma estriol levels in normal and abnormal pregnancies, an index of fetal welfare, *Am. J. Obstet. Gynecol.*, 101, 638, 1968.
33. **Nachtigall, L., Bassett, M., Hogsander, U., Slagle, S., and Levitz, M.,** A rapid method for the assay of plasma estriol in pregnancy, *J. Clin. Endocrinol. Metab.*, 26, 941, 1966.
34. **Nygren, K. G., Johansson, E. D. B., and Wide, L.,** Evaluation of the prognosis of threatened abortion from the peripheral plasma levels of progesterone, estradiol and human chorionic gonadotropin, *Am. J. Obstet. Gynecol.*, 116, 916, 1973.
35. **Odell, W. D. and Ross, G. T.,** Correlation of bioassay and immunoassay potencies for FSH, LH, TSH and HCG, in *Principles of Competitive Protein Binding Assays*, Odell, W. D. and Daughaday, W. H., Eds., J. B. Lippincott, Philadelphia, 1971, 401.
36. **Pulkkinen, M. D. and Aldercreutz, H.,** Effects of ampicillin treatment on the sulphation and glucuronidation of oestriol in pregnancy, *Acta Endocrinol.*, Suppl. 177, 239, 1973.
37. **Ratansopa, V., Schindler, A. E., Lee, T. Y., and Herrmann, W. L.,** Measurement of oestriol in plasma by gas liquid chromatography, *Am. J. Obstet. Gynecol.*, 99, 295, 1967.
38. **Reynold, W. and Mirkin, B. L.,** Urinary steroid levels in newborn infants with intra-uterine growth retardation, *J. Clin. Endocrinol. Metab.*, 36, 576, 1973.
39. **Roy, E. J. and Brown, J. B.,** A method for the estimation of oestriol oestrone and oestradiol-17β in the blood of the pregnant woman and the foetus, *J. Endocrinol.*, 21, 9, 1960.
40. **Roy, E. J., Harkness, R. A., and Kerr, M. G.,** Concentration of oestrogens in blood and urine of patients suffering from pre-eclampsia, *J. Obstet. Gynaecol. Br. Commonw.*, 70, 597, 1963.
41. **Ryan, K. J.,** *Principles and Management of Human Reproduction*, Reid, D. E., Ryan, K. J., and Benirschke, K., Eds., W. B. Saunders, Philadelphia, 1972, 4.
42. **Saxena, B. N., Emerson, K., and Selenkow, H. A.,** Serum placental lactogen (HPL) levels as an index of placental function, *N. Engl. J. Med.*, 281, 225, 1969.
43. **Schwers, J., Gurpide, E., Vande Wiele, R. L., and Lieberman, S.,** Urinary metabolites of estradiol and estriol administered intra-amniotically, *J. Clin. Endocrinol. Metab.*, 27, 1403, 1967.
44. **Selenkow, H. A., Saxena, B. N., Dana, C. L., and Emerson, K., Jr.,** Measurement and pathophysiologic significance of human placental lactogen, in *The Foeto-Placental Unit*, Pecile, A. and Finzi, C., Eds., Excerpta Medica Foundation, Amsterdam, 1969, 340.
45. **Siiteri, P. K. and MacDonald, P. C.,** Placental estrogen biosynthesis during human pregnancy, *J. Clin. Endocrinol. Metab.*, 26, 751, 1966.
46. **Simmer, H. H.,** *Biology of Gestation*, Vol. 2, Assali, N. S., Ed., Academic Press, New York, 1968, 246.
47. **Singer, W., Desjardins, P., and Friesen, H. G.,** Human placental lactogen. An index of placental function, *Obstet. Gynecol.*, 36, 222, 1970.
48. **Smith, O. W.,** Free and conjugated estrogens in blood and urine before and during parturition in normal human pregnancy, *Acta Endocrinol.*, 104, 1, 1966.
49. **Smith, O. W. and Arai, K.,** Blood estrogens in late pregnancy: an evaluation of methods with improved recovery, *J. Clin. Endocrinol. Metab.*, 23, 1141, 1963.
50. **Spellacy, W. N.,** Human placental lactogen: a laboratory aid in the obstetrical management of complicated pregnancy, presented at the International Symposium on Clinical Application of Hormone Assays in Pregnancy, Foundation De Recherche En Hormonologie, Fresnes, France, October 1973.
51. **Spellacy, W. N., Teoh, E. S., and Buhi, W. C.,** Human chorionic somatomammotropin (HCS) levels prior to fetal death in high risk pregnancies, *Obstet. Gynecol.*, 35, 685, 1970.
52. **Spona, J. and Janisch, H.,** Serum placental lactogen (HPL) as index of placental function, *Acta Endocrinol.*, 68, 401, 1971.
53. **Svendsen, R. and Sorensen, B.,** The concentration of unconjugated oestrone and 17β-estradiol in plasma during pregnancy, *Acta Endocrinol.*, 47, 237, 1964.
54. **Sybulski, S. and Maughan, G. B.,** Maternal plasma estradiol levels in normal and complicated pregnancies, *Am. J. Obstet. Gynecol.*, 113, 310, 1972.
55. **Taylor, E. S., Hagerman, D. D., Betz, G., Williams, K. L., and Grey, P. A.,** Estriol concentrations in blood during pregnancy, *Am. J. Obstet. Gynecol.*, 108, 868, 1970.
56. **Teoh, E. S., Spellacy, W. N., and Buhi, W. C.,** Human chorionic somatomammotrophin (HCS): new index of placental function, *J. Obstet. Gynaecol. Br. Commonw.*, 78, 673, 1971.
57. **Townsley, J. D., Dubin, N. H., Grannis, G. F., Garman, L. J., and Crystle, C.D.,** Circadian rhythms of serum and urinary estrogens in pregnancy, *J. Clin. Endocrinol. Metab.*, 36, 289, 1973.

58. **Townsley, J. D., Gartman, L. J., and Crystle, C. D.,** Maternal serum 17β-estradiol levels in normal and complicated pregnancies: comparison with other estrogen indices of fetal health, *Am. J. Obstet. Gynecol.,* 115, 830, 1973.
59. **Tulchinsky, D.,** unpublished data.
60. **Tulchinsky, D.,** Placental secretion of unconjugated estrone, estradiol, and estriol into the fetal and maternal circulation, *J. Clin. Endocrinol. Metab.,* 36, 1079, 1973.
61. **Tulchinsky, D. and Abraham, G. E.,** Radioimmunoassay of plasma estriol, *J. Clin. Endocrinol. Metab.,* 33, 775, 1971.
62. **Tulchinsky, D. and Hobel, C. J.,** Plasma human chorionic gonadotropin, estrone, estradiol, estriol, progesterone and 17α-hydroxy progesterone in human pregnancy. III. Early normal pregnancy, *Am. J. Obstet. Gynecol.,* 117, 884, 1973.
63. **Tulchinsky, D., Hobel, C. J., and Korenman, S. G.,** A radio-ligand assay for plasma estriol values during normal and abnormal pregnancies, *Am. J. Obstet. Gynecol.,* 111, 311, 1971.
64. **Tulchinsky, D., Hobel, C. J., Yeager, E., and Marshall, J. R.,** Plasma estrone, estradiol, estriol, progesterone and 17-OH-progesterone in human pregnancy. I. Normal pregnancy, *Am. J. Obstet. Gynecol.,* 112, 1095, 1972.
65. **Tulchinsky, D., Hobel, C. J., Yeager, E., and Marshall, J. R.,** Plasma estradiol, estriol and progesterone in human pregnancy. II. Clinical applications in Rh isoimmunization disease, *Am. J. Obstet. Gynecol.,* 113, 766, 1972.
66. **Tulchinsky, D. and Korenman, S. G.,** The plasma estradiol as an index of feto-placental function, *J. Clin. Invest.,* 50, 1490, 1971.
67. **Viatukaitis, J. L., Braunstein, G. D., and Ross, G. T.,** A radioimmunoassay which specifically measures human chorionic gonadotropin in the presence of human luteinizing hormone, *Am. J. Obstet. Gynecol.,* 113, 754, 1972.
68. **Westphal, U., Ed.,** *Steroid-Protein Interactions,* Springer-Verlag, Berlin, 1971, 356.
69. **Wilkinson, M., Effer, S. B., Younglai, E. V., and Gupta, K.,** Free estriol in human pregnancy plasma, *Am. J. Obstet. Gynecol.,* 114, 867, 1972.
70. **Wotiz, H. H., Charransol, G., and Smith, I. N.,** Gas chromatographic measurement of plasma estrogens using an electron capture detector, *Steroids,* 10, 127, 1967.
71. **Wu, C. H. and Lundy, L. E.,** Radioimmunoassay of plasma estrogens, *Steroids,* 18, 91, 1971.
72. **Yalow, R. S. and Berson, S. A.,** Immunoassay of endogenous plasma insulin in man, *J. Clin. Invest.,* 39, 1157, 1960.
73. **Yoshimi, T., Strott, C. A., Marshall, J. R., and Lipsett, M. B.,** Corpus luteum function in early pregnancy, *J. Clin. Endocrinol. Metab.,* 29, 225, 1969.

BIOCHEMICAL CHANGES AT THE MENOPAUSE

P. Wilding

TABLE OF CONTENTS

I. INTRODUCTION

Cessation of menstruation associated with decreased estrogen secretion by the ovaries in the healthy woman may occur between the ages of 35 and 60. This change (the menopause or climacteric) generally takes place between 45 and 50 years of age, although some women may experience premature menopause before the age of 40.[1] The hormonal changes that occur at this time and their relationship to the menopausal symptoms have been described and the subject has been reviewed by numerous workers.[2-4] It has been known for many years that the decreased estrogen secretion that occurs at the menopause is associated with atrophic changes in the ovaries and the well-known and well-recognized symptoms associated with the menopausal woman, namely flushing, menstrual irregularities, and atrophic changes in the gonads, are usually attributed to this diminution of estrogen production. An alternative theory of "gonadotrophin excess" secondary to estrogen deficiency has been postulated[2] as the primary cause of many of the changes and symptoms occurring at the menopause, but this theory is generally discounted. However, it is very probable that the increases in levels of certain hormones, relative to the premenopausal period, cause or influence some of the physiological or psychological changes which are noted.

Recently, controversy has grown over the value of estrogenic therapy as a method of alleviating the unpleasant symptoms observed at the menopause or halting the atrophic changes associated with the postmenopausal woman.[5,6] There is little doubt that the widespread use of estrogens, in the form of oral contraceptives, may well modify, in some small way, the patterns of change which are observed.

A certain proportion of women (8%) do experience early menopause which is usually associated with estrogen deficiency. However, a relatively high incidence (7%) of women experience the menopause after the age of 55; this is not fully explained by maintenance of ovarian function.

Knowledge and understanding of the changes occurring over the menopausal period have undoubtedly been aided by the development and introduction of simple and convenient hormone assay procedures. It is inevitable that new and more convenient methods that allow frequent assay will aid the understanding even more.

Although the alteration in hormone production and excretion in menopausal women has been described, there are relatively few reports that discuss alteration in commonly assayed serum constituents. This chapter will attempt to discuss some of the quantifiable changes which occur in plasma over the menopausal period and to relate them, where possible, to other progressive or physiological changes due to age and to changes in hormonal secretion.

II. ENDOCRINE CHANGES AT THE MENOPAUSE

A. Estrogens

In the fertile woman, estrogens (estriol, estrone, and estradiol) are produced in the ovaries in the stromal and theca cells. The fine balance of estrogen production and other gonadal steroids (androgens, progesterone, etc.) and their control of the menstrual cycle have been extensively studied. Recently, it has been established[7] that the gonadotrophic excretion may be twice as high as usual in the period immediately prior to the menopause. This increased excretion, predominantly follicle stimulating (FSH) and luteinizing hormone (LH), parallels the diminution in production of the estrogens and the atrophic changes in the ovaries. The regular production of estrogen assumes an irregular pattern which is thought to account in part for the menstrual irregularities. The study of cervicovaginal smears, as an index of the estrogen status, has established that an atrophic appearance is commonly associated with the menopause. It is significant, however, that some 40% of women below 60 years of age who are postmenopausal demonstrate no atrophic smears. Furthermore, 9% of women over 75 years of age continue to demonstrate cervical or vaginal smears with little atrophic appearance.[8] In general, however, it can be said that approximately 50% of women who are postmenopausal by at least a year will demonstrate an atrophic pattern in their vaginal smears. The apparent persistence of estrogenic activity in postmenopausal women is difficult to associate with the other well-quantified changes occurring elsewhere; this only makes interpretation of the postmenopausal syndrome more difficult. There are undoubtedly pitfalls in attempting to assess the estrogen status of smears and, hence, the figures quoted above should be regarded with some caution.[9] The precise role of estrogen on tissues, other than gonads, is not fully understood. However, several noticeable changes, such as postmenopausal osteoporosis, have been directly attributed to the marked decrease in the production of this hormone. Several studies have been reported on the effect on calcium and phosphate homeostasis;[10] these will be discussed later.

Approximately 25% of all women experience some degree of discomfort at the menopause which necessitates a visit to their general practitioner. As a result, there have been several studies to determine the therapeutic value of estrogen therapy to alleviate these uncomfortable symptoms. Work by Kaufman[6] demonstrated that problems associated with flushing and atrophic vaginitis were markedly improved by estrogen therapy. However the protagonists of estrogen therapy[6,11] are opposed by the more conservative school[5] who suggest that as estrogen therapy cannot halt atrophic changes in the ovaries, such therapy is of minimal value. The impressive report by Kaufman[6] suggests that many women may well benefit from treatment of this type.

B. Follicle Stimulating Hormone (FSH)

Increased excretion of FSH has been reported in women as they approach the menopause.[12] However, in other patients studied, the same author has reported little or no change in excretion of the hormone. The role of this hormone in premature menopause has been studied and it appears that imbalance of this hormone with luteinizing hormone may be a factor.[13] Certainly it would appear that the regular cyclic production of this hormone, which is demonstrated during the fertile period, cannot be maintained in the postmenopausal period. This involvement of the pituitary hormone is difficult to interpret, but it may well explain some of the common problems of hirsutism, which in the fertile woman are associated with hyperfunction of the pituitary. It has been suggested that hyperfunction of the pituitary is induced by reduced levels of circulating estrogen.[14]

C. Luteinizing Hormone (LH)

This hormone is also produced in a regular cycle in the fertile woman. However, approaching the menopause the regular cycle is no longer maintained and large increases in excretion of the hormone have been detected.[10] The extent of this increased production may reach 50 to 100 times higher than that normally demonstrated in younger women.

III. CHANGES IN PLASMA CONSTITUENTS

The clinical chemist is frequently faced with the problems of defining the reference values related to age and sex for many constituents in serum. Within the last decade it has become apparent that many of the constituents commonly assayed in the clinical laboratory change markedly with age and are significantly different between sexes. The improved precision and accuracy demonstrated in clinical laboratories have meant that the ability to define these reference values has improved and large-scale studies have illustrated marked changes in plasma concentrations for many constituents at the menopause. The need to carry out studies of this type has considerable justification, as there is little doubt that assessment of biochemical results in the elderly postmenopausal woman must not be related to groups of healthy male medical students.

Changes in biochemical constituents occurring at the menopause have been quantified and the average period over which the climacteric occurs has been clearly defined. Many reports[15-18] have illustrated such changes occurring in many serum and plasma constituents. However, not all these reports are in full agreement with respect to the degree or pattern of changes; there is little doubt that they are not related to age, but are associated with physiological changes occurring simultaneously with estrogen insufficiency. As part of a large-scale study of serum biochemical constituents in the healthy population, the present author has studied patterns of change for various biochemical constituents. The changes quantified during this study are summarized in Table 1. The following section will attempt to discuss some of the changes which have been noted.

A. Calcium

The problems of osteoporosis in the postmenopausal woman have been noted for many years;[19] this topic will be discussed in greater detail in a later chapter. However, it is to be expected that the plasma levels in the postmenopausal woman will alter where there is severe mobilization of bone calcium. The problem of postmenopausal osteoporosis and the influence of estrogen on calcium and phosphate metabolism has been examined[20] and there is little doubt that there is a link between estrogen deficiency and calcium deficiency in the etiology of postmeno-

TABLE 1

Serum Constituents with Significant Patterns of Change at the Menopause

Constituent	Changes in mean values from 5th to 6th decades	Level of significance (relative to males)
Urea	+ 2.26 mg/100 ml	$p < 0.025$
Uric acid	+ 0.45 mg/100 ml	$p < 0.001$
Inorganic phosphate	+ 0.22 mg/100 ml	$p < 0.001$
Calcium	+ 0.13 mg/100 ml	$p < 0.001$
Alkaline phosphatase	+ 2.12 KA Units	$p < 0.001$
Cholesterol	+26.61 mg/100 ml	$p < 0.001$
Total lipids	+ 0.21 g/100 ml	$p < 0.001$
Sodium	+ 0.74 mmol/l	$p < 0.025$

(From Wilding, P., Rollason, J. G., and Robinson, D., *Clin. Chim. Acta*, 41, 375, 1972. With permission.)

pausal osteoporosis. The work of Caniggia and Gennari[20] suggests that osteoporosis may well derive from hormonal imbalance, causing a negative calcium balance. An elevation of mean serum calcium levels of 0.13 mg/100 ml has been shown to occur between the 5th and 6th decades in women, whereas the corresponding change in males is not significant ($p < 0.001$).[15] The failure to demonstrate a significant increase in men over the similar period is thought to indicate that the change in wômen was menopausal rather than due to age. Work by several authors[11,21] supports this finding. The levels of plasma calcium in women who have undergone bilateral oophorectomy have been investigated[11] and compared with values observed in premenopausal women of similar age. Markedly higher values were observed in those women subjected to oophorectomy. The precise effect, direct or indirect, of estrogen deficiency or gonadotrophin excess in bone resorption in the postmenopausal woman still requires elucidation.

B. Inorganic Phosphate

It is to be expected that some significant change would be expected at the menopause if alterations in calcium metabolism were noted. The studies referred to above[20] certainly indicate that alterations in inorganic phosphate are detectable and probably derive from increased absorption. Increases in the mean values of inorganic phosphate have been demonstrated in the postmenopausal woman and in women who developed or had artificial menopause following oophorectomy. Wilding and colleagues[15] demonstrated an increase in the mean value of inorganic phosphate of 0.22 mg/100 ml in women between the 5th and 6th decade ($p < 0.001$). This alteration in phosphate level over this period has not been a constant finding.[16,21] It is, of course, conceivable that the increases in this constituent are not due to primary alterations in the absorptive mechanism, but are secondary to bone resorption. Because the changes in inorganic phosphate and calcium probably derive directly or indirectly from bone resorption, it is inevitable that some changes in the alkaline phosphatase activity associated with the metabolism of these constituents would be noted.

C. Alkaline Phosphatase

Changes in this enzyme between the 5th and 6th decade in women probably represent one of the most significant changes in any serum constituent, relative to males, that has been reported. It is certainly one of the most dramatic, as illustrated by Figure 1 here. The mean activity of the enzyme in the postmenopausal woman appears to increase by over 2 King Armstrong units ($p < 0.001$).[15] This finding has been confirmed by several workers.[17,22,23] The total alkaline phosphatase activity present in serum is known to originate from several sources, including bone, liver, and intestine. However, although it would appear that the increase in this enzyme following the menopause would probably derive from bone, because of the documented occurrence of osteoporosis in the postmenopausal woman, it has yet to be established. Many techniques are now available for establishing the tissue source of alkaline phosphatase, including heat inactivation, phenyl alanine sensitivity, and inhibition by urea; thus, the true cause of elevations in this enzyme over the menopausal period should soon be determined.

The changes in calcium, inorganic phosphate, and alkaline phosphatase are obviously interrelated and the fact that they increase between the 5th and 6th decade would tend to indicate an association with calcium metabolism, probably associated with bone turnover.

D. Cholesterol

It is well established that the concentration of cholesterol in plasma increases markedly at the menopause (see Figure 2). In the premenopausal woman, cholesterol levels are markedly lower than in men and yet a postmenopausal woman frequently demonstrates levels higher than in males. The reasons for this marked elevation at the menopause are not clearly established, but may well relate to changes in other plasma lipid compartments. The results, such as those demonstrated in Figure 2, are not fully supported by some workers,[17] but several large population studies do indicate that this increase is real.[16,18] Several workers have attempted to relate the increase in serum cholesterol to other parameters, such as arterial blood pressure;[24] it would appear that the postmenopausal woman not only demonstrates increased serum cholesterol values, but also increases in diastolic blood pressure.

The cholesterol in plasma exists predominantly as a β-lipoprotein. Estimation of β-lipoprotein has been difficult, but more modern specific methods, mostly immunological techniques, have enabled

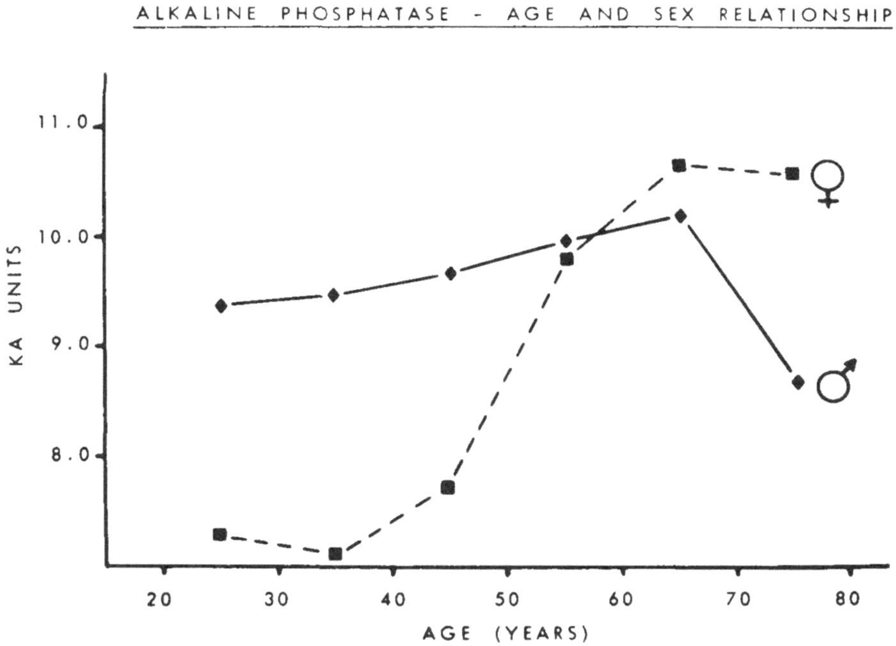

FIGURE 1. The effect of age and sex on serum alkaline phosphatase activity (■ = males and ▲ = females). (From Wilding, P., Rollason, J. G., and Robinson, D., *Clin. Chim. Acta,* 41, 375, 1972. With permission.)

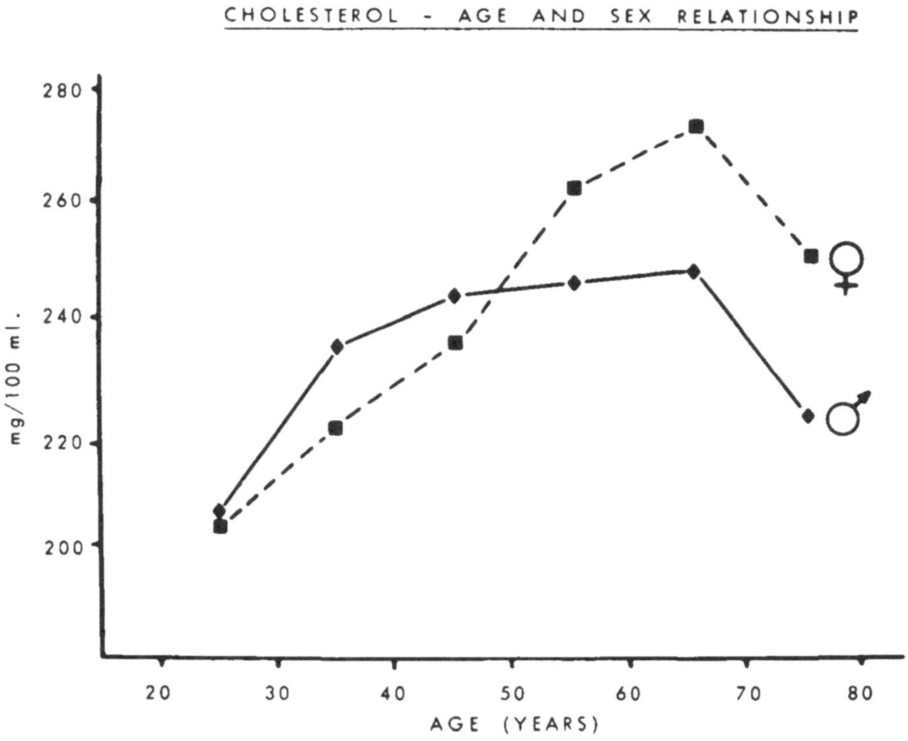

FIGURE 2. The effect of age and sex on serum cholesterol concentration (■ = males; ▲ = females).

this parameter to be assayed. Turbidimetric assays for β-lipoprotein, based on precipitation with dextran sulfate,[25] would tend to indicate that concentrations of β-lipoprotein do not follow the same pattern as that for cholesterol. Wilding and his colleagues[15] have demonstrated that β-lipoprotein, when assayed by turbidimetric methods, tends to be higher in adult women at all ages. If this technology measures true β-lipoprotein, then one must postulate an alteration in the transport mechanism of the lipoprotein over the menopausal period. It is known that β-lipoprotein and pre β-lipoprotein have other transport functions and it is conceivable that cholesterol competes for the lipoproteins transport availability in the premenopausal period. If estrogen is one of these competing components, then it is conceivable that diminution in its concentration would facilitate the transport of more cholesterol. This highly simplified explanation is by no means totally acceptable, but may have some element of truth. The markedly lower levels in plasma cholesterol in the premenopausal woman are reflected in the decreased incidence of ischemic heart disease and, likewise, the higher levels of plasma cholesterol found in postmenopausal women are associated with increased formation of atherosclerotic deposits.

Determination of total lipids has demonstrated[15] that the patterns of change throughout adult life in women mirror those for total cholesterol.

There is little evidence that the thyroid function, which markedly influences the metabolic control of synthesis and catabolism of cholesterol, is altered in the postmenopausal woman. Population studies have been carried out to establish the patterns of change in protein-bound iodine, but these would certainly not indicate changes in thyroid function sufficient to explain the marked increase in plasma cholesterol.

E. Urea

Several published studies have established that urea levels are generally lower in women, although in both sexes the mean values increase with age.[15,16,18,21] The level of urea in blood appears to increase in both sexes between the 5th and 6th decades, but the increase in women is significantly greater than the corresponding change in men.[15] Improved technology has allowed urea to be determined with greater precision and accuracy; thus, it is possible to differentiate significant abnormalities in urea concentration. Like many other of the alterations in plasma concentrations at the menopause, the urea increase is difficult to explain. Parameters that reflect liver function tend to demonstrate no marked elevation, other than alkaline phosphatase; thus, it is tempting to postulate that some alteration in renal clearance of urea may be the reason. Unlike alkaline phosphatase and cholesterol, the changes in urea concentration at the menopause are not so marked, making it more difficult to establish the direct cause.

F. Uric Acid

Uric acid concentrations in plasma are markedly lower in women at all ages, but a marked and highly significant increase in plasma concentration does occur at the menopause.[15,16] This particular plasma constituent has been studied for over a century and still the factors which determine its concentration, catabolism, and excretion have yet to be elucidated. Numerous studies have established that dietary intake, specifically purines, height, intelligence, and renal function may all contribute toward the elevation of uric acid in blood. However, it is unlikely that one of these alone is the sole reason for elevation of uric acid at the menopause. Hyperuricemia is thought to be due to increased tubular reabsorption rather than to decreased excretion by the glomerulus. Certainly, the elevated levels of uric acid noted in toxemia of pregnancy may be explained in this way. It is conceivable that the tubular reabsorption mechanism is altered in the postmenopausal period, but the failure to show marked alterations in other constituents probably indicates that other factors are involved. Physical activity is known to increase the level of uric acid and the concentration in the serum correlates with lactic acid levels.

The increase in uric acid in females between the 5th and 6th decade has been shown to approximate 0.5 mg/100 ml.[15] This increase is highly significant and surely similar findings would be demonstrated if this were solely due to alterations in renal function.

Hyperuricemia is known to vary with social strata and some evidence also exists for different genotypes. Since the changes occurring over the menopausal period cannot be explained in this way, an explanation based on renal function is probably more plausible.

G. Sodium

A slight, but significant increase appears to occur in plasma sodium concentration at the menopause.[15] This alteration may well be associated with the alteration in hormonal homeostasis and related to fluid retention. Minor alterations in plasma sodium may not fully depict the alterations in residual sodium over this period. Certainly, some of the psychological changes occurring at this time might be explained in this way. It is postulated that depression may derive from changes in residual sodium levels,[26] which could in some way explain the psychological symptoms in some women at the menopause.

H. General

Many other plasma constituents that are commonly determined in the clinical laboratory have been studied to determine their patterns of change in the adult. These include glucose, potassium, serum enzymes, bilirubin, and proteins, but there is little evidence to indicate that significant changes occur over the menopausal period. However, recent work on the alteration in levels of plasma zinc following estrogen therapy indicate that trace metals may be implicated in menopausal symptoms.[27] The plasma compartment is frequently a poor reflection of the true body state of the environment existing in other fluid or cellular compartments. The histopathologist may well detect marked cellular changes in an organ when there is little or no alteration in a plasma constituent whose metabolism is markedly affected by that organ. Certainly, the plasma constituents discussed so far are rarely changed significantly in the early stages of cancer.

The influence of change in hormonal status at the menopause on the development of breast carcinoma and other cancers has been studied by several workers,[28,29] but as the etiology of cancer is so complex, interpretation of the findings is difficult. There is little doubt that the postmenopausal woman has an increased risk of breast cancer; estrogen insufficiency would appear to be implicated. It has been demonstrated in rats that estrogen therapy can halt or inhibit the development of mammary carcinoma.[30] On the other hand, evidence has been presented which indicates that prolonged estrogen therapy may increase the risk of cancer.[31] In a recent study, treatment with anti-estrogenic agents caused regression of tumors in 10 out of 46 women with late or recurrent carcinoma of breast.[32] These contradicting findings coupled with the failure to demonstrate any real association of cancer with long-term administration of oral contraceptives have caused many physicians to hesitate before allowing themselves to treat the timeless problem of aging women who seek youth via hormone therapy.

Recent studies[33] on the management of patients suffering from hyperstimulation of the ovaries demonstrate that considerable changes in various plasma constituents can result from prolonged treatment with gonadotrophins. Thus, it is essential that the patterns of change with age for many plasma constituents be quantitated if the menopausal syndrome and the role of hormone imbalance in the etiology of cancer are to be understood and before the treatment of unpleasant menopausal symptoms is to be carried out successfully.

REFERENCES

1. **Cali, R. W.,** Management of the climacteric and post menopausal women, *Med. Clin. North Am.,* 56, 789, 1972.
2. **Fluhmann, C. F.,** Hormonal relations of menopausal symptoms, *J. Clin. Endocrinol. Metab.,* 4, 586, 1944.
3. **Heller, C. G., Chandler, R. E., and Myers, G. B.,** Effect of small and large doses of diethyl-stilbestrol upon menopausal symptoms, vaginal smear and urinary gonadotrophins in 23 oophorectomized women, *J. Clin. Endocrinol. Metab.,* 4, 109, 1944.
4. **Goldberg, M. B.,** *Medical Management of the Menopause,* Grune and Stratton, New York, 1959.
5. **Curzen, P.,** Eliminating menopause, *Practitioner,* 208, 387, 1972.
6. **Kaufman, S. A.,** Limited relationship of maturation index to oestrogen therapy for menopausal symptoms, *Obstet. Gynecol.,* 30, 399, 1967.
7. **Kaiser, R. and Daume, E.,** On a unified nomenclature for the climacterium and its symptoms, *Geburtshilfe. Frauenheilkd.,* 25, 974, 1965.

8. **McLennan, M. T. and McLennan, C. E.,** Estrogen status of menstruating and menopausal women assessed by cervicovaginal smears, *Obstet. Gynecol.,* 37, 325, 1971.
9. **Lin, T. J. and So-Bosita, J. L.,** Pitfalls in the interpretation of estrogenic effect in postmenopausal women, *Am. J. Obstet. Gynecol.,* 114, 929, 1972.
10. **Young, M. M. and Nordin, B. E. C.,** Effects of natural and artificial menopause on plasma and urinary calcium and phosphorus, *Lancet,* 2, 118, 1967.
11. **Maddison, J.,** Hormone replacement therapy for menopausal symptoms, *Lancet,* 1, 1507, 1973.
12. **Adamopoulos, D. A., Loraine, J. A., and Dove, G. A.,** Endocrinological studies in women approaching the menopause, *J. Obstet. Gynaecol. Br. Commonw.,* 78, 62, 1971.
13. **Adamopoulos, D. A., Loraine, J. A., and Ginsburg, J.,** Endocrinological findings in two patients with premature ovarian failure, *Lancet,* 1, 161, 1971.
14. **Israel, S. L.,** *Diagnosis and Treatment of Menstrual Disorders and Sterility,* Harper and Row, New York, 1967, 638.
15. **Wilding, P., Rollason, J. G., and Robinson, D.,** Patterns of change for various biochemical constituents detected in well population screening, *Clin. Chim. Acta,* 41, 375, 1972.
16. **Roberts, L. B.,** The normal ranges with statistical analysis for seventeen blood constituents, *Clin. Chim. Acta,* 16, 69, 1967.
17. **Reed, A. H., Cannon, D. C., Winkelman, J. W., Bhasin, Y. P., Henry, R. J., and Pileggi, V. J.,** Estimation of normal ranges from a control sample survey. 1. Sex- and age-related influence on the SMA 12/60 screening group of tests, *Clin. Chem.,* 18, 57, 1972.
18. **O'Kell, R. T. and Elliott, J. R.,** Development of normal values for use in multitest biochemical screening of sera, *Clin. Chem.,* 16, 161, 1970.
19. **Albright, F. and Reifenstein, E. C., Jr.,** The parathyroid glands and metabolic bone disease, *Williams & Wilkins,* Baltimore, 1948, 150.
20. **Caniggia, A. and Gennari, C.,** Sites and modes of action of an estrogen-gestagen combination on calcium and phosphate metabolism in postmenopausal osteoporosis, *Clin. Orthop.,* 85, 184, 1972.
21. **Werner, M., Tolls, R. E., Hultin, J. V., and Mellecker, J.,** Influence of sex and age on the normal range of eleven serum constituents, *Z. Klin. Chem. Klin. Biochem.,* 8, 105, 1970.
22. **Tolls, R. E., Werner, M., Hultin, J. V., and Mellecker, J. V.,** in *Advances in Automated Analysis, Technicon International Congress 1969,* Vol. 3, Barton, C. E. et al., Eds. Mediad, Inc., White Plains, N. Y., 1970, p. 9.
23. **Keating, F. R., Jones, J. D., Elveback, L. R., and Randall, R. V.,** The relation of age and sex to distribution of values in healthy adults of serum calcium, inorganic phosphorus, magnesium, alkaline phosphatase, total proteins, albumin and blood urea, *J. Lab. Clin. Med.,* 73, 835, 1969.
24. **Weiss, N. S.,** Relationship of menopause to serum cholesterol and arterial blood pressure. The United States Health Examination Survey of Adults, *Am. J. Epidemiol.,* 96, 237, 1972.
25. **Burstein, M. and Samaille, J.,** Nouvelle méthode séparation et de dosage des lipoprotéines de faible densité (New method of separation and determination of low density lipoproteins), *Ann. Biol. Clin. (Paris),* 17, 23, 1959.
26. **Gibbons, J. L.,** Biochemistry of depressive illness, in *Recent Development in Affective Disorders – A Symposium,* Copen, A. and Walk, A., Eds., Royal Medico-Psychological Association, London, 1968, Ch. 5.
27. **Aitken, J. M., Linsay, R., and McKay Hart, D.,** Plasma zinc in pre- and post-menopausal women: Its relationship to oestrogen therapy, *Clin. Sci.,* 44, 91, 1973.
28. **Gusberg, S. B.,** Precursors of corpus carcinoma estrogens and adenomatous hyperplasia, *Am. J. Obstet. Gynecol.,* 54, 905, 1947.
29. **Wilson, R. A.,** The roles of estrogen and progesterone in breast and genital cancer, *J.A.M.A.,* 182, 327, 1962.
30. **Meites, J., Cassell, E., and Clark, J.,** Estrogen inhibition of mammary tumor growth in rats; counteraction of prolactin, *Proc. Soc. Exp. Biol. Med.,* 137, 1225, 1971.
31. **Bromberg, Y. M., Liban, E., and Laufer, A.,** Early endometrial carcinoma following prolonged estrogen administration in an ovariectomized woman, *Obstet. Gynecol.,* 14, 221, 1959.
32. **Cole, M. P., Jones, C. T. A., and Todd, I. D. H.,** A new anti-oestrogenic agent in late breast cancer, *Br. J. Cancer,* 25, 270, 1971.
33. **Tyack, A. J. and Ellis, J. D.,** Management of severe ovarian hyperstimulation, *Br. Med. J.,* 2, 263, 1971.

DRUGS IN PREGNANCY AND LACTATION

D. V. Parke

TABLE OF CONTENTS

I. DRUGS IN PREGNANCY AND LACTATION

The beliefs that pregnancy and early childhood are periods of particular susceptibility to the toxic effects of drugs and other chemical substances have long been held and probably have their origin in folklore. The treatment of infants with the new wonder drug chloramphenicol in the 1946–47 European epidemic of infantile diarrhea is believed to have contributed largely to the high mortality of that tragic incident and confirmed medical and public opinion that infants were especially vulnerable to the side effects of drugs. Some 20 years later, this view was reaffirmed by the thalidomide disaster, the tragic personal and social consequences of which will remain for a generation or more to remind us of our medical and social inadequacies. More recently, in the U.S., it has been suggested that the synthetic estrogenic drug stilbestrol, given in high dosage to pregnant women to maintain their pregnancies in the fear of threatened abortion, has been the cause of carcinoma of the vagina occurring in their female offspring, now that the young ladies are reaching sexual maturity. Teratogenesis, transplacental carcinogenesis, abortion, and infantile mortality may thus be the occasional tragic legacies of drug administration to infants or the results of incautious taking of drugs during pregnancy, the puerperium, or the postnatal period of lactation.

Medical research during the past two decades has shown that there is much substance to this belief that the fetus and neonate are particularly vulnerable to the toxic effects of drugs. Now there is accumulated scientific evidence that the deactivation and detoxication of drugs and other chemicals may be impaired during the later stages of pregnancy and that drugs may cross the placenta from the maternal blood into the fetus, effecting damage that may result in the death or maldevelopment of the unborn infant or give rise to latent malignancy. Furthermore, it has been shown that the fetus and the neonate have little natural protection against the toxic effects of drugs, which might reach the child not only by direct administration but also indirectly through secretion in the mother's milk.

II. DETOXICATION OF DRUGS

In most cases modern drugs are selectively toxic compounds, exercising their therapeutic benefits by the selective poisoning of some infectious or invasive agent or by the selective inhibition of some enzyme system, subcellular membrane, hormone, or endogenous substrate in the body. Modern medicines are, therefore, generally highly efficacious compounds, but also relatively toxic, and need to be prescribed and administered only where there is a specific need and then with considerable caution and under medical care.

The body, however, has a natural defense mechanism which protects it against the toxic effects of drugs and terminates the pharmacological and therapeutic action by progressive metabolism. This metabolic deactivation of drugs is brought about by enzymes of the liver and other body tissues. It is possibly purely fortuitous and is the result of the action of enzymes whose normal function is presumably to protect the organism against the accumulation and undesirable effects of non-nutrient chemicals which are naturally present in the food and the environment.[1,2] Consequently, the actions of these enzymes are the same toward drugs as to the natural anutrient chemicals and, although leading ultimately to the deactivation and enhanced excretion of drugs, they may result in enhanced therapeutic activity (activation), to a change in the nature of the activity, or even to increased toxicity (intoxication). As an example of this, prontosil, an inactive chemical substance, undergoes metabolism in the body to form sulfanilamide, an antibacterial drug (activation); iproniazid, an antidepressant drug, undergoes metabolism in the body to form isoniazid an antitubercular drug (change of activity); and cyclophosphamide undergoes metabolism to form hydroxycyclophosphamide, an effective cytotoxic agent used in the treatment of malignant disease.

Most drugs are lipid-soluble compounds which readily enter the body's tissues by absorption from the gastrointestinal tract or through the skin, but once inside a cell they do not readily leave until made more polar and less lipid-soluble by metabolism. Those drugs that are themselves highly polar molecules, e.g., barbital and acetazolamide, are readily excreted unchanged, without prior metabolism.

The metabolism of drugs takes place especially in the liver but also in other tissues, such as the blood, the gastrointestinal tract, kidneys, lungs, and skin.[1,2] The metabolic reactions that occur have been classified into two major classes, namely, *biotransformations*, in which the drug molecule undergoes a chemical change (oxygenation, oxidation, reduction, or hydrolysis), and *conjugations*, in which the synthesis of a derivative of the drug, or of its metabolite, occurs (acetylation, methylation, glucuronide, sulfate, glycine, and glutathione formation). For example, the hypnotic drug phenobarbital undergoes oxygenation to form *p*-hydroxyphenobarbital, which is subsequently conjugated to yield *p*-hydroxyphenobarbital glucuronide, which is then more rapidly excreted from the body (see Figure 1). The hepatic enzyme system that is responsible for the hydroxylation of drugs is located in the endoplasmic reticulum (microsomes) and comprises the carbon monoxide-binding pigment, cytochrome P-450, which together with NADPH-cytochrome c reductase, cytochrome P-450 reductase, and possibly other enzymic entities, form the electron transport chain catalyzing the O_2- and NADPH-dependent incorporation of oxygen into the substrate (see Figure 2).

The enzyme systems responsible for the metabolism of drugs are readily inducible and metabolism may be enhanced by pretreatment with

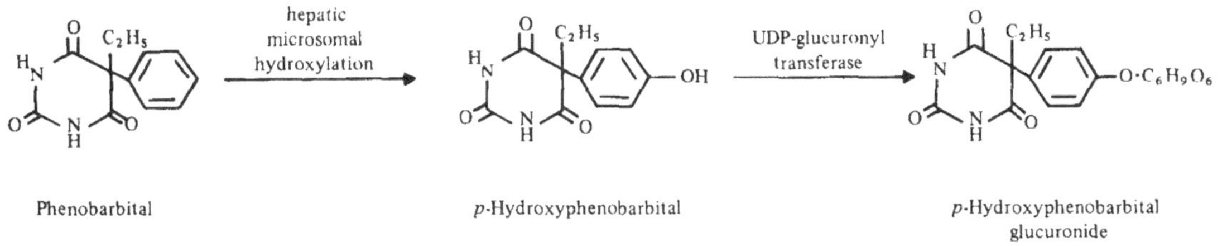

FIGURE 1. Metabolism of phenobarbital.

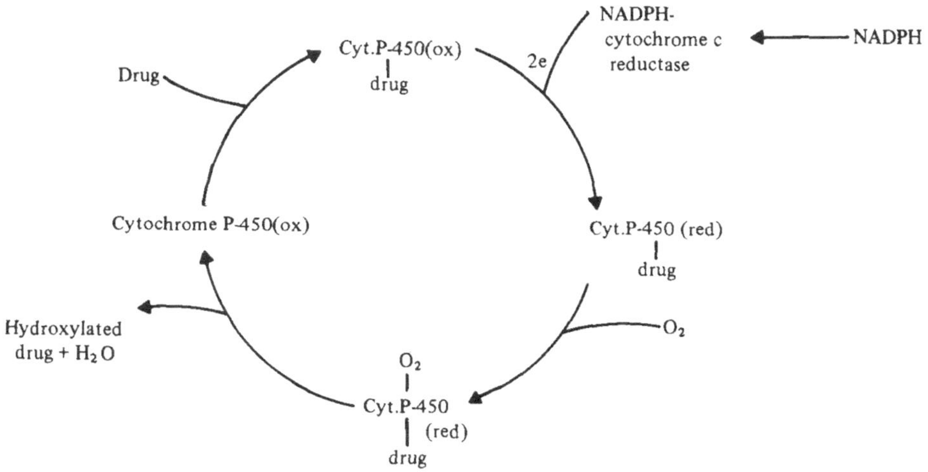

FIGURE 2. The hepatic microsomal drug-hydroxylating enzyme system.

certain drugs (e.g., phenobarbital) and other xenobiotic chemicals (e.g., the food antioxidant, BHT, and the insecticide, DDT) which consequently, gives rise to decreased pharmacological activity of some drugs. Conversely, inhibition of metabolism may occur, as by the simultaneous administration of other drugs and certain steroid hormones; this may lead to enhancement and prolongation of pharmacological activity.[1]

Metabolism of drugs may not always result in the formation of more polar, inactive metabolites, which are rapidly eliminated from the tissues and excreted from the body, but may, in certain circumstances, give rise to highly reactive metabolic intermediates, such as the arene oxides which subsequently interact with proteins, nucleic acids, and other cellular components and become covalently bound to the tissues.[3] Such metabolism has been suggested to account for the retention and accumulation of metabolites of the synthetic estrogen, mestranol, in the liver and other tissues of women taking this oral contraceptive drug.[4]

III. EFFECTS OF ORAL CONTRACEPTIVES

"It is common experience among gynaecologists, clinicians and general practitioners that once a woman has confidence in a method of contraception she will adhere to it for the whole of her active sexual life. This could mean that a woman would remain under the influence of this type of pill for 35 to 40 years. It would seem extremely unlikely to me, thinking as a biologist, that one could submit the delicately balanced hormone system of a woman to this violent alteration for 40 years without something serious happening."[5,6]

A variety of methods of oral contraception are now available which may be categorized as follows: (1) the combination estrogen-progestogen pill; (2) the sequential method, using estrogen (day 5 to 25) to inhibit ovulation and progestogen (the last 5 days) to initiate shedding of the endometrium with normal menstruation; and (3) low-dosage progestogen administered continuously. These various methods employ the synthetic steroids mestranol or ethynylestradiol as estrogens and chlormadinone acetate, medroxyprogesterone acetate, norethisterone, or norethynodrel, etc. as progestogens (see Figure 3).

Dodd's predictions,[5] made in 1961, have to some extent been borne out and a number of adverse side effects of oral contraceptive steroids have been observed. The more serious of these are thromboembolic incidents, such as deep-vein thrombosis, pulmonary embolism, and coronary and cerebral thrombosis. The occurrence has, fortunately, been low, but the risk of incidence incurred by oral contraception is probably three to four times normal.[7,8] The estrogens, especially mestranol, seemed to be the causative agent and with the introduction of reduced dosage of this component, it is to be hoped that these adverse reactions will diminish. Other reported side effects are relatively minor and include a few cases of cholestatic jaundice,[10] mild metabolic disturbances, weight gain, skin pigmentation, etc. The more serious fears of carcinogenesis and mutagenesis have, fortunately, not been substantiated. A recent extensive study in rodents of the potential carcinogenicity of a wide range of oral contraceptives, steroids, and combinations of these found no evidence of carcinogenesis of significance to man.[11] Similarly, a comprehensive study of women taking "the pill" and their babies showed that there was no increase in chromosomal aberrations of the circulating lymphocytes of mothers or their offspring.[12]

The effects of continuous administration of these synthetic steroids bear comparison with continuous medication with drugs and result in a number of biochemical changes which have been compared to the state of pregnancy. Since steroids and drugs are known to be metabolized by the same enzyme system in the liver,[13] it is perhaps not surprising to find that the taking of oral contraceptives may have consequences on the metabolism of drugs.

The progestational steroids, norethynodrel and progesterone, inhibit a variety of drug-metabolizing enzyme activites of rat liver homogenates in vitro, but pretreatment of animals with norethynodrel leads to an increase in the activities of a number of these enzymes.[14] Therefore, in rats, short-term treatment with steroids inhibits microsomal enzymes and prolongs the action of certain drugs, but, in general, when administered long-term these chemicals have an enzyme-inducing effect. However, the doses of steroids causing induction were much higher than those used in oral contraception and in recent animal studies the effects of steroids on the drug-metabolizing enzymes were found to be dose-related. Thus, although in high dosage, these steroids may induce hepatic drug metabolism; at the low dosage used for oral contraception, inhibition of these enzymes occurs.[15] Studies in women taking various oral contraceptive preparations, mainly mixtures of estrogen and progestogen, showed that the pill had an inhibitory effect on drug metabolism resulting in a modest, but significant increase in the plasma half-life of antipyrine[15] and a slower rate of metabolism of meperidine (pethidine) and promazine.[16] Thus, it may be expected that women taking the pill would exhibit an enhanced pharmacological response to drugs, although there is but little clinical evidence, as yet, to substantiate this prediction.

Oral contraceptives are known to give rise occasionally to a variety of hepatic abnormalities, in particular cholestatic jaundice, similar to the intrahepatic cholestasis of pregnancy. In a study of the effects of oral contraceptive steroids on hepatic function, it has been shown that the estrogen ethynylestradiol exhibits inhibition of the glucuronide conjugation of bilirubin and ortho-aminophenol by rat liver preparations in vitro, which is proportional to the concentration of the steroid added. Mestranol, another estrogen, and a variety of progestogens did not show any similar inhibition of this enzyme system.[17]

IV. METABOLISM OF DRUGS DURING PREGNANCY

There is now much evidence that during pregnancy significant changes may occur in both the rate and pattern of metabolism of drugs. The conjugation of drugs and other xenobiotics with glucuronic acid to form glucuronides was shown to be inhibited in experimental animals by Cessi.[18] This was subsequently attributed to inhibition of the glucuronyl transferase enzyme(s) by the high circulating levels of pregnanediol and other progestational and estrogenic hormones which occur in both experimental animals and humans during pregnancy.[19-22] More recent work, using 4-methyl-umbelliferone as substrate and a new fluorimetric method of determination, has confirmed that in both rats and rabbits the specific activities of hepatic glucuronyl transferase are inhibited to

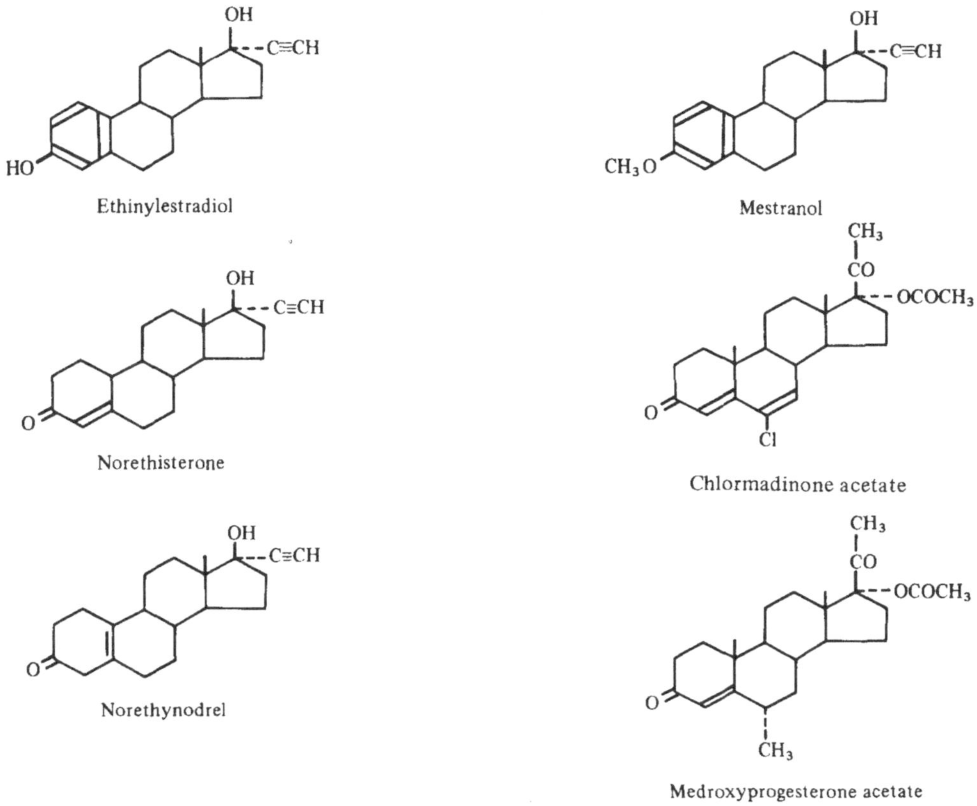

FIGURE 3. Oral contraceptives and progestogens.

the extent of 20 to 30% of normal during late pregnancy.[23] Using another new procedure involving the dialysis of the liver homogenate, Halac and Sicignano[24] have found that there is a progressive decrease of the specific activity of rat hepatic bilirubin and *p*-nitrophenol glucuronyl transferase(s) which shows an immediate recovery to normal following delivery. The inhibition was obviously not due to the presence of a dialyzable inhibitor and thus may not, in this case at least, be the result of high plasma levels of circulating progestational steroids, unless these were tenaciously bound to the enzyme. Conjugation of drugs with sulfate to form sulfate esters has also been shown to be decreased during pregnancy in the rat and is attributed to the inhibitory effect of high tissue levels of estrogens.[25]

It was further shown that the biotransformations of drugs involving hydroxylation were also inhibited during pregnancy.[26] This impairment of the biotransformation of drugs has been confirmed by other workers and shown not to be due to the presence of inhibitors,[27] but is associated with a decrease in the hepatic cytochrome P-450.[28,29] The inhibitory effect in both rats and rabbits is maximal toward the termination of the pregnancy and involves an inhibition of about 30% of biphenyl-hydroxylase and *p*-nitrophenol reductase in the rat, and in rabbit an inhibition of coumarin-7-hydroxylase of some 60%.[23]

In experimental animals, such as the rat and rabbit, the specific activities of the hepatic oxidative drug-metabolizing enzymes are substantially reduced, but due to the fact that there is a corresponding increase in the size of the liver in pregnancy, the total enzyme activities are the same in the pregnant as in the nonpregnant animals. Therefore, if drugs are administered on a body weight basis, the increased dosage corresponding to the increased body weight in pregnancy gives rise to a substantial increase in pharmacological response, but if a dose corresponding to the nonpregnant body weight is administered to a pregnant animal, the pharmacological response is normal. The significance to human therapeutics of this impairment of drug metabolism observed in pregnant animals may only be speculated upon, but it would seem reasonable to assume that provided dosage is not increased to allow for the increased body weight of pregnancy, drugs should be metabolized at normal rates and thus give rise to normal levels of pharmacological response.[23]

V. PLACENTAL TRANSFER OF DRUGS

The placenta is no longer regarded as a simple semipermeable membranous system, but a highly complex tissue with a unique metabolic pattern. Nevertheless, with respect to drugs and xenobiotics, it does behave similarly to a semipermeable filter which is readily penetrated by lipid-soluble molecules of quite large molecular dimension. Consequently, it is penetrated by numerous drugs and chemicals which find their way into the tissues of the fetus. In fact, no longer is the foremost consideration whether or not a drug may cross the placenta, but rather at what rate it does so and the mechanisms involved. These new concepts have led to the view that any drug or chemical that is found in the maternal blood will cross the placenta unless it is metabolized during its passage.[30]

Although the different forms of placentation with the consequent different physiological relationships between maternal and fetal circulations make studies of the placental transport of drugs in experimental animals difficult to relate to man, this is valid only in terms of the kinetics of transport and it is generally supposed that where placental transfer of a drug occurs in a laboratory animal, it will also cross the human placenta. The following mechanisms have been considered to be involved in the transport of substances across the placenta: (1) ultrafiltration, (2) simple and facilitated diffusion, (3) active transport, (4) pinocytosis, and (5) breaks in the placental villi. The most important as far as drugs are concerned is "simple diffusion," although "active transport" may also be involved in the transfer of those compounds, such as antimetabolites, which are closely related to endogenous substrates.

The concept of the placenta as a simple lipoid barrier has been widely accepted as a criterion of the permeability of the placenta to drugs. Therefore, drugs that are highly dissociated into lipid-insoluble ionic forms at physiological pH, or in their nonionized forms are relatively lipid-insoluble, such as succinylcholine and d-tubocurarine, will penetrate the placenta to only a minor degree, whereas highly lipid-soluble compounds such as thiopental and antipyrine will readily cross this barrier. However, as simple diffusion is the major mechanism involved, the rate of transfer will also be dependent on the concentration gradient, that is, essentially the concentration in the maternal blood. If administered at sufficiently high dosage, even the relatively lipid-insoluble succinylcholine and tubocurarine will cross the placenta in sufficient amounts to be demonstrable in the fetal blood. Other factors may modify this concept of simple diffusion across a lipoid barrier, and binding to the plasma proteins of the mother and to the proteins of the placenta is one of the major considerations which have to be borne in mind. Other factors, such as the kinetics of plasma concentration, determined by the rates of maternal absorption, metabolism, and excretion, and the possibility of placental metabolism, will also affect the rate and extent of transfer of drugs across the placenta.

In the mechanism of simple diffusion, the rate of transfer is to a large degree dependent on the molecular weight of the substance being transferred. In general, the lower the molecular weight, the greater the diffusibility, so that drugs of a molecular weight of less than 600 are readily transferred if sufficiently lipid-soluble, whereas at a molecular weight of over 1,000, diffusion becomes almost impossible. Recent studies in the author's laboratories on the simple diffusion of drugs across the gastrointestinal mucosa have shown that an optimal lipid-solubility exists for the absorption of drugs from the gastrointestinal tract and that extremely lipid-soluble drugs may not be so readily transferred as moderately lipid-soluble ones.[31] However, no studies have yet been undertaken with the placenta to ascertain if a similar situation exists with this organ, but since all mammalian membranes are similar in that they are dynamic, biphasic aqueous/lipid systems, some optimum of lipid-solubility would be expected. Certain other factors would also be expected to vary the rate of simple diffusion of drugs across the placenta; these include conditions such as pre-eclampsia and diabetes mellitus, which alter the permeability of the placenta, and changes in maternal and fetal blood flow conditioned by anesthesia, compression of the cord, etc.

Because of the concept of the placenta as a lipid-permeable barrier to drugs, analogies have frequently been drawn between the placenta and the blood-brain barrier, in the assumption that those drugs which cross the placenta into the fetus would also find their way into the brain tissues. This hypothesis is not strictly true and it has been recently shown that the parasympatholytic drug, phenylglutarimide hydrochloride, crosses the placenta of the rat to reach higher concentrations in the fetus and amniotic fluid than occur in the maternal circulation, whereas no significant amounts of this drug were detected in the brain.[32]

In two comprehensive reviews of this field, Moya and his colleagues[30,33] have shown that numerous drugs, including pratically all those used in obstetrics, with the possible exception of the muscle relaxants, cross the placenta to some degree. Administration of a number of barbiturates, including barbitone, a drug of relatively low lipid-solubility, resulted in equilibrium being established between maternal and fetal blood within minutes; this was maintained for several hours. The highly lipid-soluble thiopental, like other thiobarbiturates, was rapidly transferred across the placenta and was detected in cord blood only 45 sec after injection to the mother. Chlorpromazine and other tranquilizers; ethanol, chloral hydrate, paraldehyde, and other hypnotics; the gaseous anesthetics such as trichlorethylene, cyclopropane, halothane, and nitrous oxide; and a variety of sulfonamides and antibiotics have been shown to pass readily across the placenta into the fetus. However, the highly polar, relatively lipid-insoluble muscle relaxant drugs, such as succinylcholine, decamethonium, and tubocurarine, cross the placenta with difficulty and only at high maternal plasma concentrations, so that the usual clinical doses have no demonstrable effects on the newborn.

On the basis of these findings that a number of sedatives, hypnotics, and anesthetics pass readily from the maternal circulation into the fetus, it became common obstetric practice to deliver the baby rapidly, within a minute or two of administration of the drug to the mother, to minimize the transfer of drug to the baby with the attendant risk of respiratory depression. However, subsequent workers have shown that the duration of time elapsing between sedation or anesthesia of the mother and delivery of the infant is of little consequence in determining the condition of the infant, and indeed the longer the interval, the greater is the maternal metabolism and the less is the likelihood of neonatal respiratory depression. Furthermore, as Flowers demonstrated,[34] although there is equilibrium in the concentration of a number of barbiturates in the maternal vein and the umbilical artery and vein at delivery, the deep anesthesia of the mother contrasts greatly with the alertness at birth of the infants.

More recent studies have shown the placental transfer of diphenylhydantoin,[27,35] diazepam;[36] pethidine and pentazocine;[37] imipramine;[38] digoxin;[39] nortriptyline;[40] the synthetic progestin, norethynodrel, which is used in the maintenance of pregnancy;[41] and the chlorinated hydrocarbon pesticides, DDT, gammexane (BHC), dieldrin, and heptachlor epoxide.[42] The extensive use of diazepam in labor and cesarian section has been associated with transient hypoactivity and toxicity in the neonate.[43] The rapid transfer of the drug and its metabolites across the placenta with a pronounced uptake and long retention in the fetal cerebellum, spinal cord, and peripheral nerves has been shown to occur in monkeys;[44] a similar accumulation of diazepam in human fetuses[36] may explain this neonatal hypoactivity. The slow diminution of diazepam in fetal tissues and cord blood may indicate inhibition of the reverse transfer of the drug back across the placenta to equilibrate with the declining maternal plasma levels, possibly due to an enhanced binding of the drug to fetal blood cells, plasma, or tissue proteins.[36] However, in a study of the plasma protein binding of diphenylhydantoin, imipramine, and a variety of other drugs, it was found that binding to cord plasma is significantly lower than binding in adult plasma.[45]

It is probable that all narcotics are capable of traversing the placenta. Pinpoint pupils can be seen in babies of mothers given morphine during labor, and infants born to mothers addicted to morphine or heroin are now a major social problem in some areas of the U.S. These babies of addicted mothers have been often reported to show symptoms of withdrawal;[46] indeed, withdrawal may be diagnosed in the fetus in utero.[47] Morphine, meperidine, and methadone have been shown to be present in cord blood of neonates at approximately the same concentrations as found in the maternal blood, the mothers having been given the drug from 0.6 to 3.5 hr prior to delivery.[48] There was, however, no correlation between cord blood levels and the condition of the infants at birth. Nicotine has been shown to cross the placenta in several species; in the rat, nicotine and its metabolites rapidly attain high levels in the fetus which for unchanged nicotine are higher in the fetal than in the maternal circulation. Concentrations of nicotine in the rat decidua and placenta were also found to be high.[49]

VI. PLACENTAL DRUG METABOLISM

The role of the placenta in the metabolism and development of the fetus has been likened to the role of the liver in the adult animal. It has been suggested that in the same way that the liver is responsible for the deactivation of drugs and toxic chemicals in the adult, so may the placenta be similarly responsible for the detoxication of drugs and xenobiotics in the fetoplacental unit. Consequently, there has recently developed a considerable interest in the possible metabolism of drugs and foreign compounds in placental tissues.[50-52]

Electron microscopy of the placental tissue has demonstrated that significant quantities of smooth endoplasmic reticulum are present in the trophoblastic cells[53] and a carbon monoxide-binding pigment is present in both the microsomes and mitochondria; this has been identified as cytochrome P-450.[54] The presence of cytochrome P-450 in placental tissue, although resembling that associated with the hepatic microsomal drug-metabolizing enzymes (see Figure 2), is only indicative of potential capacity for the biological oxygenation of drugs. Indeed, conflicting results have been reported, both for the amount of cytochrome P-450 present in the placental tissues and for the placental metabolism of drugs and other foreign compounds, but these discrepancies may possibly be explained in terms of species differences, gestational age, or both of these. Homogenates of human placental tissue, of 9 to 12 weeks' gestation, exhibited minimal activity for the *N*-demethylation of aminopyrine, para-hydroxylation of aniline, O-demethylation of codeine, oxidative deamination of amphetamine, sulfoxidation of chlorpromazine, oxidation of hexobarbital, and hydroxylation of benzo[a]pyrene and zoxazolamine.[55] *N*-demethylation of paramethylaniline in human placentas occurred from early pregnancy,[56] but no measurable *N*-demethylation of 3-methyl-4-monomethylaminoazobenzene was observed except where the placentas were obtained from mothers with histories of regular cigarette smoking.[57] Aromatic hydroxylase activity was studied by Creaven and Parke[26] in several animal placentas, but no oxidative metabolism of coumarin or biphenyl was detectable. Other investigators[56,58] have, however, reported negative results with respect to drug oxygenation in the human placenta. Juchau and Symms[59] investigated the mechanism of the para-hydroxylation of aniline in human placenta material; comparisons of the soluble-fraction of human placenta with a model tissue free system suggested that hemoglobin was a principal factor responsible for the observed hydroxylation of aniline. The study further suggested that the placental trophoblasts per se were extremely low in, or possibly devoid of, aniline para-hydroxylating activity.

Benzo[a]pyrene hydroxylase (also known as aryl hydrocarbon hydroxylase) activity of human placental tissue has been widely studied by a number of workers.[57,60,61] This enzyme, which may be cytochrome P-450 dependent, is nevertheless quite distinct from the enzyme responsible for hydroxylating monocyclic aromatic rings and carrying out other hydroxylating reactions of drugs. Little or no aryl hydrocarbon hydroxylase was found in human placentas from nonsmokers, but this enzymic activity was found in significant amounts in placentas obtained from women who smoked 10 to 40 cigarettes a day.[62] The induction of this benzopyrene hydroxylase in hepatic and extrahepatic tissues is well known and may reach very high levels.[63] The variation in the extent of induction of this enzyme between different individuals is as great for the placenta as it is for other tissues, and may reflect an abnormal carcinogenic risk in those individuals who show a poor response. Steroids synthesized by human placental tissues, that is, β-estradiol, estrone, and progesterone, markedly inhibited the placental-catalyzed hydroxylation of benzpyrene and it is possible that the low drug-metabolizing activity exhibited by placental tissues may be attributed to the high concentration of these steroids present in the placenta.[52,64]

The reduction of a wide variety of drugs and other substrates has been shown to occur in the presence of placental subcellular fractions.[55,65,66] Reduction of para-nitrobenzoic acid is reported to proceed via both enzymic and nonenzymic pathways involving NADPH- and NADH-generating systems.[66] Human placenta also contains in the soluble fraction an NADPH-dependent enzymic/nonenzymic system capable of the reduction of the azo linkage of neoprontosil to form sulfanilamide. This enzyme system was shown to be different from the azo-reductase system(s) present in hepatic tissues.[65]

The placenta is largely deficient in its capacity for conjugating drugs and other chemicals. Glucuronide conjugation in the placenta has been reported, but many other workers have been unable to find any glucuronyl transferase activity. Berte et al.[67] have demonstrated glucuronide formation with oxazepam which was present in the rabbit placental 9,000 g supernatant fraction; conjugating activity with the coenzyme uridine diphosphate glucuronic acid (UDPG) was higher at day 14 of the gestation than at day 30 and was increased by pretreatment with phenobarbital but not with the substrate oxazepam. Glycine conjugation is catalyzed in the presence of human placental mitochondria,[50] but the major conjugation present in the placenta seems to be with sulfate.[68] It may be that in having a limited ability to form conjugates with drugs and other xenobiotics, the placenta functions protectively by allowing these chemicals and endogenous substrates, such as estrogens and bilirubin, to pass out from the placenta and the fetus back into the maternal circulation, where they are conjugated and thus unable to return to the fetoplacental unit.

In a recent survey with human full-term placentas, hexobarbital oxidase and p-nitroreductase activities were found to be present, but cytochrome P-450, ethylmorphine-*N*-demethylase, and UDP-glucuronyl transferase were not detected.[69] Treatment of the pregnant women with barbiturates increased the placental hexobarbital oxidase but had no effect on any other of the enzymes.

In the light of this very limited ability of the placenta to conjugate and biotransform drugs and other xenobiotics, particularly the very limited ability to hydroxylate drugs, it is surprising to find that the placenta has a very efficient enzyme system for the biosynthesis of estrogens. This biosynthesis of estrogens in the placenta, involving the aromatization of the steroid ring A of C_{19} steroids, has been demonstrated to occur in the placentas of sheep, cows, horses, pigs, monkeys, and primates, as well as in the human.[52] This biosynthesis of estrogens involves three separate oxygenations of the molecule and it has been suggested[70] that one or more of these reactions may be mediated by the NADPH-dependent enzyme system cytochrome P-450. In view of the very high level of steroid aromatizing activity present in the placenta, it is somewhat surprising that the cytochrome P-450-dependent hydroxylation of drugs is not demonstrable similarly in placental tissue. In a recent extensive survey of the drug-metabolizing activity of human and animal placentas, and a detailed study of the placental enzymes responsible for the aromatization of steroids in estrogen biosynthesis, it was shown that estrogen biosynthesis was independent of cytochrome P-450 and that the placentas contained very limited amounts of this cytochrome in either the microsomal or mitochondrial fractions and, consequently, also had very limited ability to metabolize drugs.[52]

The placental metabolism of drugs may be related to the subsequent effects of these drugs on the fetus and the newborn infant. Cigarette smoking, which produces induction of the placental aryl hydroxylase, may produce low-weight fetuses and placentas, increase the risk of spontaneous abortion,[71] and also possibly produce teratogenic effects.[72] The situation is somewhat complicated by the presence of carbon monoxide in the cigarette smoke, for carbon monoxide binds avidly to hemoglobin, and hypoxia of the fetus may be the cause of toxicity rather than nicotine, benzpyrene, and other chemicals in the smoke.[73]

VII. DRUG METABOLISM IN THE FETUS AND NEONATE

Considerable evidence has been accumulated that drugs are only slowly metabolized in fetal and newborn animals and that the activities of the drug-metabolizing enzymes of the fetus and neonate are considerably less than those of the normal adult. It has been shown[74,75] that the liver microsomes of newborn rabbits, mice, and guinea pigs are either lacking in certain of the hepatic drug-metabolizing enzyme activities or exhibit low levels of these enzymes. However, these enzyme activities developed rapidly after birth and eventually achieved adult levels at varying times, depending on the particular enzyme concerned and the animal species. The causes of the low enzyme activities have been attributed to the presence of inhibitory substances in the neonate,[74] to the acquisition of inhibitory gonadal hormones from the mother,[76] or to lack of stimulation of the system responsible for the formation of these enzyme systems.[77] At least one anomaly to this pattern of impaired neonatal drug-metabolism exists, namely, the 2-hydroxylation of the model substrate biphenyl, for although the 4-hydroxylation of biphenyl follows the normal pattern of development, that is, it is low at birth but progressively increases with age, hydroxylation at the 2-position generally shows the converse pattern and decreases progressively after birth.[78] Undoubtedly two different enzyme systems are involved in the hydroxylation of biphenyl and it is likely that the 2-hydroxylase is not an enzyme normally concerned with drug metabolism.

Evidence for impaired ability to metabolize drugs in the newborn human infant includes a slower rate of excretion of phenylbutazone metabolites in neonates compared to older children, the slower appearance in the plasma of acetanilide metabolites after administration to newborn infants, and the prolonged plasma half-lives of sulfobromophthalein, *p*-aminobenzoic acid, and bilirubin, which reflect deficiencies of the conjugating systems;[79] however, human neonates do appear to be able to metabolize nortriptyline to 10-hydroxynortriptyline.[40] Even antibiotics, which are known not to be metabolized and are excreted unchanged, have been shown to have a longer plasma half-life in neonates, which decreases with age. This is attributed to the development of the glomerula filtration rate and the tubular secretory capacity of the kidney, which reach maturity at about 3 months of age.[80] The plasma protein binding of drugs in the neonate may also be less than that which occurs in adults and, hence, the fraction of unbound drug in the neonate plasma is higher; yet, plasma concentrations of diphenylhydantoin,[81] nortriptyline, and other drugs are much lower in the neonate than in the adult when administered at a similar dosage on a body weight basis. As metabolism is believed to be less efficient in the infant, this lower plasma concentration most probably indicates a greater volume of distribution, with increased proportions of the drugs being bound to proteins of the tissues.

The conjugation of drugs, xenobiotics, and their metabolites with glucuronic acid to form the more polar, glucuronide conjugates is known to be impaired in the neonates of a number of different species. The deficiency seems to be due to a deficiency in the activities of the hepatic microsomal glucuronyl transferases, enzymes which transfer the glucuronic acid moiety from uridine diphosphate glucuronic acid (UDPG) to the drug or its metabolite.[1] An alternative hypothesis is that there is a defective uptake of the various drugs by the liver due to deficiencies of the intracellular organic-anion binding proteins. This view is supported by evidence that the major of these binding proteins, acceptor protein Y, is deficient in fetal and neonatal guinea pigs.[82] This deficiency in glucuronide conjugation may give rise to high serum bilirubin levels in the newborn, leading to kernicterus, and may be exacerbated by the administration of drugs such as chloramphenicol, which compete with bilirubin for glucuronide conjugation and biliary excretion, or by drugs such as the sulfonamides, which are highly bound to plasma proteins and competitively displace bilirubin from these binding sites. A number of substances may also act as inhibitors of the glucuronyl transferases with consequent impairment of glucuronide conjugation. Among these inhibitors are a variety of steroid hormones including progesterone, testosterone, and pregnanediol, the latter being the most active producing 50% inhibition of glucuronyl transferase at a concentration of 0.4×10^{-4} *M* by competitive inhibition.[21] Pregnanediol glucuronide is an even more effective inhibitor of glucuronide syntheses than the free steroid and possibly exerts its inhibitory action by interfering with the secretory mechanism of bilirubin glucuronide and other glucuronides. Thus, the low levels of glucuronyl transferase activity in the neonate may be attributable to high tissue levels of progesterone, pregnanediol, and other metabolites transferred across the placenta from the mother before birth, and via the milk during lactation. Other nonsteroidal inhibitors of glucuronyl transferase, such as the antibiotic novobiocyin and the prothrombin-activator vitamin K, which are noncompetitive inhibitors of the enzyme, are known and may also contribute to hyperbilirubinemia when administered to infants.[20] In a more comprehensive study of this problem, using a variety of substrates, Dutton[83] has shown that there are probably several different glucuronyl transferase enzymes in any one species and that the perinatal development of these enzymes show individual differences within one animal species; furthermore, the development of any one of these enzymes may vary in different animal species and, in some instances, may exhibit the highest activity at birth. Although glucuronyl transferase activity is absent from all early-fetal livers, the time of its appearance and the rate of its development can vary from species to species and substrate to substrate.

Similar postnatal development has been shown to occur for the formation of glycine conjugates,[84] but, in contrast, neonates and fetuses seem readily capable of forming sulfate ester conjugates.[85]

The low activities in the neonate of enzymes that effect the conjugation of drugs and their metabolites is paralleled by a similar reduction in the activities of the enzymes responsible for the biotransformations of drugs. In all mammalian species examined, the enzymes that carry out the various hydroxylation reactions, or reductions, of drugs and chemicals have low activities at birth and increase progressively with age, reaching adult levels within a few weeks of birth. The metabolism of hexobarbital, pentobarbital, meprobamate, carisoprodol, and strychnine were all found to be metabolized in the newborn rat at rates of less than 10% of the adult values, with the consequent increase in barbiturate hypnosis, carisoprodol paralysis, and the toxicity of strychnine.[86] The metabolism of these drugs increased rapidly with the development of the animal and reached normal adult levels at 30 days after birth, after which they slowly declined again. The metabolism of chlorpromazine, hexobarbital, meperidine, *N*-methylaniline, 3,4-benzpyrene, and *p*-nitrobenzoic acid in the human fetus are all effected at much lower rates than in the adult. Most of these drugs undergo hydroxylation, which is mediated by the hepatic microsomal enzyme, cytochrome P-450; this has recently been demonstrated to occur in human fetal liver.[87,88]

In a study of the metabolism of the tranquilizing drug, chlorpromazine, by liver preparations from fetal, neonatal, and adult rabbits, it was found that at 4 days before term only sulfoxidation could occur; at 3 days after birth, sulfoxidation, 7-hydroxylation, and mono-demethylation occurred; and in adult animals, sulfoxidation, hydroxylation, mono- and di-demethylation, and combinations of these were observed, showing that even different hydroxylation reactions may undergo different rates of development (see Table 1).[89]

The impaired metabolic hydroxylation of drugs in the neonatal rat has been correlated with a reduction in several of the components of the hepatic microsomal electron transport system responsible for the hydroxylation of drugs. Cytochrome P-450 and cytochrome P-450 reductase are about 10%; NADPH-cytochrome c reductase is some 30 to 40% of normal adult levels at birth and reaches adult levels in about 4 to 5 weeks.[90,91] Fouts and Devereux[92] made similar findings with the neonatal rabbit, with liver concentrations of cytochrome P-450 being 20%, cytochrome b_5 50%, and NADPH-cytochrome c reductase 60% of normal adult values. For the neonatal rabbit lung the corresponding values were higher, 30% for cytochrome P-450, 80% for cytochrome b_5, and 50% of NADH-cytochrome c reductase. All components, both in the liver and the lung, reached adult levels at 30 days after birth. Impairment of drug metabolism is, therefore, probably not confined to the liver, but also affects extrahepatic metabolism; metabolism in vitro by neonatal rabbit lung preparations showed only 10 to 20% the adult rate of demethylation of benzphetamine and hydroxylation of benzpyrene.

In a recent study, the concentrations of NADPH-cytochrome c reductase and cytochrome P-450 in human fetal liver were found to be similar to those for adults, which is in marked contrast to that found previously for other animals. The distribution of these enzymes in human fetal liver homogenates is atypical because the cells of fetal liver are more resistant to homogenization than those of adult liver.[93] The consequence is that the drug-metabolizing activity was found to be associated with the cell debris instead of with the microsomal fraction. This may account for the low levels of drug-metabolizing activities found in fetal liver microsomal preparations in vitro, but does not account for the reduced activities observed in vivo.

The ability of fetal tissues to carry out the enzymic reduction of drugs develops in the human fetus at about 12 weeks gestation and nitrore-

TABLE 1

The Development of Chlorpromazine Metabolism in the Rabbit

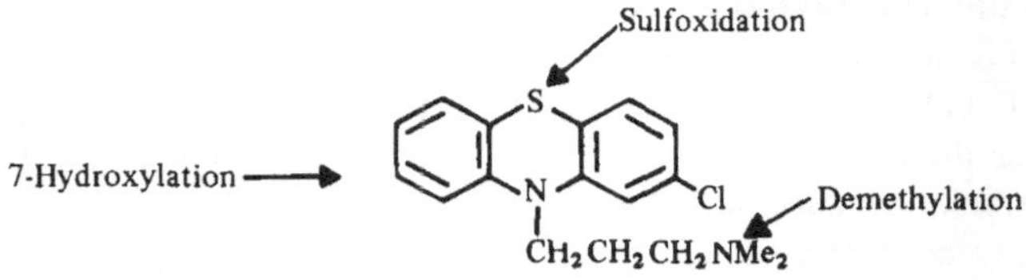

Chlorpromazine

	Incubation time (min)	Metabolites formed
Adult	15	I, II, III
	30	I – VII
	45	I – X
3-day-old	45	I, II, III
Fetus	45	I

The metabolites formed were as follows: I, CP-sulfoxide; II, 7-hydroxy CP; III, desmethyl-CP; IV, desdimethyl-CP; V, desmethyl-CP-sulfoxide; VI, desmethyl-7-hydroxy-CP; VII, 7-hydroxy-CP-sulfoxide; VIII, desdimethyl-CP-sulfoxide; IX, desdimethyl-7-hydroxy-CP; X, desmethyl-7-hydroxy-CP-sulfoxide.

ductase activity is present in the adrenal and the liver, with lesser activity in the lung. Human fetal livers will carry out the reduction of aromatic nitro compounds, but the requirement of high coenzyme levels for maximum activity in vitro indicates that the activity in vivo may well be slow.[94]

Kardish and Feuer[95] have recently suggested that the factors thought to be responsible for the impaired conjugation of drugs with glucuronic acid may also be the casue of the impaired development of the fetal and neonatal hydroxylating enzyme systems. They have shown that a number of reduced progesterone metabolites, namely, 5α- and 5β-pregnanediol and pregnanolone and 5α-isomers of pregnanedione, when administered to newborn rats inhibited the metabolic hydroxylation of drugs in liver preparations to 40 to 80% of controls. Furthermore, premature weaning of the rats increased the hepatic drug-metabolizing activity by as much as 120%, which was reversed by administration of the reduced progesterone derivatives. In the human, as in other species, the levels of these progesterone derivatives during pregnancy are very high both in maternal tissues and in the fetus and neonate; they are also secreted in the milk. It is probable that the presence of these steroids would inhibit the metabolic hydroxylation of drugs in the fetus and neonate, but it has not been demonstrated unequivocally that they lead to the impaired development of cytochrome P-450 and the other components of the microsomal enzyme system. Impaired drug metabolism in the fetus and neonate has been attributed to a variety of possible causes, including the presence of steroid inhibitors, the absence of the enzyme proteins, decreased permeability of cell membranes, and the absence of intracellular transporting mechanisms.

Administration of drugs during pregnancy may have a stimulatory effect on the low levels of the drug-metabolizing enzymes found in the fetus and neonate. Treatment of pregnant rabbits or rats near term with known inducing agents of the hepatic drug-metabolizing enzyme system, such as phenobarbitone, results in an increase of these enzymes in the fetus or the neonate.[96,97] Treatment of neonatal animals with phenobarbitone and other inducing chemicals also increases the drug-metabolizing activities and enhances the rate of development of cytochrome P-450 and other components of the microsomal enzyme system.[90] Similarly, administration of phenobarbitone and other enzyme inducers, such as chlorcyclizine, chloroquine, and 3,4-benzpyrene, has been shown to induce hepatic glucuronide formation in newborn rats.[98] Administration of phenobarbitone to pregnant women[99] or to young babies[100] results in an enhanced activity of glucuronyl transferase with reduction in serum bilirubin concentrations; liver function as measured by sulfobromophthalein clearance was also enhanced.[101] Phenobarbitone has, therefore, been quite widely used with success in the treatment of neonatal hyperbilirubinemia, and can reduce the probable need for replacement transfusions.[100] On the other hand, neonatal infants with hyperbilirubinemia resulting from erythroblastosis fetalis showed a much reduced ability to conjugate bilirubin when the mothers had previously been treated with the immunosuppressant drugs, amethopterin, or 6-mercaptopurine. These antimetabolites, which may cross the placenta and are known to inhibit protein synthesis, probably produce this inhibitory effect by blocking the *de novo* synthesis of the glucuronyl transferase and other hepatic enzymes.[102]

VIII. DRUG-INDUCED TERATOGENESIS

Despite public concern over the teratogenic potential of drugs, present information would indicate that medicines account for only a very small percentage of the total burden of human developmental defects, although it is possible that synergistic effects of drugs and chemicals, perhaps with other agents, may account for some of the substantial percentage (75%) of developmental defects which are of unknown etiology. Despite the costly lesson of the thalidomide catastrophe, drugs are still frequently prescribed and widely taken during pregnancy. A recent survey in Scotland showed that in a sample of over 1,000 mothers, 97% took prescribed drugs and 65% self-administered drugs during pregnancy.[103] Furthermore, it was found that mothers (458) who gave birth to infants with developmental defects took a significantly greater number of medicines of one kind or another (aspirin, amphetamine, barbiturates, sulfonamides, etc.), than did the control group (911). Nevertheless, a direct causal relationship of teratogenicity with a given drug is difficult to establish because of possible contributions to the abnormal development by other environmental agents, genetic factors, and infectious diseases. The time during gestation at which the drug is taken is often critical both for the appearance of drug-induced embryotoxicity and for the nature of the effect, whether it be intrauterine death, teratogenesis, or carcinogenesis.[104] Where a drug gives rise to a characteristic pattern of deformities, particularly if produced by taking the drug only at a certain time during gestation, it is possible, although not easy, to establish a causal relationship.

In a recent review of this problem, Wilson[105] classifies drugs under four headings, namely (1) those positively implicated as being teratogenic, (2) those suspected of some teratogenic potential, (3) those possibly teratogenic under some conditions, and (4) those with little or no teratogenic potential. Some drugs representative of the first three categories are classified in Table 2. Drugs that have been studied for their teratogenic potential in some detail and found to be relatively free from risk include the sulfonamides, cortisone, meclizine, and lysergic acid diethylamide; although a weak mutagen, the last-named drug is suspected of causing malformations in humans only when taken in amounts excessive to therapeutic dosage and possibly in association with other drugs of abuse.[105] Thus, apart from teratological catastrophes such as thalidomide, few drugs have been shown to have teratogenic potential in man and, in general, it is likely that drugs used therapeutically are not a major factor in causing congenital malformations. Nevertheless, the interactions of drugs with other environmental chemicals may represent a hazard which is difficult to evaluate: in this context, it has been shown that the widely used food perservative, benzoic acid, may potentiate the teratogenic potential of aspirin in rats.[106]

Studies of the modes of action of teratogenic drugs have revealed several mechanisms apart from direct toxicity to the fetus. These include indirect toxicity to the fetus; physiological and nutritional effects to the fetoplacental unit; metabolic, nutritional, and hormonal effects in the mother; and effects on the sperm.[107] Most drugs probably act at more than one of these sites. Thalidomide, cytotoxics and immunosuppressive drugs, and antibiotics, such as the tetracyclines, probably exert their toxic effect directly on the fetus; folic acid antagonists such as aminopterin and methotrexate, and possibly anti-epileptic drugs such as diphenylhydantoin, ptobably act indirectly by

TABLE 2

Status of Drugs as Teratogens[105]

Drug	Type of defects	Species affected
1. **Positive implication of teratogenicity**		
Thalidomide	Musculo-skeletal	Human, baboons, macaque monkeys, and rabbits
Steroid hormones:		
Androgens and progestogens	Masculinization of female fetuses	Human, monkeys
Diethylstilbestrol	Vaginal adenocarcinoma	Human
Corticosteroids	Cleft palate	Mice, rabbits, and humans
Folate antimetabolites: Aminopterin	Intrauterine death and various deformities	Human, rodents
2. **Suspected teratogenic potential**		
Anticonvulsants:		
Phenylhydantoin Phenobarbitone Trimethadione	Cleft lip or palate, cardiac malformations	Human, mice
Neurotropic – anorexogenic agents:		
Amphetamine	Heart malformations	Human, mice
Phenmetrazine	Biliary tract atrasia	
Oral hypoglycemics:		
Tolbutamide	Bone deformities	Human, rodents
Alkylating agents	Intrauterine death and multiple malformations	Human, rodents
3. **Possible teratogens**		
Antibiotics:		
Chloramphenicol	Deafness	Rodents, human?
Streptomycin	Urinary tract anomalies	
Tetracyclines	Teeth and bone defects	
Antituberculous drugs:		
Ethionamide	Defects of central nervous system	Human?
Streptomycin-isoniazid-*p*-aminosalicylate combination	General malformations	
Imipramine	Limb deformities	Rabbits, rats, human?
Insulin	Fetal death, general deformities	Human?
Salicylates (aspirin)	Growth retardation, general malformations	Rats, mice, monkeys, humans?

affecting fetal nutrition and metabolism; catecholamines and vasopressin probably cause the loss of fetal limbs through production of fetal vasospasm and the resulting gangrenous necrosis; and steroids, antithyroid, and hypoglycemic drugs probably result in congenital abnormalities because of interference with fetal hormonal balance. In the fetoplacental unit, serotonin may adversely affect placental blood flow, depriving the fetus of adequate nutrition and causing all types of congenital malformations. Cortisone, a potent teratogen causing cleft palate in some species (mouse and rabbit), results in a reduction in the volume of amniotic fluid in susceptible animals which, it is suggested, causes flexion of the fetal head so that the tongue prevents closure of the palate. Drugs such as the phenothiazines or monoamine oxidase inhibitors, which alter the hormonal status of the mother, or drugs which affect the maternal nutrition or metabolism may consequently lead to malformations in the fetus;[107] similarly, the male gametes may contribute to congenital malformations and a number of drugs, including thalidomide, have been shown to affect fetal development when administered to the male parent.[108]

The classical and most tragic example of drug-induced teratogenicity is that of thalidomide; an estimated 10,000 babies born with rudimentary stumps instead of normal limbs are grim testimony of the now well-established relationship between the maternal use of this drug and fetal malformations.[109] Some 5,000 to 7,000 children in West Germany alone, the country of origin of the drug, were deformed by thalidomide and the incidence of malformation was nearly 100% when the drug was taken between the 34th to 45th day after the last menstrual period, even if taken only in small doses or for short periods of time.[110] The malformations were very varied in type, involving abnormalities of the limbs, the heart, the gastrointestinal and urinogenital systems, and sensory organs.[111] The development of the limb buds is affected in such a way that the hands and feet arise from the trunk similar to the flippers of a seal, hence, the term "phocomelia" (Greek: *phoke* = seal and *melos* = limbs). Although, at first, it appeared difficult to reproduce these malformations in experimental animals, this has now been achieved in a variety of species, including rat, mouse, rabbit, monkey, and chicken. Using the radioactively labeled ^{14}C-thalidomide, Koransky and Ullberg[112] showed that the drug readily crosses the mouse placenta, becoming evenly distributed in the fetus where it may be demonstrated in the developing bones. ^{14}C-Thalidomide, together with a number of its spontaneous hydrolysis products, has also been demonstrated to occur in the blastocyst of the rabbit after oral administration to the pregnant doe.[113]

A number of published reports have appeared during recent years suggesting possible effects of anticonvulsant drugs on human fetal development. The anticonvulsant phenytoin has been shown to be highly teratogenic to mice even in single doses, producing cleft palate, skeletal abnormalities, and other deformities;[114] this toxicity is dose-dependent, is affected by the extent of metabolism of the drug, and is also influenced by the timing of administration of the drug in relation to the pregnancy. In a study in rhesus monkeys given diphenylhydantoin at a dosage higher than recommended for humans, the only evidence of embryotoxicity was a minor urinary tract anomaly.[105] Dilantin® has been shown to cross the placental barrier in rats and mice and fetal distribution studies in these animals have suggested that the teratology may be due to interference with folate utilization[27] or with steroid hormone metabolism.[35] Placental transfer of the drug has also been demonstrated in human subjects[115,116] and a recent report from Melbourne, Australia, presents seven cases of babies with skeletal and other abnormalities born to mothers taking phenytoin and phenobarbitone throughout their pregnancy.[117] Although the etiology of the role of phenytoin in human malformation is still far from clear and Loughnan and his co-workers have suggested impairment of folate metabolism, collagen synthesis, nucleic acid dysfunction, or chromosomal damage as possibilities, an association is strongly indicated and it is advised that maternal serum levels of phenytoin should be monitored and controlled during pregnancy.[117]

Several reports have linked anencephaly and spina bifida with some dietary factor associated with the eating of "blighted" potatoes during pregnancy. In a study of the effects of a "blighted" *(Phytophthera infestans)* potato concentrate administered to Wistar rats and cotton-eared marmosets, no abnormalities were observed in the rat, but cranial osseous defects were observed in 4 out of 11 progeny of the marmosets, suggesting a causal relationship between blighted potato and cranial osseous defects in primates.[118a] However, more recent experiments have failed to substantiate this.[118b]

The chlorinated weed killers, 2,4-dichlorophenoxyacetic acid (2,4-D) and 2,4,5-trichlorophenoxyacetic acid (2,4,5-T), have also been used as defoliants in the Vietnam War and allegations of teratogenic effects in humans have been made. Experiments have shown them to be teratogenic when administered at high dosage (ca. 100 mg/kg) to hamsters,[119] but the incidence of developmental abnormalities, mostly associated with fetal head development, were markedly increased when the content of 2,3,7,8-tetrachlorodibenzo-*p*-dioxin, an impurity sometimes present in 2,4,5-T, was also increased. The possible hazard of congenital abnormalities from human exposure to the low levels of these chemicals when used as herbicides is difficult to assess, but the high level of dosage required for teratogenic effects in experimental animals would indicate a wide margin of safety. Similarly, the insecticides DDT and carbaryl (Sevin®) have been shown to have embryopathic and teratogenic potential at very high dosage in experimental animals; tissue accumulation of DDT appears to be greater in intrauterine life than in the neonate, but it is concluded that normal human exposure constitutes only a very low level of risk of embryotoxicity.[120]

IX. FETAL DRUG-TOXICITY

Although drugs given after the first trimester of pregnancy are unlikely to be teratogenic they may nevertheless adversely affect the growth or function of normally formed tissues or organs of the fetus. Tetracyclines form chelate complexes with calcium and, as a consequence, may become concentrated in fetal bones and teeth, producing yellow discoloration and enamel hypoplasia of the teeth and reversible impairment of skeletal growth. Streptomycin and other ototoxic drugs may damage the fetal labyrinth and auditory nerve; chloramphenicol can be toxic to the fetus in late pregnancy, as it is to the neonate, because of the inability of the fetus to conjugate the antibiotic. Sulfonamides administered to the mother in late pregnancy will be transferred to the fetus where, because of their high degree of binding to plasma proteins, displacement of protein-bound bilirubin may occur, giving rise to jaundice and kernicterus in the neonate. The fetus is particularly sensitive to coumarin anticoagulant drugs which may cross the placenta to cause fetal or neonatal hemorrhage and should be discontinued 4 weeks before term.[121] Salicylates given in late pregnancy may also result in fetal and neonatal hemorrhage due to hypoprothrombinemia. Antithyroid drugs can result in goiter in the fetus and iodides in proprietary cough medicines taken by the mother may give rise to goiter and hypothyroidism.[122]

Drugs may also adversely affect the fetus by affecting uterine and placental circulation; prednisolone has been found to produce fetal distress and an increased incidence of stillbirths, possibly because of placental insufficiency; and noradrenaline and other sympathomimetic drugs may cause fetal anoxia through vasoconstriction of the uterus. Similarly, drugs may result in adverse changes in the fetal circulation and the local anesthetics lidocaine and mepivacaine used in paracervical block in the mother may result in bradycardia.[123]

Clinical studies have shown that women who smoke heavily during pregnancy experience a higher incidence than nonsmokers of abortion, premature births, delayed delivery, low infant birth weight, and loss of birth weight with decreased survival potential. These observations, generally poorly controlled, have been confirmed by more precise studies of the effects of acute doses of nicotine to rats, when similar adverse effects to smoking were observed in pregnancy and in neonates.[124]

Attention has recently been focused on a hitherto rare form of fetal drug toxicity, namely, transplacental carcinogenicity. In the period 1966–69, 8 patients aged 15 to 22 in Boston, presented with clear-cell adenocarcinoma of the vagina (mesonephroma), a rather rare form of malignancy,[125] and a further 5 cases were subsequently documented in the U.S.[126] All 13 patients were born in the period 1946–53, when estrogens were being widely used for the treatment of threatened abortion and a high degree of significance has been established between the treatment of mothers with stilbestrol during pregnancy and the subsequent development of these malignant tumors in their daughters. More than 60 cases of this rare malignancy have now been reported and other nonsteroidal estrogens are under suspicion.[127] Although other causative factors cannot be completely excluded, it is reasonable to advise that stilbestrol should not be given during pregnancy and that it should not be used as a post-coital contraceptive because of the risk to the fetus if the prophylaxis were unsuccessful.

Similar transplacental carcinogenesis has been demonstrated in experimental animals treated with urethane and other carcinogens.[104,129] A single injection of urethane to mice administered on the 9th day of gestation gave rise to lung and tail deformities and malignant teratomas, to hepatomas when administered after day 11, and to lung tumors after day 13, showing a correlation between organogenesis, teratogenesis, and carcinogenesis.[129] Druckrey[104] has shown a similar correlation between transplacental carcinogenesis and teratogenesis induced by nitrosamines and triazenes and the development of fetal drug-metabolizing enzymes.

X. EFFECTS OF DRUGS IN PARTURITION

Drugs administered to the mother during labor carry the risk of depressing respiration and other vital functions in the neonate. The gaseous anesthetics and the short-acting barbiturates quickly pass across the placenta into the fetus, whereas narcotic analgesics attain equilibrium between maternal and fetal circulations more slowly. There is a general belief that it is the safest practice to carry out delivery within the first hour of administration of narcotics to the mother, although frequently the neonate shows a markedly lower response than the mother to the same plasma level of barbiturates and other drugs.[34]

Muscle relaxants cross the placenta with difficulty, but if administered in very high dosage, can reach the fetus and result in paralysis at birth.[130] Local anesthetics used in epidural and paracervical anesthesia, such as lidocaine and mepivacaine, usually have no adverse effects on the neonate, but occasionally may produce fetal bradycardia.[131] Of the gaseous anesthetics widely used in labor, nitrous oxide with oxygen causes little, if any neonatal depression; cyclopropane produces no ill effects to the neonate if administered at analgesic concentrations, although it may cause neonatal depression if used to produce deeper planes of anesthesia; trichloroethylene (trilene) rapidly gives rise to high plasma levels in the fetus and after 15 min may be higher than in the maternal circulation, thus resulting in neonatal depression if used at too great a concentration or for too long a period; and halothane rapidly reaches the fetus and produces a greater incidence of neonatal depression than nitrous oxide and is not advised for routine obstetric practice.[132] Meperidine has been claimed to be the safest narcotic for use in obstetrics, but similar to other narcotics, such as morphine, methadone, levorphanol (Dromoran®), and phenazocine, may frequently give rise to depression in the neonate, particularly when the infant is born between 1 and 4 hr after administration of the drug. Babies born to mothers who are dependent on narcotics, such as heroin and pethidine, may exhibit the typical signs of withdrawal syndrome at birth.

It has been suggested that prostaglandins (especially $PGF_{2\alpha}$) may have a natural uterotonic function mediating the expulsion of the fetus in parturition; drugs that inhibit synthesis of endogenous prostaglandins might, therefore, prolong parturition. Aspirin and indomethacin, both inhibitors of prostaglandin synthesis, have been shown in the rat to reduce uterine motility, delay delivery, and substantially increase fetal mortality;[133] these and other anti-inflammatory drugs that are inhibitors of prostaglandin synthesis should, consequently, be avoided in the later stages of pregnancy.

XI. TOXICITY OF DRUGS TO NEONATES

Newborn infants, especially if premature, are particularly sensitive to the toxic actions of drugs and other xenobiotics, primarily because the hepatic enzymes that are concerned with the metabolic deactivation of these compounds are insufficiently developed. Impaired renal excretion because of the incomplete development of glomerular filtration and tubular secretion mechanisms at birth exacerbate the situation with the consequence that drugs are slowly metabolized, slowly deactivated, and slowly excreted.

Hyperbilirubinemia and kernicterus may be produced by several drugs by: (1) competitive displacement of bilirubin from plasma protein binding sites, (2) competition with the bilirubin for hepatic conjugation with glucuronic acid and with bilirubin glucuronide for biliary excretion,

(3) noncompetitive inhibition of the hepatic enzymes which form the bilirubin glucuronide, viz., the glucuronyl transferases, and (4) oxidation of hemoglobin to methemoglobin with the consequent hemolysis of the erythrocyte and excessive formation of bilirubin. Sulfonamides and many nonsteroidal anti-inflammatory drugs are known to affect the protein-binding of bilirubin; paracetamol and salicylates are competitive inhibitors and novobiocin and vitamin K are noncompetitive inhibitors of bilirubin glucuronyl transferase; nitrofurantoin, phenacetin, nitrites, and sulfonamides may cause methemoglobinemia and hemolysis. Moreover, as has been previously considered, the hepatic glucuronyl transferases in the neonate are normally at suboptimal levels of activity because of impaired development or inhibition by progestational steroids. Deficiency of the ability to synthesize glucuronide conjugates undoubtedly leads to the enhanced susceptibility of the neonate to chloramphenicol, paracetamol, salicylates, and many other drugs.

The acute toxicities of a number of different groups of drugs have been shown to be higher in newborn rats than in adult animals.[134] This is attributed to deficiencies in the hepatic drug-metabolizing enzymes, which do not generally become manifest until after birth, and then increase to reach adult levels at about 30 days in rats and 8 weeks in humans.[86] Consequently, because of these enzyme deficiencies, the newborn is particularly susceptible to the toxic effects of drugs. Deficiencies of the hepatic microsomal mixed-function oxidases lead to impaired oxidative metabolism and increased toxicity from paracetamol, phenylbutazone, and many other drugs, although others would appear to be metabolized at normal rates.[79] Administration of chloramphenicol to newborn infants gives rise to the "grey syndrome" consisting of cyanosis, vomiting, and vascular collapse[135] and was one of the major contributory causes of thousands of infant deaths which occurred in the epidemics of infantile diarrhea that spread across Europe immediately following World War II. Impaired glucuronide conjugation, and possibly other metabolic pathways, of this antibiotic no doubt contribute greatly to its high toxicity to the neonate.

Hexachlorophene 2,2′-methylenebis (3,4,6-trichlorophenol), is an antiseptic which is widely used as a solution in the bathing of babies, especially to prevent nursery staphylcoccus infections, and is also contained in soaps and baby toilet powders. In 1972, the U.S. Food and Drug Administration saw fit to restrict the use of hexachlorophene and restrictions were also subsequently imposed by the Committee on Safety of Medicines of the United Kingdom. The reason for this was a number of reports of convulsions occurring in neonates and young children bathed with strong solutions of this antiseptic. Hexachlorophene has been shown to be absorbed from the normal skin of humans and rats, and in large doses may affect the central nervous systems of both species, giving rise to edema of the white matter or "status spongiosus."[136,137] It may also produce paralysis in rats, rabbits, cats, and pigs and blindness in sheep.[138] Hexachlorophene, both in vivo and in vitro, stimulates the mitochondrial respiration of rat liver and calf brain, followed by inhibition of respiration, indicative that uncoupling of oxidative phosphorylation may be the primary biochemical lesion of hexachlorophene intoxication.[139,140]

Most of the organophosphate insecticides are more toxic to weanling animals than to adults, the esterase enzyme system that is responsible for the detoxication of malathion and malaoxon being depressed or absent at birth like many other hepatic microsomal enzymes.[141] Weanling rats are also more susceptible than adults to the toxicity of parathion and it has been shown that both parathion and its more toxic metabolite paraoxon are metabolized less effectively by the weanling animals.[142]

In 1968–69 several hundred cases of poisoning by cooking oils contaminated with tetrachlorobiphenyl were reported in the Fukuoka prefecture of Japan. The symptoms were acne-like skin eruptions characterized by hyperkeratosis and atrophy of the skin. Several pregnant women and their fetuses were also affected by this contaminated oil; at delivery nine infants had dark-brown staining of the skin and two more were stillborn, their abnormal skin containing chlorobiphenyls. The chlorinated hydrocarbon apparently passes across the placenta, like so many drugs and foreign chemicals.[143]

The ingestion of lead, a cumulative poison, is widespread among young children. It may be ingested in the food, water, or air; from contaminated food, milk, and water conducted in lead pipes; from air contaminated with lead from industrial lead-smelting plants; and possibly from vehicle exhausts, the result of alkyl lead additives present in petrol, but especially from lead-containing paints. The chronic toxic effects of lead include decreased hemopoiesis, porphyria, renal tubular dysfunction, and damage to the central nervous system including encephalopathy and hyperactivity; acute symptoms include vomiting, anorexia, ataxia, and convulsions. The extent of poisoning of children by this metal is probably not fully realized and should be suspected in young children showing signs of lassitude, vomiting, tremors, or hyperactivity.[144] Inorganic forms of lead have a greater potential for toxic injury to the embryo, fetus, and neonate than to the adult, but in contrast, the organic forms of lead, while more toxic than inorganic lead to adults, have a lesser effect on embryo, fetus, and neonate.[145]

XII. DRUGS AND LACTATION

Although the popularity of breast-feeding fluctuates according to its social acceptability, it is generally believed by nutritionists that any alternative method of infant feeding is a mere substitute for nature's ideal diet of breast milk. The current realization that drugs and foreign chemicals may be secreted in breast milk and, of course, also in cow's milk further complicates the problem of infant nutrition. Drugs may also affect the secretion of milk and its composition. Chlorpromazine and other phenothiazines have been found to have mammogenic side effects, producing galactorrhea in females and gynecomastia in male patients receiving these drugs. Contraceptive estrogen-progestogen preparations may diminish lactation and ethynylestradiol can result in increased protein content of the milk.[146]

Drugs and chemicals that are present in the circulating blood of the mother will pass into the alveolar cells of the mammary gland and will be excreted into the milk. Those drugs that are present in the maternal blood in high concentration for long periods of time will be secreted in the milk in the most significant amounts. Lipid-soluble drugs are secreted in association with the fatty components of the milk and aqueous-soluble drugs may be bound to lactalbumin, casein, and other proteins of milk. Nevertheless, the mammary gland is, on the whole, a relatively unimportant route for the excretion of drugs.

Although the presence of drugs in milk does not necessarily imply adverse effects on the infant, most reports in the literature have been directed to those instances where the consequences were unfavorable. These aspects of neonatal nutrition and pharmacology have recently been comprehensively reviewed.[147] Alcohol is reported to be secreted in breast milk in only very small amounts, but infant intoxication following the excessive consumption of alcohol by the mother has been recorded. Chloroform anesthesia of the mother may similarly produce deep sleep in her breast-fed infant. Anticoagulants may be secreted in the milk and a recent report cites the case of an infant developing a huge hematoma following the repair of an inguinal hernia whose mother was breast-feeding him while taking phenindione.[148] Male infants may develop gynecomastia following the mothers taking estrogens or excessive dosage of oral contraceptives.[149] Thiouracil is secreted in breast milk in higher concentrations than occur in the plasma and the neonate could develop goiter and hypothyroidism; iodides may have similar consequences.[150] It has, therefore, become apparent that many drugs should not be prescribed for nursing mothers, or if such drugs are essential to the mother's health, the infant should not be breast-fed. Among the more noted of these drugs are the anticoagulants, antimetabolites, antithyroid drugs, narcotics, iodides, and bromides, tetracyclines, sulfonamides, barbiturates, and steroids.

In many instances, the concentrations of drugs secreted in the milk are low and would have no adverse effects upon the child. This is known to be generally valid for insulin, aspirin, alcohol, caffeine,[147] for diphenylhydantoin,[116] and even methotrexate, which because of its low pK values is present in blood plasma in the lipid-insoluble ionized form and does not readily pass into the milk.[151] The secretion into milk of drugs which are the inducers of the hepatic microsomal drug-metabolizing enzymes, e.g., phenobarbitone, may result in substantial increases in these enzymes in the suckling offspring[152] and enhance the rates of metabolism of drugs in the neonate.

Chlorinated hydrocarbons, such as DDT and its metabolite DDE, being highly lipid-soluble and not readily metabolized, tend to accumulate in lipid tissues and are widely occurring in the milk of most species including human. In recent studies, the milk of lactating women examined in various areas of the U.S. was shown to contain chlorinated pesticides, such as DDT, DDE, benzene hexachloride (BHC), dieldrin, and heptachlor epoxide in every case.[153,154] The levels of total pesticides found in human milk in the latter study ranged from two to ten times the legal limit established in the U.S. for commercial cow's milk.

Breast milk sometimes is the causative agent of neonatal unconjugated hyperbilirubinemia, probably because of the presence of inhibitors of hepatic glucuronyl transferase. It has been suggested that the inhibiting substance is the steroid hormone 3α,20β-pregnanediol, but a more recent study of this problem has shown that another inhibitory substance is also likely to be present since, whereas the inhibitory milk decreases the glucuronide conjugation of bilirubin in rat liver slices, pregnanediol appears to exert its inhibitory effect on the secretion of the bilirubin glucuronide from the liver slices.[155]

XIII. CONCLUSIONS

The introduction of toxic chemicals into our society in the form of drugs, pesticides, food additives, etc. has conferred great benefits in terms of health, amenity, and prosperity, but in the past this has been achieved in an empirical manner with little thought for safety, an attitude of mind from which we are only very slowly emerging. The ignorance and complacency of society concerning the potential hazards of unrestricted usage of toxic chemicals have been equaled only by the scientific naïveté of the medical profession and the diffident reluctance of chemists, biologists, and other scientists to assume responsiblity for matters concerning health. The result has been a number of human tragedies over the use of drugs such as thorotrast, chloramphenicol, and thalidomide; pesticides such as parathion and paraquat; and toxic industrial chemicals such as the bladder carcinogen, β-naphthylamine, and the toxic lubricant additive tri-ortho-cresyl phosphate, which too often has found its way into human food.

That the most prominent of these tragedies have affected the newborn is perhaps in itself the most significant indication of the particular vulnerability of the pregnant woman, her fetus, and the neonate to the toxic effects of drugs and chemicals. Neither is it insignificant that major regulatory legislation has usually been introduced following each of these tragedies, for the health of the young is a highly emotive area that quickly arouses public feeling.

The present testing of drugs for potential teratogenicity is generally of adequate standard, but few drug regulatory authorities require that this evaluation of safety be extended to include monitoring of the whole perinatal period or request reproductive and neonatal studies. However, where these requirements are not already implemented, they are currently under consideration. Furthermore, there is no reason, save expense, why chemicals other than drugs, such as cosmetics and industrial chemicals, pesticides, and agricultural chemicals, should not be subjected to the same rigorous testing as is applied to drugs. Indeed, there is a case for even more stringent standards being imposed as a greater section of the community will come into contact with these chemicals more frequently, often involuntarily, and without the same degree of personal benefit.

It is to be hoped that society has learned from its past mistakes, which have exacted such a great price from the very young, that drugs and other chemicals are potential perdition as well as blessing, and must be studied and evaluated with extreme care to ensure their safe usage and freedom from future catastrophes.

REFERENCES

1. **Parke, D. V.**, in *Recent Advances in Pharmacology*, Robson, J. M. and Stacey, R. S., Eds., Churchill, London, 1968, Ch. 2 and 3.
2. **Parke, D. V.**, The biochemistry of foreign compounds, in *International Series of Monographs in Pure and Applied Biology*, Vol. 5, Pergamon Press, Oxford, 1968.
3. **Udenfriend, S.**, Arene oxide intermediates in enzymatic hydroxylation and their significance with respect to drug toxicity, in *Drug Metabolism in Man*, Vessell, E. S., Ed., *Ann. N.Y. Acad. Sci.*, 179, 295, 1971.
4. **Bolt, H. M. and Remmer, H.**, Retention and metabolism and elimination of 17α-ethynyloestradiol-3-methyl ether (mestranol), *Xenobiotica*, 2, 77, 1972.
5. **Dodds, E. C.**, Rime and reason in endocrinology, *J. Endocrinol.*, 23, 3, 1961.
6. **Dodds, E. C.**, Oral contraceptives: the past and future, *Clin. Pharmacol. Ther.*, 10, 147, 1969.
7. Anon., Oral contraceptives and thromboembolism, *Lancet*, 1039, 1967.
8. **Drill, V. A.**, Oral contraceptives and thromboembolic disease. *J.A.M.A.*, 206, 77, 1968.
9. **Bolton, C. H., Hampton, J. R., and Mitchell, J. R. A.**, Effect of oral contraceptive agents on platelets and plasma-phospholipids, *Lancet*, 1336, 1968.
10. **Metreau, J. M., Dhumeaux, D., and Berthelot, P.**, Oral contraceptives and the liver, *Digestion*, 7, 318, 1972.
11. Committee on Safety of Medicines, Tests on the pill for carcinogenicity, *Br. Med. J.*, 4, 5834, 1972.
12. **Bishun, N. P., Williams, D. C., Mills, J., Lloyd, N., Raven, R. W., and Parke, D. V.**, Chromosome damage induced by chemicals, *Chem. Biol. Interact.*, 6, 375, 1973.
13. **Parke, D. V.**, The effect of drugs and steroid hormones on the enzymes of the endoplasmic reticulum, in *Effects of Drugs on Cellular Control Mechanisms*, Rabin, B. R. and Freedman, R. B., Eds., Macmillan, London, 1971.
14. **Juchau, M. R. and Fouts, J. R.**, Effects of norethynodrel and progesterone on hepatic microsomal drug metabolizing enzyme systems, *Biochem. Pharmacol.*, 15, 891, 1966.
15. **O'Malley, K., Stevenson, I. J., and Crooks, J.**, Impairment of human drug metabolism by oral contraceptive steroids, *Clin. Pharmacol. Ther.*, 13, 552, 1972.
16. **Crawford, J. J. and Rudofsky, S.**, Some alterations in the pattern of drug metabolism associated with pregnancy, oral contraception and the newly born, *Br. J. Anaesth.*, 38, 446, 1966.
17. **Hargreaves, T., Piper, R. F., and Cam, J.**, Effect of oral contraception on glucuronyl transferase, *Nature*, 234, 110, 1971.
18. **Cessi, C.**, Comportamento della glucuronoconingazione nella cavia gravida, *Boll. Soc. Ital. Biol. Sper.*, 28, 1857, 1952.
19. **Lathe, G. H. and Walker, M.**, Inhibition of bilirubin conjugation in rat liver slices by human pregnancy and neonatal serum and steroids, *Quart. J. Physiol.*, 43, 257, 1958.
20. **Hsia, D. Y. Y., Riabov, S., and Dowben, R. M.**, Inhibition of glucuronyl transferase by steroid hormones, *Arch. Biochem.*, 103, 181, 1963.
21. **Hsia, D. Y. Y., Dowben, R. M., and Riabov, S.**, Inhibitors of glucuronyl transferase in the newborn, *Ann. N.Y. Acad. Sci.*, 111, 326, 1964.
22. **Hartiala, K., Pulkkinen, M., and Rauramo, L.**, Resorption and glucuronide conjugation of salicylamide during pregnancy, *Nature*, 196, 678, 1963.
23. **Neale, M. G. and Parke, D. V.**, Effects of pregnancy on the metabolism of drugs in the rat and rabbit, *Biochem. Pharmacol.*, 22, 1451, 1973.
24. **Halac, E., Jr. and Sicignano, C.**, Re-evaluation of the influence of sex, age, pregnancy and phenobarbital on the activity of UDP-glucuronyl transferase in rat liver, *J. Lab. Clin. Med.*, 73, 677, 1969.
25. **Pulkkinen, M. O.**, Sulphate conjugation during development in human, rat and guinea pig, *Acta Physiol. Scand.*, 66, 120, 1966.
26. **Creaven, P. J. and Parke, D. V.**, The effect of pregnancy on the microsomal metabolism of foreign compounds, *Fed. Eur. Biochem. Soc. Sec. Meet. Abst.*, 128, 88, 1965.
27. **Gabler, W. L. and Falace, D.**, The distribution and metabolism of dilantin in non-pregnant, pregnant and foetal rats, *Arch. Int. Pharmacodyn. Ther.*, 184, 45, 1970.
28. **Neale, M. G. and Parke, D. V.**, The effect of pregnancy on the hydroxylation and reduction of drugs and the cytochrome P-450 content of rat liver microsomes, *Biochem. J.*, 113, 12, 1969.
29. **Guarino, A. M., Gram, T. E., Schroeder, D. H., Call, J. B., and Gillette, J. R.**, Alterations in kinetic constants for hepatic microsomal aniline hydroxylase and ethylmorphine *N*-demethylase associated with pregnancy in rats, *J. Pharmacol. Exp. Ther.*, 168, 224, 1969.
30. **Moya, F. and Thorndike, V.**, Passage of drugs across the placenta, *Am. J. Obstet. Gynecol.*, 84, 1778, 1962.
31. **Houston, J. B., Upshall, D. G., and Bridges, J. W.**, The absorption of aliphatic carbamates by everted gut sacs of rat, *Proc. Biochem. Soc.*, 128, 141P, 1972.
32. **Douglas, J. S. and Nicholls, P. J.**, The fate of phenglutarimide hydrochloride in the rat and man, *Xenobiotica*, 3, 605, 1973.
33. **Moya, F. and Smith, B. E.**, Uptake, distribution and placental transport of drugs and anesthetics, *Anesthesiology*, 26, 465, 1965.

34. **Flowers, C. E.**, Diamox as a diuretic in pregnancy, *Am. J. Obstet. Gynecol.*, 78, 730, 1959.
35. **Waddell, W. J. and Mirkin, B. C.**, Distribution and metabolism of diphenylhydantoin-^{14}C in fetal and maternal tissues of pregnant mouse, *Biochem. Pharmacol.*, 21, 547, 1972.
36. **Idänpään-Heikkilä, J. E., Jouppila, P. U., Puolakka, J. O., and Vorne, M. S.**, Placental transfer and fetal metabolism of diazepam in early human pregnancy, *Am. J. Obstet. Gynecol.*, 109, 1011, 1971.
37. **Beckett, A. H. and Taylor, J. F.**, Blood concentrations of pethidine and pentazocine in mother and infant at time of birth, *J. Pharm. Pharmacol.*, 19, Suppl. 50S, 1967.
38. **Douglas, B. H. and Hume, A. S.**, Placental transfer of imipramine, a basic lipid-soluble drug, *Am. J. Obstet. Gynecol.*, 99, 573, 1967.
39. **Rogers, M. C., Willerson, J. T., Goldblatt, A., and Smith, T. W.**, Serum digoxin concentrations in the human fetus, neonate and infant, *N. Engl. J. Med.*, 287, 1010, 1972.
40. **Sjoqvist, F., Bergfors, P. G., Borga, O., Lind, M., and Ygge, H.**, Plasma disappearance of nortriptyline in a newborn infant following placental transfer from an intoxicated mother: evidence for drug metabolism, *J. Pediatr.*, 80, 496, 1972.
41. **Freudenthal, R. I., Martin, J., Cook, C. E., and Wall, M. E.**, Metabolism and distribution of norethynodrel in pregnant and non-pregnant mice, *Toxicol. Appl. Pharmacol.*, 24, 125, 1973.
42. **Curley, A., Copeland, F., and Kimbrough, R. D.**, Chlorinated hydrocarbon insecticides in organs of stillborn and blood of newborn babies, *Arch. Environ. Health*, 19, 628, 1969.
43. **Flowers, C. E., Rudolph, A. J., and Desmond, M. M.**, Diazepam (Valium) and as adjunct in obstetric analgesia, *Obstet. Gynecol.*, 34, 68, 1969.
44. **Idänpään-Heikkilä, J. E., Taska, R. J., Allen, H. A., and Schoolar, J. C.**, Placental transfer of diazepam-^{14}C in mice, hamsters and monkeys, *J. Pharmacol. Exp. Ther.*, 176, 752, 1971.
45. **Pruitt, A. W. and Dayton, P. G.**, A comparison of the binding of drugs to adult and cord plasma, *Eur. J. Clin. Pharmacol.*, 4, 59, 1971.
46. **Perlmutter, J. F.**, Drug addiction in pregnant women, *Am. J. Obstet. Gynecol.*, 99, 569, 1967.
47. **Goodfriend, M. J., Shey, I. A., and Klein, M. D.**, The effect of maternal narcotic addiction on the newborn, *Am. J. Obstet. Gynecol.*, 71, 29, 1956.
48. **Agpar, V., Burns, J. J., Brodie, B. B., and Papper, E. M.**, The transmission of meperidine across the human placenta, *Am. J. Obstet. Gynecol.*, 64, 1368, 1952.
49. **Mosier, H. D., Jr. and Jansons, R. A.**, Distribution and fate of nicotine in the rat fetus, *Teratology*, 6, 303, 1972.
50. **Juchau, M. R. and Yaffe, S. J.**, in *Foeto Placental Unit*, Pecile, A. and Finzi, C., Eds., Excerpta Medica Foundation, Amsterdam, 1969, 260.
51. **Juchau, M. R.**, Mechanisms of drug biotransformation reactions in the placenta, *Fed. Proc.*, 31, 48, 1972.
52. **Chakraborty, J. P., Hopkins, R., and Parke, D. V.**, Biological oxygenation of drugs and steroids in the placenta, *Biochem. J.*, 125, 15P, 1971.
53. **Strauss, L., Goldenburg, N., Hirota, K., and Okudaira, Y.**, Symposium on the placenta, *Birth Defects, Original Article Series*, 1, 13, 1965.
54. **Meigs, R. A. and Ryan, K. J.**, Cytochrome P-450 and steroid biosynthesis in the human placenta, *Biochim. Biophys. Acta*, 165, 476, 1968.
55. **Juchau, M. R., Niswander, K. R., and Yaffe, S. J.**, Metabolizing systems in homogenates of human immature placentas, *Am. J. Obstet. Gynecol.*, 100, 348, 1968.
56. **Pelkonen, O., Vorne, M., Jouppila, P., and Karki, N. T.**, Metabolism of chlorpromazine and *p*-nitrobenzoic acid in the liver, intestine and kidney of the human fetus, *Scand. J. Clin. Lab.*, 27 (Suppl. 116), 7, 1971.
57. **Welch, R. H., Harrison, Y. E., Gommi, B. W., Poppers, P. J., Finster, M., and Conney, A. H.**, Stimulatory effect of cigarette smoking on the hydroxylation of 3,4-benzpyrene and the *N*-demethylation of 3-methyl-4-monomethylaminoazobenzene by enzymes in human placenta, *Clin. Pharmacol. Ther.*, 10, 100, 1969.
58. **Hagerman, D. D.**, Enzymology of the placenta, in *Foetus and Placenta*, Blackwell, Oxford, 1969, 421.
59. **Juchau, M. R. and Symms, K. G.**, Aniline hydroxylation in human placenta: mechanistic aspects, *Biochem. Pharmacol.*, 21, 2053, 1972.
60. **Nebert, D. W., Winkler, J., and Gelboin, H. V.**, Aryl hydrocarbon hydroxylase activity in human placenta from cigarette-smoking and non-smoking women, *Cancer Res.*, 29, 1763, 1969.
61. **Conney, A. H., Welch, R., Kuntzman, R., Change, R., Jacobson, M., Munro-Faure, A. D., Peck, A. W., Bye, A., Poland, A., Poppers, P. J., Finster, M., and Wolf, J. A.**, Effects of environmental chemicals on the metabolism of drugs, carcinogens and normal body constituents in man, *Ann. N.Y. Acad. Sci.*, 179, 155, 1971.
62. **Pelkonen, O., Jouppila, P., and Karki, N. T.**, Effect of maternal cigarette smoking on 3,4-benzpyrene and *N*-methylaniline metabolism in human foetal liver and placenta, *Toxicol. Appl. Pharmacol.*, 23, 399, 1972.
63. **Gelboin, H. V., Wiebel, F., and Coon, H. G.**, Analysis of microsomal hydroxylase induction in cell cultures, *Biochem. J.*, 130, 48P, 1972.
64. **Juchau, M. R.**, Human placental hydroxylation of 3,4-benzpyrene during early gestation and at term, *Toxicol. Appl. Pharmacol.*, 18, 665, 1971.
65. **Juchau, M. R., Krasner, J., and Yaffe, S. J.**, Reduction of azo linkages in human placental homogenates, *Biochem. Pharmacol.*, 17, 1969, 1968.

66. **Juchau, M. R.,** Reduction of aromatic nitro groups in human and rodent placental homogenates, *J. Pharmacol. Exp. Ther.,* 165, 1, 1969.
67. **Berte, F., Manzo, L., De Bernardi, M., and Benzi, G.,** Ability of the placenta to metabolise oxazepam and aminopyrine before and after drug stimulation, *Arch. Int. Pharmacodyn. Ther.,* 182, 1, 1969.
68. **Troen, P., De Miguel, M., and Alonso, C.,** Perfusion studies of the human placenta. IV. Conjugation of estriol-16-^{14}C, *Biochemistry,* 5, 332, 1966.
69. **Kyegombe, D., Franklin, C., and Turner, P.,** Drug-metabolizing enzymes in the human placenta, their induction and repression, *Lancet,* 1, 405, 1973.
70. **Akhtar, M. and Skinner, S. J. M.,** The intermediary role of a 19-oxo androgen in the biosynthesis of oestrogen, *Biochem. J.,* 109, 318, 1968.
71. **Kullander, S. and Kallen, B.,** Prospective study of smoking and pregnancy, *Acta Obstet. Gynaecol. Scand.,* 50, 83, 1971.
72. **Fedrick, J., Alberman, E. D., and Goldstein, H.,** Possible teratogenic effect of cigarette smoking, *Nature,* 231, 529, 1971.
73. **Longo, L. D.,** Carbon monoxide in the pregnant mother and fetus and its exchange across the placenta, *Ann. N.Y. Acad. Sci.,* 174, 313, 1970.
74. **Fouts, J. R. and Adamson, R. H.,** Drug metabolism in the newborn rabbit, *Science,* 129, 897, 1959.
75. **Jondorf, W. R., Maickel, R. R., and Brodie, B. B.,** Inability of newborn mice and guinea pigs to metabolize drugs, *Biochem. Pharmacol.,* 1, 352, 1959.
76. **Feuer, G. and Liscio, A.,** Effects of drugs on the hepatic drug metabolism of the foetus and newborn, *Int. J. Clin. Pharmacol.,* 3, 30, 1970.
77. **Fouts, J. R. and Hart, L. G.,** Hepatic drug metabolism during the perinatal period, *Ann. N.Y. Acad. Sci.,* 123, 245, 1965.
78. **Creaven, P. J., Parke, D. V., and Williams, R. T.,** A fluorimetric study of the hydroxylation of biphenyl in vitro by liver preparations of various species, *Biochem. J.,* 96, 879, 1965.
79. **Rane, A. and Sjoqvist, F.,** Drug metabolism in the human foetus and newborn infant, *Pediatr. Clin. North Am.,* 19, 37, 1972.
80. **West, J. R., Smith, H. W., and Chasis, H.,** Glomerular filtration rate, effective renal blood flow and maximal tubular excretory capacity in infancy, *J. Pediatr.,* 32, 10, 1948.
81. **Jalling, B., Boreus, L. O., Rane, A., and Sjoqvist, F.,** Plasma concentrations of diphenylhydantoin in young infants, *Pharmacol. Clin.,* 2, 200, 1970.
82. **Levi, A. J., Gatmaitan, Z., and Arias, I. M.,** Deficiency of hepatic organic anion-binding protein as a possible cause of non-haemolytic, unconjugated hyperbilirubinaemia in the newborn, *Lancet,* 2, 139, 1969.
83. **Dutton, G. J.,** Variations in glucuronide formation by perinatal liver, *Biochem. Pharmacol.,* 15, 947, 1966.
84. **Brandt, I. K.,** The development of the hippuric acid-synthesising system in the rat, *Dev. Biol.,* 10, 202, 1964.
85. **Wengle, B.,** Studies on ester sulphates. 21. On sulphate conjugation in human foetal tissue extracts, *Acta Soc. Med. Ups.,* 69, 105, 1964.
86. **Kato, R., Vassanelli, P., Frontino, G., and Chiesara, E.,** Variation in the activity of liver microsomal drug-metabolizing enzymes in rat in relation to the age, *Biochem. Pharmacol.,* 13, 1037, 1964.
87. **Yaffe, S. J., Rane, A., Sjoqvist, F., Boreus, L. O., and Orrenius, S.,** The presence of mono-oxygenase system in human fetal liver microsomes, *Life Sci.,* 9, 1189, 1970.
88. **Pelkonen, O. and Karki, N. T.,** A demonstration of cytochrome P-450 in human foetal liver microsomes in early pregnancy, *Acta Pharmacol. Toxicol.,* 30, 158, 1971.
89. **Bhatnagar, S. P.,** The metabolism of chlorpromazine by developing rabbit liver, *Can. J. Physiol. Pharmacol.,* 49, 649, 1971.
90. **Basu, T. K., Dickerson, J. W. T., and Parke, D. V.,** Effect of development on the activity of microsomal drug-metabolizing enzymes in rat liver, *Biochem. J.,* 124, 19, 1971.
91. **MacLeod, S. M., Renton, K. W., and Eade, N. R.,** Development of hepatic microsomal drug-oxidizing enzymes in immature male and female rats, *J. Pharmacol. Exp. Ther.,* 183, 489, 1972.
92. **Fouts, J. R. and Devereux, T. E.,** Developmental aspects of hepatic and extrahepatic drug-metabolizing enzyme systems: microsomal enzymes and components in rabbit liver and lung during the first month of life, *J. Pharmacol. Exp. Ther.,* 183, 458, 1972.
93. **Ackermann, E., Rane, A., and Ericsson, J. L. E.,** The liver microsomal mono-oxygenase system in the human foetus: distribution in different centrifugal fractions, *Clin. Pharmacol. Ther.,* 13, 652, 1972.
94. **Juchau, M. R.,** Drug biotransformation in the human fetus: nitro group-reduction, *Arch. Int. Pharmacodyn. Ther.,* 194, 346, 1971.
95. **Kardish, R. and Feuer, G.,** Relationship between maternal progesterones and the delayed drug metabolism in the neonate, *Biol. Neonate,* 20, 58, 1972.
96. **Hart, L. G., Adamson, R. H., Dixon, R. L., and Fouts, J. R.,** Stimulation of hepatic microsomal drug metabolism in the newborn and fetal rabbit, *J. Pharmacol. Exp. Ther.,* 137, 103, 1962.
97. **Pantuck, E., Conney, A. H., and Kuntzman, R.,** Effect of phenobarbital on the metabolism of pentobarbital and meperidine in fetal rabbits and rats, *Biochem. Pharmacol.,* 17, 1441, 1968.

98. **Arias, I. M., Gartner, L., Furman, M., and Wolfson, S.,** Effect of several drugs and chemicals on hepatic glucuronide formation in newborn rats, *Proc. Soc. Exp. Biol. Med.*, 112, 1037, 1963.
99. **Maurer, H. M., Wolff, J. A., Finster, M., Poppers, P. J., Pantuck, E., Kuntzman, R., and Conney, A. H.,** Reduction in concentration of total serum bilirubin in offspring of women treated with phenobarbitone during pregnancy, *Lancet,* 2, 122, 1968.
100. **Yeung, C. Y. and Field, C. E.,** Phenobarbitone therapy in neonatal hyperbilirubinaemia, *Lancet.* 2, 135, 1969.
101. **Yeung, C. Y. and Yu, V. Y. H.,** Phenobarbitone enhancement of bromsulphothalein clearance in neonatal hyperbilirubinaemia, *Pediatrics,* 48, 556, 1971.
102. **Rubaltelli, F. F.,** Serum conjugated bilirubin in newborns of mothers treated with immunosuppressant drugs, *Acta Pediatr. Scand.*, 61, 606, 1972.
103. **Nelson, M. M. and Forfar, J. O.,** Association between drugs administered during pregnancy and congenital abnormalities of the fetus, *Br. Med. J.*, 1, 523, 1971.
104. **Druckrey, H.,** Specific carcinogenic and teratogenic effect of indirect alkylating methyl and ethyl compounds and their dependency on stages of ontogenic development, *Xenobiotica,* 3, 271, 1973.
105. **Wilson, J. G.,** Present status of drugs as teratogens in man, *Teratology,* 7, 3, 1973.
106. **Kimmel, C. A., Wilson, J. G., and Schumacher, H. J.,** Studies on metabolism and identification of the causative agent in aspirin teratogenesis in the rat, *Teratology,* 4, 15, 1971.
107. **Sullivan, F. M.,** Mechanisms of action of teratogenic drugs, *Proc. R. Soc. Med.*, 63, 1252, 1970.
108. **Lutwak-Mann, C., Schmid, K., and Keberle, H.,** Thalidomide in rabbit serum, *Nature,* 214, 1018, 1967.
109. **Sjöström, H. and Nilsson, R.,** Thalidomide and the Power of the Drug Companies, *Penguin Books,* Harmondsworth, England, 1972.
110. **Lenz, W. and Knapp, K.,** Thalidomide embryopathy, *Arch. Environ. Health,* 5, 100, 1962.
111. **Sutherland, J. M. and Light, I. J.,** The effects of drugs upon the developing fetus, *Pediatr. Clin. North Am.*, 12, 781, 1965.
112. **Koransky, W. and Ullberg, S.,** Autoradiographic investigations of ^{14}C-labelled thalidomide and glutethimide in pregnant mice, *Proc. Soc. Exp. Biol. Med.*, 116, 512, 1964.
113. **Fabro, S., Schumacher, H., Smith, R. L., and Williams, R. T.,** Identification of thalidomide in rabbit blastocysts, *Nature,* 201, 1125, 1964.
114. **Harbison, R. D. and Becker, B. A.,** Relation of dosage and time of administration of diphenylhydantoin to its teratogenic effect in mice, *Teratology,* 2, 305, 1969.
115. **Mirkin, B. L.,** Diphenylhydantoin: placental transport, fetal localisation, neonatal metabolism and possible teratogenic effect, *J. Pediatr.*, 78, 329, 1971.
116. **Mirkin, B. L.,** Placental transfer and neonatal elimination of diphenylhydantoin, *Am. J. Obstet. Gynecol.*, 109, 930, 1971.
117. **Loughnan, P. M., Gold, H., and Vance, J. C.,** Phenytoin teratogenicity in man, *Lancet,* 1, 70, 1973.
118a. **Poswillo, D. E., Sopher, D., and Mitchell, S.,** Experimental induction of foetal malformation with "blighted potato": a preliminary report, *Nature,* 239, 462, 1972.
118b. **Poswillo, D. E., Sopher, D., Mitchell, S. J., Coxon, D. T., Curtis, R. F., and Price, K. R.,** Further investigations into the teratogenic potential of imperfect potatoes, *Nature,* 244, 367, 1973.
119. **Collins, T. F. X. and Williams, C. H.,** Teratogenic studies with 2,4,5-T and 2,4-D in the hamster, *Bull. Environ. Contam. Toxicol.*, 6, 559, 1971.
120. **Khera, K. S. and Clegg, D. J.,** Perinatal toxicity of pesticides, *Can. Med. Assoc. J.*, 100, 167, 1969.
121. **Bloomfield, D. K.,** Fetal deaths and malformations associated with the use of coumarin derivatives in pregnancy. A critical review, *Am. J. Obstet. Gynecol.*, 107, 883, 1970.
122. **Carswell, F., Kerr, M. H., and Hutchinson, J. H.,** Congenital goitre and hyperthyroidism produced by maternal ingestion of iodides, *Lancet,* 1, 1241, 1970.
123. **Heymann, M. A.,** Inter-relations of foetal circulation and the placental transfer of drugs, *Fed. Proc.*, 31, 44, 1972.
124. **Becker, R. F. and King, J. E.,** Studies on nicotine absorption during pregnancy. II. The effects of acute heavy doses on mother and neonates, *Am. J. Obstet. Gynecol.*, 95, 515, 1966.
125. **Herbst, A. L., Ulfelder, H., and Poskanzer, D. C.,** Adenocarcinoma of the vagina – association of maternal stilboestrol with tumour appearance in young women, *N. Engl. J. Med.*, 284, 878, 1971.
126. **Greenwald, P., Barlow, J. J., Nasca, P. C., and Burnett, W. S.,** Vaginal cancer after maternal treatment with synthetic oestrogens, *N. Engl. J. Med.*, 285, 390, 1971.
127. **Heinonen, O. P.,** Diethylstilboestrol in pregnancy. Frequency of exposure and usage patterns, *Cancer,* 31, 573, 1973.
128. **Kuchera, L. K.,** Postcoital contraception with diethylstilboestrol, *J.A.M.A.*, 218, 562, 1971.
129. **Nomura, T. and Okamoto, E.,** Transplacental carcinogenesis by urethan in mice: teratogenesis and carcinogenesis in relation to organogenesis, *Gann,* 63, 731, 1973.
130. **Older, P. O. and Harris, J. M.,** Placental transfer of tubocurarine, *Br. J. Anaesth.*, 40, 459, 1968.
131. **Asling, J. H., Schnider, S. M., Margolis, A. J., Wilkinson, G. R., and Way, E. L.,** Paracervical block anaesthesia in obstetrics. II. Etiology of fetal bradycardia following paracervical block anaesthesia, *Am. J. Obstet. Gynecol.*, 107, 626, 1970.

132. **Moya, F. and Thorndike, V.**, The effects of drugs used in labour on the foetus and newborn, *Clin. Pharmacol. Ther.*, 4, 628, 1963.
133. **Aiken, J. W.**, Aspirin and indomethacin prolong parturition in rats: evidence that prostaglandins contribute to expulsion of foetus, *Nature*, 240, 21, 1972.
134. **Yeary, R. A., Benish, R. A., and Finkelstein, M.**, Acute toxicity of drugs in newborn animals, *J. Pediatr.*, 69, 663, 1966.
135. **Lischner, H., Seligman, S. J., Krammer, A., and Parmelee, A. H., Jr.**, An outbreak of neonatal deaths among term infants associated with administration of chloramphenicol, *J. Pediatr.*, 59, 21, 1961.
136. **Pines, W. L.**, The hexachlorophene story, *F.D.A. Papers*, 6, 11, 1972.
137. **Lockhart, J. D.**, How toxic is hexachlorophene? *Pediatrics*, 50, 229, 1972.
138. **Kimbrough, R. D.**, Review of recent evidence of toxic effects of hexachlorophene, *Pediatrics*, 51, 391, 1973.
139. **Cammer, W., Moore, C. L., and Korey, S. R.**, The effect of hexachlorophene on the respiration of brain and liver mitochondria, *Biochim. Biophys. Res. Commun.*, 46, 1887, 1972.
140. **Caldwell, R. S., Nakaue, H. S., and Buhler, D. R.**, Biochemical lesion in rat liver mitochondria induced by chlorophene, *Biochem. Pharmacol.*, 21, 2425, 1972.
141. **Brodeur, J. and Dubois, K. P.**, Studies on factors influencing the acute toxicity of malathion and malaoxon in rats, *Can. J. Physiol. Pharmacol.*, 45, 621, 1967.
142. **Gagné, J. and Brodeur, J.**, Metabolic studies on the mechanisms of increased susceptibility of weanling rats to parathion, *Can. J. Pharmacol.*, 50, 902, 1972.
143. **Miller, R. W.**, Cola-coloured babies. Chlorobiphenyl poisoning in Japan, *Teratology*, 4, 211, 1971.
144. **Hardy, H. L., Chamberlin, R. I., Malof, C. C., Boylen, G. W., and Howell, M. C.**, Lead as an environmental poison, *Clin. Pharmacol. Ther.*, 12, 982, 1971.
145. **McClain, R. M. and Becker, B. A.**, Effects of organolead compounds on rat embryonic and fetal development, *Toxicol. Appl. Pharmacol.*, 21, 265, 1972.
146. **Toaff, R., Ashkenazi, H., Schwartz, A., and Hetzberg, M.**, Effects of estrogen and progestogen on the composition of human milk, *J. Reprod. Fertil.*, 19, 477, 1969.
147. **Giacoia, G. P.**, Drugs and breast milk, *Pediat. Clin. North Am.*, 19, 151, 1972.
148. **Eckstein,.H. and Jack, B.**, Breast feeding and anticoagulant feeding, *Lancet*, 1, 672, 1970.
149. **Curtis, E. M.**, Oral contraceptive feminization of a normal male infant, *Obstet. Gynecol.*, 23, 295, 1964.
150. **Knowles, J. A.**, Excretion of drugs in milk – a review, *J. Pediatr.*, 66, 1068, 1965.
151. **Johns, D. G., Rutherford, L. D., Leighton, P. C., and Vogel, C. L.**, Secretion of methotrexate into human milk, *Am. J. Obstet. Gynecol.*, 112, 978, 1972.
152. **Darby, F. J.**, Changes in drug metabolizing activities in the livers of suckling rats as a result of the treatment of lactating mothers with phenobarbitone and chlorpromazine, *Biochem. J.*, 122, 41, 1971.
153. **Curley, A. and Kimbrough, R.**, Chlorinated hydrocarbon insecticides in plasma and milk of pregnant and lactating women, *Arch. Environ. Health*, 18, 158, 1969.
154. **Hagyard, S. B., Brown, W. H., Stull, J. W., Whiting, F. M., and Kemberling, S. R.**, DDT and DDE content of human milk in Arizona, *Bull. Environ. Contam. Toxicol.*, 9, 169, 1973.
155. **Hargreaves, T. and Piper, R. F.**, Effect of inhibitory breast milk and $3\alpha,20\beta$-pregnanediol on glucuronyl transferase, *Arch. Dis. Child.*, 46, 195, 1971.

BIOCHEMICAL CHANGES IN CARCINOMA

R. W. Morgan

TABLE OF CONTENTS

I. BIOCHEMICAL CHANGES IN CARCINOMA

Carcinoma may create several biochemical changes in the body, either through increased use of metabolites, interference with normal functioning, or, in the case of functioning tumors, the production of unusual or excessive amounts of biochemical agents, including hormones. In addition to the effects directly related to the primary tumor, secondary tumors may create even more bizarre problems by their local actions, especially if they invade other endocrine organs. It is highly likely that some changes in the biochemical milieu do not result from, but may precede or even perhaps initiate, malignant transformation of previously normal cells.

In studying biochemical parameters of malignancy, we must ask ourselves if abnormal values are due to

1. Changes predating and likely contributing to carcinogenesis
2. Production of metabolites or hormones by the tumor
3. Interference with normal production due to organ damage
4. Excess metabolism or even binding of metabolites, enzymes, and hormones
5. Combinations of the above.

The numerous changes occurring after the onset of malignancy make it difficult to infer much about the premalignant biochemical or endocrine milieu, especially since not all tumors of a single organ exhibit the same behavior, nor do they likely all arise through the same pathway.

Some assistance in sorting the *causes* from the *effects* may come from the application of epidemiologic methods to questions of biochemical concern. The epidemiologist studies the distribution and determinants of disease *in populations.* The prime interest is in exposing the causal web of a disease; that web may include noxious agents (i.e., irradiation), infections, genetic factors, behavior, etc. The basic epidemiologic model of disease production is shown in Figure 1. Each of the three interacting corners usually represents a multitude of variables.

In many instances, the epidemiologist can identify, before malignancy, a population at high risk of later cancer. By turning our attention to the biochemical examination of that population and comparing their characteristics with either a low-risk or normal-risk group, one may make inferences about the biochemical determinants of the disease under study.

The following diseases are examples of female cancers in which endocrine/biochemical factors appear to be important and where epidemiologic methods are playing a considerable role in elucidating the disease etiology.

II. CARCINOMA OF THE BREAST

Carcinoma of the breast is the most common cause of female cancer mortality in Western, predominantly white countries. Annually, it is

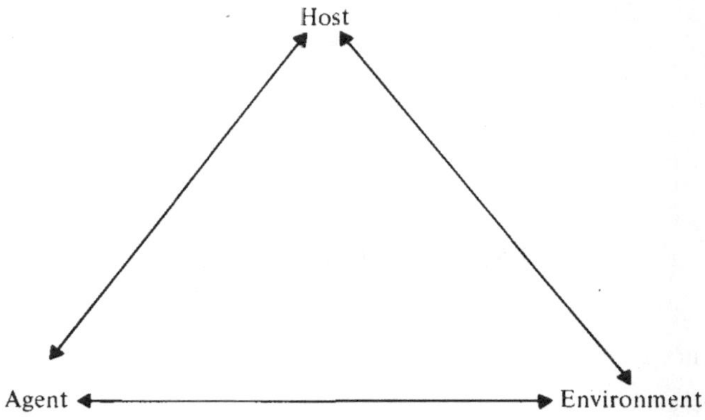

FIGURE 1. Disease: an epidemiologic model.

responsible for about 12,000 deaths in Great Britain and 28,000 deaths per year in the U.S. There has been little, if any, change in incidence or survival in most areas in the past 50 years. It is one of the most extensively studied human tumors. References 1 to 3 are recent comprehensive reviews of the topic.

Certain epidemiologic observations about carcinoma of the breast date back to Ragoni Stern in the 17th century. He noted the high frequency of carcinoma of the breast in nuns and its relative scarcity among the parous women of Verona. This observation has been substantiated a number of times since, and has, in fact, served as the basis for endocrine hypotheses currently being examined in many centers.

Geographic variability in breast cancer rates has been known for a long time, probably over 100 years. The reason for this extreme variability has never been clear; however, since high cancer rates tended to coincide with industrial and cultural development, at least in the Western physicians' views, there was considerable speculation about why the more "primitive" societies seem to have less of this cancer. One of the explanations was that they were more likely to breast-feed their children. On this slender evidence, generations of medical students were told (and unfortunately are still told in some centers) that breast cancer would be diminished in incidence (if not abolished) if women would only use their breasts for the purpose that nature intended. Unfortunately, different cultures and races have many differences other than breast-feeding. The breast-feeding explanation for geographic differences fails to consider that the decline of breast-feeding in Western civilizations is comparatively recent. The old doctors' tale was finally put to rest by MacMahon and his colleagues[23-25] who, in an international collaborative study, interviewed breast cancer patients and controls from the same hospitals. They found that within areas of low, medium and high rates of breast cancer, there was almost no difference in breast-feeding habits of cases and controls. The study did, however, indicate a strong association of breast cancer with late age of first pregnancy, as well as with the nulliparity previously noted by many authors since the time of Stern. The apparent protective effect of early pregnancy has thus led investigators to more closely examine the marked hormonal changes coincident with pregnancy.

The association of breast cancer with a late age of first pregnancy gave a new interest to ovarian function as a possible risk factor in breast cancer. Earlier epidemiologic studies demonstrated that surgically castrated women had, in later years, a better experience than had a similar group of noncastrated females. These observations may have been responsible, in part, for the surgical vogue of oophorectomy for premenopausal patients with breast cancer. The lack of success for oophorectomy in these patients may indicate that the estrogen production is not responsible for tumor growth as much as tumor initiation or maintenance.

In 1966, Lemon et al. hypothesized that the protective effects of pregnancy could be due to an increase of estriol production which exceeded the increase in estradiol and estrone during pregnancy. They suggested that since estriol had never been shown to be carcinogenic but was suspected of being protective, that this could in fact be the key to the pregnancy question.[4] In 1970, Cole and MacMahon extended the Lemon hypothesis to speculate that if the relative levels of estriol, estrone, and estradiol determined the risk of breast cancer, then there should be major geographic variation in the distribution of these hormones in populations.[5] Their method of testing the hypothesis was relatively simple: Groups of normal women of various ages were asked to donate urine specimens in both luteal and follicular phases. Areas of Japan, Taiwan, Hong Kong, Canada, and the U.S. participated, with volunteers ranging in ages from 15 to 40. Excluded from the study were any women taking oral contraceptives, other steroids, etc. The pilot study for that project demonstrated significant differences in the ratio of estriol/estrone + estradiol (Table 1).[6] The larger study now being completed by that group is confirming most of the original findings. The marked geographic variation in estrogen profiles in apparently normal women reinforces the idea that "normal values" for any biochemical determination must be specific for age, sex, and race. In the international study quoted above, variability in methods of specimen collection and analysis cannot account for the geographic differences since all collection methods were standardized and determinations done in one central laboratory.

The assay of urine for estrogen compounds in nonpregnant women presents certain problems. For a start, the laboratory method of Brown et al.[7] is technically demanding and performed in few centers. Although serum methods for fractionated estrogen determination have been reported,[8] they are not currently available nor has there been any great stimulus to develop such assays. Serum determinations would have a number of advantages: Measurement is not dependent on kidney/-bladder function and patient cooperation and serum assay would likely include, besides the traditional "big three" (estrone, estradiol, estriol), other fractions, especially labile.

The marked geographic differences in hormone profiles invariably lead to the question: "Why?" Are there genetic factors at work? Diet? Other environmental influences? Currently, one group of investigators is examining the estrogen profiles of teenaged daughters from high-risk families and comparing their estrogen profiles to other teenagers of the same age within the same population. A difference between these two groups of young women could strengthen the hypothesis that an endocrine profile is inherited rather than acquired. The same group is participating in a large scale dietary study in which cases and controls will be interviewed concerning past dietary history; current dietary habits will be assessed with special attention paid to caloric and saturated fat intake. The interest in fat intake is derived from earlier observations that geographic distribution of breast cancer resembles coronary artery disease. Hill et al.[9] have suggested that dietary fat could be converted to estrogen-like compounds by the action of bile components and gut bacteria. Such a hypothesis is attractive inasmuch as it links diet, geography, and endocrine factors for the disease.

Also following on the interest in estrogens, some groups are now doing studies of estrogen profiles in cases of breast cancer and suitable controls. Although such studies are certainly needed, they all have the same difficulty: Does the breast tumor produce or metabolize substances

TABLE 1

Concentrations (μg/liter) of Urinary Estrogens in Asian and North American Women

	Follicular phase		Luteal phase	
Estrogen or quotient	Asian	North American	Asian	North American
Estrone (El)	5.3	7.0	7.7	13.0
Estradiol (E2)	2.5	3.4	3.6	7.0
Estriol (E3)	8.2	8.3	15.1	14.6
E3 / (El + E2)	1.27	.83	1.64	.89
E3 / (E1 + E2 + E3)	.53	.41	.56	.44

(From MacMahon, B. et al., *Lancet,* 2, 900, 1971. With permission.)

which might interfere with hormone levels or their measurement? Also, endocrine profiles late in life may not accurately reflect earlier levels which might be so important to tumor induction.

There is evidence that some breast carcinomas are capable of binding estrogens.[10] This finding has a number of implications concerning both etiology and treatment. Until we can quantitate the effect of the tumor upon levels of circulating estrogens, the determination of urinary or even serum estrogen fractions may be a grossly inaccurate estimation of either total body or mammary gland estrogen levels. Such uncertainty makes etiologic inferences impossible. Further work on estrogen-binding may have therapeutic implication, however, in determining best candidates for either endocrine or ablative therapy. Currently, selection of such patients is usually by age or menopausal status.

There is a large body of knowledge concerning the relationship of hormones to mammary cancer induction in lower animals, especially mice and rats. The rat is a model more appropriate to human considerations since the rat, like the human, is protected by pregnancy against later development of breast cancer. The mouse, on the other hand, has an increased risk following pregnancy. These risks are likely related to the differences in changes of estriol/estradiol/estrone between species during pregnancy.

Factors studied in animal populations include estrogen binding, competition with testosterone for binding sites,[10] and effect of estrogen deprivation/treatment in animals given 7,12-dimethylbenz(a)-anthracene (DMBA).[11,12] Testosterone does not appear to compete for estrogen-binding molecules. Estrogen does appear necessary for tumor initiation, but perhaps not for maintenance or malignant transformation. Local application of diethylstilbestrol (DES) potentiates DMBA carcinogenesis, indicating that the estrogen acts directly, rather than by stimulating prolactin secretion, as has been suggested.

Naturally, one must always be concerned about the applicability of animal results to problems in humans. This would seem especially true where mammary tumor carcinogenesis is concerned: We know that the human differs greatly in endocrine physiology; we are also, as yet, not aware of any chemical carcinogen that acts, in humans, in a manner similar to DMBA.

The estrogen-binding capacity of some malignant tissue does not appear to be shared by fibrocystic disease of the breast,[13] a frequent precursor of breast cancer and thought to be a risk factor. Not only does fibrocystic disease sometimes precede malignancy, pathologists' tumor reports frequently note the coexistence of benign and malignant disease. Unfortunately, the prevalence of fibrocystic change in the general population is not known, since only symptomatic cases are likely to be biopsied. Although tissues from the two diseases differ in estrogen-binding capacity, both have higher glucose-6-phosphate dehydrogenase (G6PD) and lower α-glycerol phosphate dehydrogenase (α-GPD) activities than the normal breast.[13] The other cytoplasmic enzymes for which activities differ in normal and malignant breast tissue include isocitrate dehydrogenase (decarboxylating), glucose phosphate isomerase, phosphoglucomutase, glutamate dehydrogenase, aspartate aminotransferase, pyruvate kinase, hexokinase, and lactate dehydrogenase. For etiologic consideration, the relationship of G6PD and αGPD may be most important since perhaps the enzyme abnormalities may precede malignant transformation.

Some investigators have chosen to study androgens rather than estrogens. Bulbrook and his colleagues have reported that androsterone and etiocholanolone excretion is lower in women who develop breast cancer than women who do not.[14] This excretion appears to be genetically determined and the suggestion is that androgen abnormalities in early life are important and bear certain similarities to hypotheses centered on estrogen. Most of the data concerning androgen related to the development of breast cancer come from the study on the Isle of Guernsey. This is, in epidemiologic parlance, a "freezer study." Large groups of apparently healthy women have their urine collected and stored in a frozen state. When a woman in the study develops breast cancer, her urine is unfrozen and the appropriate metabolite measured. At the same time, one or more unaffected controls suitably matched for age can be withdrawn from the freezer for similar analysis. Although androgen studies are interesting, it is difficult to reconcile the estrogen and androgen hypothesis unless one considers that perhaps estrogens have a role for tumor induction and the androgens then maintain or stimulate tumor growth.

Patients with breast cancer are reported to have

a much greater excretion of corticosteroid sulfates and cortisol than do controls.[15] However, since the study was retrospective in nature, one cannot decide whether the corticosteroid increase reflects on a metabolic disturbance predating the tumor, steroid synthesis by the tumor, or host response to malignancy. Since tumor cells are reported to be capable, in vitro, of steroid synthesis,[16] this appears to be the most likely explanation.

In 1937, Bittner reported apparent transmission of carcinoma in mouse milk. The agent in the milk was later shown to be mouse mammary tumor virus and since that time, people have speculated as to whether or not a similar virus might be responsible for human breast cancer. The examination of the problem is hampered by the long time period between infancy and later development of breast cancer. This effectively prohibits a single investigator from following a group of women from infancy. Likewise, most women do not have accurate information as to whether or not they were breast-fed in infancy. Attempts to search for evidence of virus infection within the breast are technically difficult; even if no virus is found, this does not rule out the possibility that the tumor was originally induced by a virus. Virologists have concentrated on examining milk from high-risk and low-risk mothers; although initial reports were encouraging, they have not been substantiated. Recently, the problem has been tackled with an epidemiologic approach.[17] A large group of elderly women were interviewed about their child-bearing and child-feeding practices. The breast disease histories of the respondents and of their daughters were ascertained at the same interview. The data provided two cohorts of women, one breast-fed and the other never breast-fed. The experience in developing both benign and malignant breast disease was identical for the two groups, thus lending no support to any hypothesis concerning viral transmission via milk. Naturally, other modes of viral transmission for carcinogenesis would still be a possibility, especially in utero or later in life. An important issue in possible viral etiology of breast cancer is the interplay of biochemical factors, especially at the intracellular level with oncogenic viruses. We can only speculate on the sequence of events that relates an early age of menarche to appearance of a tumor 30 or more years later; we are certainly attracted to the concept of endocrine events creating a biochemical milieu sensitive to later oncogenic insult by irradiation, virus, or whatever.

Can a genetically controlled biochemical abnormality explain the familial nature of breast cancer? We know that women with breast cancer (especially premenopausal) pass to their daughters an appreciably increased risk of breast cancer. They may also pass on a greater likelihood of benign breast disease. Similarly, women with benign breast disease produce daughters with an excess of malignant disease. For this reason, the association of abnormal levels of G6PD and α-GPD in both benign and malignant disease may explain, in part, the apparent familial association of the two diseases. Naturally, a familial occurrence does not mean a genetic cause; the common environmental experiences (including diet) of both mother and daughter might be contributing factors.

The effect of radiation of production of breast carcinoma has also been studied by epidemiologic methods.[18] To test the hypothesis that excessive exposure to radiation might increase risk, a group of women who had previously had a therapeutic pneumothorax for tuberculosis was identified. Such patients would normally have rather high radiation exposure because of the multiple fluoroscopies involved in ascertaining the state of the pneumothorax. The former tuberculosis patients had a considerably higher breast cancer rate than would be expected from population statistics and cancers were almost invariably on the side of the previous pneumothorax where the radiation exposure was most intense. Thus, we can infer that high dosages of diagnostic irradiation increase risk of mammary cancer.

Carcinoma of the female breast is a disease of the educated and affluent. Although it occurs in all social classes, it has a marked proclivity for the upper socioeconomic strata. This may reflect a combination of known factors, including both genetics and a later age at first pregnancy in the upper classes. Or it may represent unexplored variables such as diet, environmental influences, etc. Biochemists have done little to explore the possible differences in biochemical make-up of persons of difference social class, except where problems of nutrition are concerned. Yet there is ample evidence that social class is a profound influence on development of many diseases, some of which must be mediated through biochemical mechanisms. There would also appear to be a potential biochemical contribution in explaining the epidemiologic association of breast and uterine corpus malignancies. Both diseases have similar geographic distributions and both organs are exquisitely estrogen-sensitive. Do they share similar mechanisms of malignant change?

Male and female breast cancer have certain epidemiologic similarities, including geographic and racial distribution. There seems to be a hormonal etiology for at least a portion of male disease since a number of cases have been reported in men undergoing estrogen treatment for various reasons, including sex change and hormonal therapy of prostatic cancer.[19,20] Likewise, a number of cases have occurred in males with chromosomal abnormalities, such as Klinefelter's syndrome,[21] associated with endocrine abnormality. Carcinoma of the male breast has received relatively little study, probably because of its relative rarity. Also, most researchers see little applicability to the female of research findings in the male breast, given the difference in mass, function, and hormonal milieu.

In discussing breast cancer etiology, the author has frequently referred to various agents or experiences as risk factors. A risk factor is not necessarily a cause; however we can use risk factors in an effort to control occurrence, morbidity, and mortality from breast cancer.

To prevent occurrence, we will naturally wish to limit diagnostic and therapeutic irradiation. We cannot advise a woman to select her parents carefully, nor are we likely to convince her to have a child early in her reproductive life, rather than later, when she is more likely to be educated, financially stable, and perhaps married.

We should use the risk factors to select women for participation in screening programs, whether those programs are based on thermography, mammography, clinical examination, or newer techniques yet to be developed. Breast cancer appears to be one disease where early diagnosis may improve survival; since screening programs are never effectively applied to whole populations, selection on the basis of risk factors is the logical solution. This principle could be applied to public health programs or clinical practice.

The same principles useful for public health screening programs might be applied to answer biochemical questions. The use of known risk factors to identify high- and low-risk populations would allow the biochemist to compare hormonal profiles to generate and test hypotheses concerning the biochemical precursors of malignant change.

III. ADENOCARCINOMA OF THE CERVIX AND VAGINA

Epidemiologic studies frequently begin with or arise out of the observations of astute clinicians. In 1970, a gynecologist noted an apparent increase in the number of teen-aged patients with adenocarcinoma of the cervix and/or vagina.[22] Normally, this disease is rare; fortunately, the clinical observations stimulated further inquiry. The investigators ascertained that there had been an increase in the actual number of cases of the conditions as compared with the experience from previous years. They were dealing with an epidemic – an occurrence of disease in a frequency greater than that usually expected.

An important item present in the medical history of many patients was the observation that they were born of a "high-risk" pregnancy. This is not a piece of information usually obtained in a medical history for patients this age; fortunately, history-taking in these instances was of above-average quality. After identifying further cases from other sources, cases and a group of similar controls were studied with special regard to their prenatal histories, including maternal drug treatments and radiation exposure, both of which might be expected to be more frequent in high-risk pregnancies. In investigating the patients and the controls, investigators were careful to search for prenatal factors as dilligently in the controls as in the cases. Of a number of possible risk factors, only the taking of stilbestrol by the mother appeared to be a major risk factor (Table 2). Later,

TABLE 2

Maternal History of Cases of Vaginal Adenocarcinoma and Matched Controls

	Case	Control
Estrogen given in this pregnancy	7	0
No estrogen in this pregnancy	1	32

$p < 0.00001$

From Herbst, A. L. et al., *N. Engl. J. Med.*, 284, 878, 1971. With permission.)

similar studies by other investigators confirmed this association. Although the cancers caused by stilbestrol have been apparently few in number compared to the thousands of women who received the drug, presumptive identification of stilbestrol as a potential carcinogen has greatly changed its use, especially as an abortificient or so-called "morning-after pill."

Since the publication connecting stilbestrol to cervical and vaginal carcinoma in young women, epidemiologists have attempted (so far unsuccessfully) to link stilbestrol intake with other tumors of the urogenital tract or breasts. The action of the hormone appears to be tissue-specific in persons exposed in utero. Is stilbestrol oncogenic for more mature tissues? To date there has been no publication concerning the role of stilbestrol in the production of adenocarcinoma of the cervix in later life. Even in later life, adenocarcinoma is uncommon; squamous carcinoma is relatively frequent, and the epidemiology well known. Squamous cell cervical carcinoma is easily distinguishable, readily diagnosed, and most common among the lower socioeconomic strata, the promiscuous, and the multiparous. Most of these features are the opposite of the cases of adenocarcinoma reported in the stilbestrol study.

It still remains for basic scientists to elucidate the exact mechanism by which the stilbestrol acted. Fortunately, however, epidemiologic evidence is now accepted enough so that important action has been taken to reduce medical and agricultural use of stilbestrol and its derivatives while further evidence is collected. This is an interesting departure from some previous instances (such as the relationship of smoking to lung cancer) where epidemiologic evidence was not well accepted and where little or no public action took place on the basis of early reports, even though the frequency of lung cancer is far higher than that of adenocarcinoma of the female genital tract.

Unfortunately, there are still some parts of the stilbestrol-carcinoma problem which remain unsolved. There is no simple and accurate way of ascertaining the number of women who received stilbestrol during gestation. This means there is no satisfactory way to estimate the cancer rate among the female fetuses exposed to the drug. In many ways, we are fortunate to have detected this cancer-drug relationship, since the cancer is rare and the exposure to the drug occurs many years prior to appearance of the tumor. It is predictable that this type of study will be repeated for other drugs and other tumors now that clinicians have a higher index of suspicion concerning the long-term safety of such agents. Certainly there are many questions yet to be answered concerning the eventual side effects of agents as potent and common as oral contraceptives. Many epidemiologists are now beginning to watch what amounts to the largest uncontrolled clinical trial of any hormonal agent ever introduced for therapeutic use. If, for instance, the critical ages for female breast tumor induction are 15 to 20, with clinical appearance far later, we would not yet expect any increase in breast carcinoma to be detectable. The women started on oral contraceptives at an early age have not yet entered the cancer age group. Time will tell much.

Part of the problem in assessing the safety of the oral contraceptive is pharmacological. We do not know the eventual fate of all of the estrogen and progesterone derivatives in vivo. And even assuming we were able to measure the serum levels of all of the derivatives of the compounds ingested, we do not yet know their carcinogenic potential in humans over time.

A better knowledge of steroid metabolism has implications beyond contraception or cancer etiology. There have been serious suggestions that teenage daughters with a "bad" estrogen profile and a positive family history should be treated to modify their hormonal status. Before such radical action is initiated, we need to know much more about estrogen metabolism and its long-term effects, whether harmful or beneficial. In view of the extremely long (up to 20 years) latent period between "insult" and malignancy, as in the vaginal adenocarcinoma, any new hormonal treatments should be approached with great caution.

REFERENCES

1. **MacMahon, B., Cole, P., and Brown, J.**, Etiology of human breast cancer: a review, *J. Natl. Cancer Inst.*, 50, 21, 1973.
2. **Vakil, D. V. and Morgan, R. W.**, Etiology of breast cancer – genetic aspects, *Can. Med. Assoc. J.*, 109, 29, 1973.
3. **Vakil, D. V. and Morgan, R. W.**, Etiology of breast cancer – epidemiologic aspects, *Can. Med. Assoc. J.*, 109, 201, 1973.
4. **Lemon, H. M., Wotiz, H. H., Parsons, L., et al.**, Reduced estriol excretion in patients with breast cancer prior to endocrine therapy, *J.A.M.A.*, 196, 1128, 1966.
5. **Cole, P. and MacMahon, B.**, Estrogen fractions during early reproductive life in the etiology of breast cancer, *Lancet*, 1, 604, 1969.
6. **MacMahon, B., Cole, P., Brown, J. B., et al.**, Estrogen profiles of Asian and North American women, *Lancet*, 2, 900, 1971.
7. **Brown, J. B., Bulbrook, R. D., and Greenwood, F. C.**, An additional purification step for a method for estimating estriol, estrone and estradiol-17β in human urine, *J. Endocrinol.*, 16, 49, 1957.
8. **Wu, C–H., Lundy, L. E., and Lee, S. G.**, A rapid radiommunoassay for plasma estrogen, *Am. J. Obstet. Gynecol.*, 115, 169, 1973.
9. **Hill, M. J., Goddard, P., and Williams, R. E. O.**, Gut bacteria and etiology of cancer of the breast, *Lancet*, 2, 1213, 1972.
10. **Korenman, S. G.**, Human breast cancer as a steroid hormone target, in *Estrogen Target Tissues and Neoplasia*, Dao, T. L., Ed., University of Chicago Press, Chicago, 1972.
11. **Beuving, L. J. and Bern, H. A.**, Hormonal influence upon normal, preneoplastic, and neoplastic mammary gland, in *Estrogen Target Tissues and Neoplasia*, Dao, T. L., Ed., University of Chicago Press, Chicago, 1972.
12. **Sinha, D. and Dao, T. L.**, Estrogen and induction of mammary cancer, in *Estrogen Target Tissues and Neoplasia*, Dao, T. L., Ed., University of Chicago Press, Chicago, 1972.
13. **Hilf, R., Wittliff, J. L., Rector, W. D., et al.**, Studies on certain cytoplasmic enzymes and specific estrogen receptors in human breast cancer and in nonmalignant disease of the breast, *Cancer Res.*, 33, 2054, 1973.
14. **Bulbrook, R. D., Hayward, J. L., and Spicer, C. C.**, Relation between urinary androgen and corticoid excretion and subsequent breast cancer, *Lancet*, 2, 395, 1971.
15. **Gosh, P. C., Lockwood, E., and Pennington, G. W.**, Abnormal excretion of corticosteroid sulfates in patients with breast cancer, *Br. Med. J.*, 1, 328, 1973.
16. **Adams, J. B. and Wong, M. S. F.**, A correlation between urinary steroid metabolites and pathways of steroidogenesis in human breast-tumor tissue, *Lancet*, 2, 1163, 1968.
17. **Morgan, R. W., Vakil, D. V., and Chipman, M. L.**, Breast feeding, family history, and breast disease, *Am. J. Epidemiol.*, in press.
18. **Myrden, J. A. and Hiltz, J. E.**, Breast cancer following multiple fluoroscopies during artificial pneumothorax treatment of pulmonary tuberculosis, *Can. Med. Assoc. J.*, 100, 1032, 1969.
19. **McClure, J. A. and Higgins, C. C.**, Bilateral carcinoma of male breast after estrogen therapy, *J.A.M.A.*, 146, 7, 1951.
20. **Symmers, W. S. C.**, Carcinoma of breast in trans-sexual individuals after surgical and hormonal interference with the primary and secondary sex characteristics, *Br. Med. J.*, 2, 83, 1968.
21. **Jackson, Q. W., Muldal, S., Ockbey, C. H., et al.**, Carcinoma of the male breast in association with the Klinefelter syndrome, *Br. Med. J.*, 1, 223, 1965.
22. **Herbst, A. L., Ulfelder, H., and Poskanzer, D. C.**, Adenocarcinoma of the vagina. Association of maternal stilbestrol therapy with tumor appearance in young women, *N. Engl. J. Med.*, 284, 878, 1971.
23. **Lin, T. M., Chen, K. P., and MacMahon, B.**, Epidemiologic characteristics of cancer of the breast in Taiwan, *Cancer*, 27, 1497, 1971.
24. **Valaoras, V. G., MacMahon, B., Trichopoulos, D., et al.**, Lactation and reproductive histories of breast cancer patients in Greater Athens, 1965–67, *Int. J. Cancer*, 4, 350, 1969.
25. **Ravnihar, B., MacMahon, B., and Lindtner, J.**, Epidemiologic features of breast cancer in Slovenia 1965–1967, *Eur. J. Cancer*, 7, 295, 1971.

CALCIUM METABOLISM AND THE MENOPAUSE

J. C. Gallagher and B. E. C. Nordin

TABLE OF CONTENTS

I. INTRODUCTION

Although it has been known for well over a century that fractures are more common in old people than in the young,[1] it is only in recent years that fracture epidemiology has received the attention it deserves and that the pathological changes in bone underlying the progressive fragility which accompanies aging have begun to be understood.

A milestone in this development of scientific thought was the definition by Pommer in 1885[2] of the histological difference between rickets and osteomalacia on the one hand and of osteoporosis on the other. According to Pommer, whose views have been largely accepted ever since, rickets and osteomalacia represented a failure to mineralize new bone, whereas osteoporosis represented a destruction of existing bone. Contrary to earlier views, Pommer considered that calcium could not be withdrawn from the skeleton without a simultaneous resorption of the collagenous matrix in which it is embedded. In his opinion, and this remains largely if not entirely true today, there was no such thing as halisteresis, that is, the removal of mineral alone from the bony tissue. A reduction in the ash content of bone must, therefore, be due to a delay in the mineralization process; removal of mineral always involves removal of matrix and it does not, therefore, affect the percentage of ash in the bony tissue as such.

Although the difference between osteomalacia and osteoporosis was placed on a sound pathological basis nearly 100 years ago, the corresponding clinical definitions were established very much later. The biochemistry of rickets was not defined until the early 1920's[3] and as late as 1939, Meulengracht[4] was still describing spinal osteoporosis with vertebral crush fractures as "osteomalacia of the spinal column," although in fact vertebral compression is relatively rare in osteomalacia.

In 1941, however, Albright et al.[5] defined a clinical entity which is now recognized as the prototype of clinical osteoporosis. In that year, they described 40 postmenopausal women with the syndrome of vertebral biconcavity and crush fractures which they attributed, correctly, to a reduced quantity of skeletal tissue. With remarkable insight, they emphasized the striking association between this condition and the postmenopausal state, noted that no fewer than ten of the cases had undergone an artificial menopause, and attributed the syndrome of spinal osteoporosis to estrogen deficiency. They insisted that the effect of estrogen deficiency on bone must operate on the bone forming rather than the bone resorbing side of the bone turnover cycle; this belief was based on animal experiments, which will now be reviewed.

In 1926 Riddle and Rheinhart[6] showed that the blood calcium of doves and pigeons rose during ovulation; this was subsequently confirmed in hens.[7] Later, other workers described ossification in the bone marrow of female pigeons which was not seen in the male species.[8] However, the important observation was that this ossification of the female marrow was cyclic and varied directly with the size of the developing ovarian follicles. An even more direct relationship between estrogenic activity and ossification of the long bones of pigeons was shown in further experiments by Pfeiffer and Gardner[9] in which injections of estrogens into male pigeons caused the appearance of numerous osteoblasts and ossification of the marrow as well as a rise in serum calcium, which is now known to be purely a rise in the protein-bound fraction.[10] The ossifying effect of estrogens on the bone marrow of mice was shown to be similar to that in pigeons, and was also shown to be inhibited by testosterone.[11] From these experiments in avians and mammals, Albright et al.[5] suggested that osteoporosis in the postmenopausal women arose from estrogen deficiency, as a result of which the osteoblasts were deficient and too little bone was laid down. However, the work of Lindquist et al.[12] showed some years later that in rats, at least, the effect of estrogens was to inhibit bone resorption rather than stimulate bone formation.

Further evidence of a link between estrogen deficiency and bone was produced in 1942 when Albright et al.[13] described 11 cases of women of short stature who had osteoporosis and who all had ovarian agenesis. This again was taken to signify that estrogen deficiency reduced the rate of new bone formation.

This was perhaps unfortunate. Nurtured in the belief that vitamin D deficiency produced rickets as a result of calcium malabsorption and that calcium deficiency in general must, therefore, have the same effect, Albright and his colleagues turned their attention from the mineral component of the skeleton to its collagenous framework and sought in this area an explanation of the link between the gonadal hormones and the skeleton. Arguing with perfect logic that a reduced quantity of bone must be a consequence of decreased bone formation or increased bone destruction, they opted for the former, influenced it seems by another brilliant speculation, namely, that the 'S' hormone of the adrenal cortex (subsequently named cortisone) interfered with protein synthesis and that postmenopausal osteoporosis resulted from the same basic fault as that seen in Cushing's syndrome.[14,15]

Many years were to elapse before actual measurements of bone turnover became possible and cast considerable doubt on certain aspects of this hypothesis.

II. AGING AND BONE

In recent years attention has turned from the rather limited problem of vertebral crush fractures in postmenopausal women to a wider study of the effects of age on bone in general and is now turning to the specific role of the menopause in the light of the wider experience which has been gained. That bone loss is an almost universal accompaniment of aging is now generally accepted and many different techniques have been employed to study this process and to understand it in greater depth.

One of the first of these studies was that of Trotter et al. in 1960.[16] They measured the weight/volume ratio (density) of a variety of bones from young and old, male and female, and white and Negro skeletons and showed that about 15% of skeletal mass was lost during aging. One of their most striking observations was that the bones of the Negro skeleton are denser than the bones of a white skeleton and that the bones of the male are denser than the those of the female skeleton. Consequently, the white female skeleton has the lowest bone density in early adult life and reaches the lowest bone density in later life.

More refined procedures have amply confirmed these early observations and increasingly incriminate the menopause as a causal factor in the accelerated bone loss of the female skeleton. These can be grouped into measurements of cortical and trabecular bone.

A. Cortical Bone

In 1960, Barnett and Nordin[17] measured the cortical (C) and total width (T) of the second metacarpal and showed that the ratio C/T fell with age in both sexes. In 1963 Meema[18] measured the thickness of the radial cortex and showed that it too fell with age. In men the loss of bone started at about the age of 60 years, while the loss in women started at about the age of 45 years; he suggested that the earlier loss in women was related to the menopause. In 1966, Nordin et al.[19] confirmed that the loss of metacarpal cortex starting at about the age of 45 years in women was related to the menopause by dividing his normal women into pre- and postmenopausal groups and also showed that loss of femoral cortex in women was only slight and generally began at about the age of 60 years. This fall in metacarpal cortex starting at about the age of 45 to 50 years in women, and between 55 and 65 years in men, has now been demonstrated in many studies, the large series by Garn[20] and Smith and Rizek[21] being particularly important. More recently the procedure of measuring metacarpal cortical thickness has been modified to express the cortical area relative to the total area (CA/TA) where it is assumed that the metacarpal is cylindrical, and thus it can be derived from the Barnett/Nordin index as follows:

$$CA/TA = 1 - (1 - \frac{C}{T})^2$$

where C is the cortical thickness and T is the total thickness. Figure 1 shows that CA/TA falls with age, starting at about 40 to 50 years in women and 10 to 15 years later in men.

B. Trabecular Bone

This is the main constituent of the vertebra but is also present and has been measured in the phalanges, radius, calcaneus, and iliac crest. Different groups of workers have come to rather different conclusions about the age at which trabecular bone loss starts. In post-mortem studies, Lindahl and Lindgren,[22] Baylink et al.,[23] and Arnold[24] all suggested that trabecular bone loss started at about the age of 20, but their data are equally compatible with the view that the loss actually starts at about the age of 40 years. The more recent data of Arnold[25] also show that loss of trabecular bone from the vertebrae in women begins at about the age of 40; other workers have come to the same conclusion.[26] In 1966 and 1968, Nordin et al.,[27] using an in vivo method of lumbar spine densitometry, showed that spinal density fell after the menopause and occurred more rapidly in women after an artificial menopause.

C. The Role of the Menopause

Despite all this evidence, some confusion still remains as to whether loss of bone in women starts at the menopause and is related to declining ovarian function or starts somewhat earlier and is to be regarded as a manifestation of "aging." One of the reasons for this confusion is that the majority of workers, using either histological or radiological techniques, have not clearly separated their cases into pre- and postmenopausal subjects, but have simply related their observations to the ages of the individuals concerned. While it is true that the average age of the menopause (as defined by the cessation of menstruation) is about 48 years, there is a considerable spread around this mean. A substantial number of women reach the menopause in their early 40's and it is not excessively rare for it to occur in the fourth decade. Thus, any unselected group of subjects in which a bone parameter is plotted against age without specific reference to menopausal status is liable to include in the younger age groups a

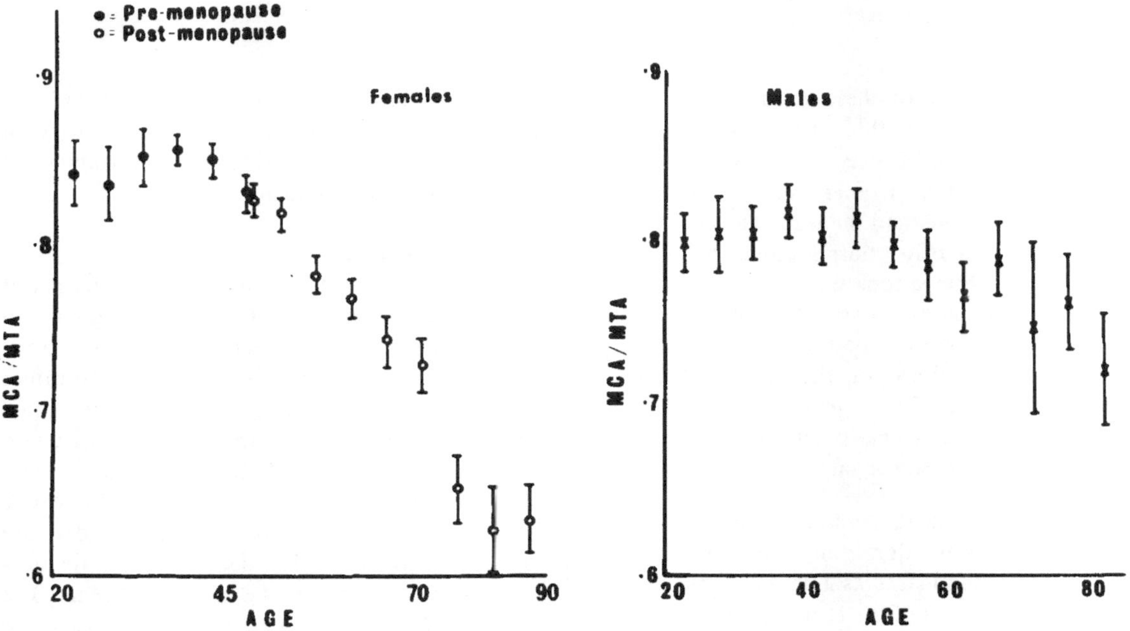

FIGURE 1. The fall with age in metacarpal cortical area/total area (MCA/MTA) measured on the second metacarpal is shown for men (right) and for pre- and postmenopausal women (left).

number of postmenopausal women whose increasing contribution to the total number of cases lowers the mean value of the group. Another reason is probably that the cessation of menstruation is only the final and visible manifestation of declining ovarian function which in fact frequently commences some years earlier.[28] It is common or even usual for menstruation to be irregular before it actually ceases and not uncommon for the typical menopausal flushes to commence during this period. Nonetheless, if the subjects of any study are clearly separated into pre-and postmenopausal groups, it becomes reasonably clear that the first loss of bone in women is related to the menopause on whatever bone the measurements are made.[19]

In order to test this point further, we have applied statistical analyses to the relative contributions of age and years since menopause to the loss of bone observed in a cross-sectional study of 82 normal postmenopausal and 61 oophorectomized women. The result of this analysis is shown in Table 1. It indicates that whereas age is the significant determinant of bone loss in normal postmenopausal women, years since menopause is the significant determinant following an artificial menopause.[29] The probable reason for this discrepancy has partly been explained above. It is that whereas the date of an artificial menopause is exact, the date of a natural menopause is not. Not only is it difficult for a woman many years after the event to remember the date of the last menstrual period, but ovarian activity declines for some years before the cessation of menstruation. This explains why the effect of oophorectomy on bone is most obvious in patients oophorectomized at an early age. Thus, Aitkin et al.[30] have shown that oophorectomy before the age of 45 years is followed by significant bone loss within 3 to 6 years after the operation, whereas if oophorectomy is performed after the age of 45, the loss of bone is no different from that after a natural menopause.

D. Longitudinal X-ray Studies

Sequential observations on the same individuals bear out the information derived from the cross-sectional studies reviewed above. The present authors' technique has been described elsewhere.[31] Briefly, it involves duplicate measurements with needle calipers of metacarpal total and medullary width on the second, third, and fourth metacarpals of the two hands. A difference in cortical width of 0.1 mm is highly significant and

TABLE 1

Analysis of Variance of CA/TA by Multiple Regression on Age and Years Since Menopause

Independent variables	Artificial menopause (61) Sum of squares or mean square	d.f.	Variance ratio F	Significance 'P'	Natural menopause (82) Sum of squares or mean square	d.f.	Variance ratio F	Significance 'P'
Age and Y.S.M.	0.083257	2			0.159275	2		
Y.S.M. alone	0.075642	1			0.065014	1		
Age after elimination of Y.S.M.	0.007615	1	2.30	<0.1	0.094261	1	17.09	<0.005
Age and Y.S.M.	0.083257	2			0.159275	2		
Age alone	0.075642	1			0.158828	1		
Y.S.M. after elimination of age	0.034430	1	10.38	<0.005	0.000447	1	0.081	<0.25
Sum of squares due to deviations from regression	0.192407	58			0.435664	79		
Deviations mean square	0.003317				0.005515			
Regression equations	CA/TA = 0.979 −0.0021		Age −0.0072 Y.S.M.*		CA/TA = 1.202 −0.0078		Age +0.0007 Y.S.M.	

*years since menopause

represents a change of about 2% in the cross-sectional area of the cortex. A significant change may be detected within a year in an individual who is losing bone rapidly, or it may take 5 years or more in one who is losing bone very slowly. In the average postmenopausal women, a significant loss of bone can be detected in approximately 2 years.

Longitudinal observations of this kind performed on normal premenopausal women show no significant loss of bone in the group as a whole, and we can now state with considerable conviction that if a normal premenopausal woman is found to be losing bone, she is probably approaching her menopause.

Longitudinal observations on 29 normal postmenopausal women are shown in Figure 2. The average decline in cortical width is 0.044 mm/years, which is very close to the mean rate of loss in the cross-sectional study. A comparable study on oophorectomized women has shown a similar rate of loss, although, as already indicated, it starts at the time of oophorectomy and, therefore, generally starts earlier than in normal women.

E. Fracture Epidemiology

Associated with the loss of bone that accompanies and follows the menopause, there is a striking increase in the lower forearm fracture rate. This has been shown by Alffram and Bauer[32] in Sweden and by Knowelden et al.[33] in the United Kingdom; their data are presented in Figure 3. It will be noted that no corresponding rise in the lower forearm fracture rate occurs in men, corresponding to the fact that there is little or no bone loss in the male sex at this time (see Figure 1).

Neither Alffram's nor Knowelden's data distinguish between pre- and postmenopausal women and we have, therefore, collected information about the menstrual history from 113 unselected cases of lower forearm fracture in women attending the Orthopaedic Department in Leeds. The result of this survey is shown in Figure 4. Only 4 of these women were premenopausal and the remainder postmenopausal; 10 of the latter had undergone an artificial menopause. There is, therefore, little room for doubt that the postmenopausal woman is peculiarly prone to lower forearm fracture (Colles') which we assume is attributable to the loss of bone in the distal radius and ulna which occurs at this time. As would be

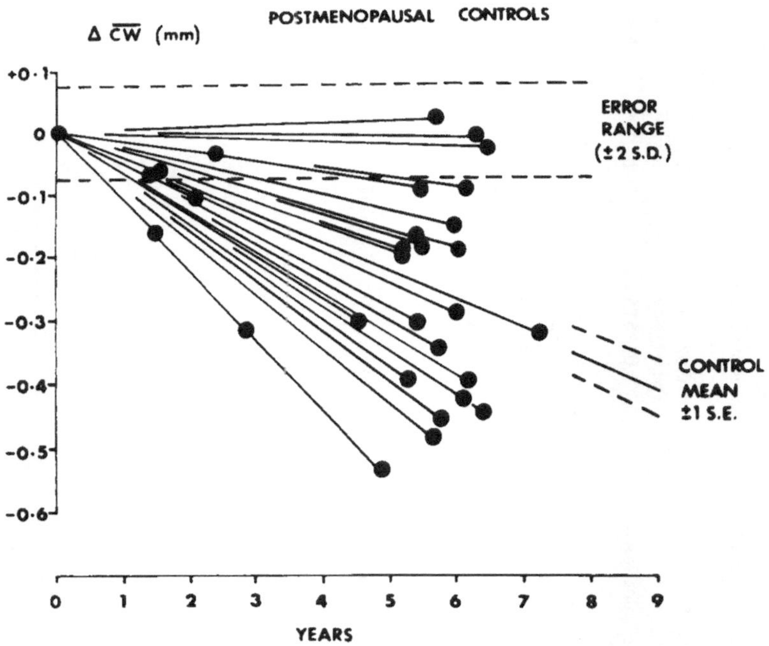

FIGURE 2. The change in mean metacarpal cortical width (mm/year) with time in 29 normal postmenopausal women.

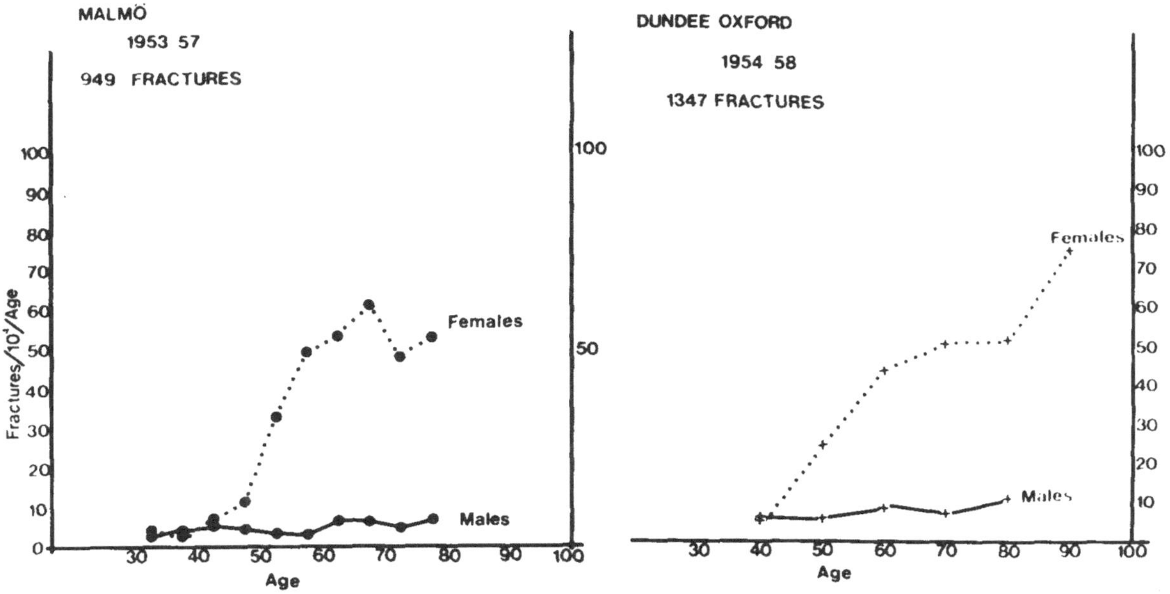

FIGURE 3. The incidence of lower forearm fracture (Colles') in relation to age is shown for two separate studies in Dundee – Oxford (Knowelden[33]) and Malmo (Alfram[32]).

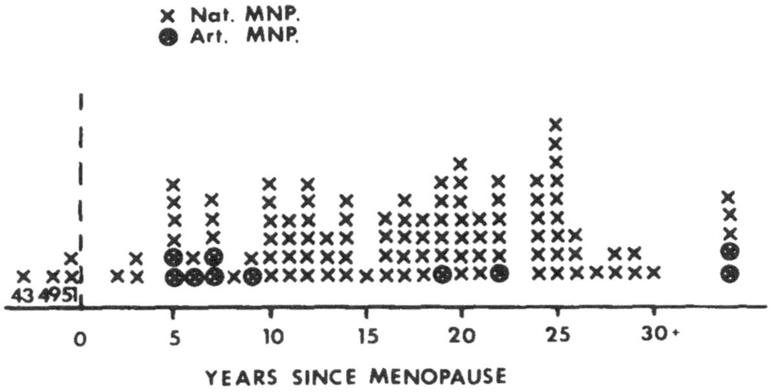

FIGURE 4. The relationship between Colles' fracture and the menopause is shown in a group of pre- and postmenopausal women in Leeds.

expected, there is a tendency for these Colles' fracture cases to have lower metacarpal CA/TA ratios than controls of the same age, just as hip fractures also have lower CA/TA ratios than controls, most of them being below the mean value and some of them below the lower limit.[34] It must be emphasized, however, that the fracture probably occurs through trabecular bone at the lower end of the radius and ulna and that the metacarpal cortex is only an indirect and possible insensitive measure of what is happening at the fracture site. It is not clear at the present time whether the fracture population represents subjects who entered the menopause with less than the average amount of bone, as suggested by Newton-John and Morgan,[35] or whether they are individuals who have lost bone more rapidly than the rest of the population since the menopause.

F. Postmenopausal Hormonal Status

Although it is clear that postmenopausal women excrete less estrogen than premenopausal women and generally have less estrogen activity on the vaginal smear assessment, there is a considerable range of estrogenic activity in the postmenopausal population. The vaginal smear maturation index[36] and total urine estrogen excretion[37] both vary during the menstrual cycle in premenopausal women, the vaginal smear index varying between about 60 and 100 and urine estrogen excretion between about 10 and 100 μg/24 hr.

In postmenopausal women the vaginal smear maturation index is almost invariably below 50 and in a substantial proportion it is zero. The actual proportion of postmenopausal women with what may be termed vaginal atrophy (no estrogenized cells on vaginal smear) varies somewhat from series to series. According to Meisels,[38] 25% of postmenopausal women show vaginal atrophy, 25% have low maturation indices, and the remainder have "moderate estrogenic activity." In our own small series of 130 postmenopausal women without fractures, approximately half have no estrogenic activity on vaginal smear (Table 2). As far as urine estrogen excretion in postmenopausal women is concerned, Brown[39] reports a range of 3.2 to 11.2 μg/24 hr; similar figures have been reported by McBride.[40] In the authors' small series of 37 normal postmenopausal women, 15 had a urinary estrogen (measured by Brown's method) below 2 μg/24 hr and 22 had values between 2 and 10 μg/24 hr (Table 3). In the present authors' experience, there is no difference in respect to vaginal smear maturation values between natural and oophorectomized postmenopausal women which suggests that the estrogenic activity of postmenopausal women does not derive from the ovary. Stimulation tests with ACTH in oophorectomized women cause an increase in total urine estrogens, while Dexamethasone® suppression causes a fall in urine estradiol and estrone but not estriol.[41] However, there is no real evidence for direct secretion of estrogens by the adrenal gland, and the present view is that postmenopausal estrogenic activity derives mainly from the conversion of androstenedione to estrone possibly in the peripheral tissues.[42]

In this connection, it is of interest to note that postmenopausal women with fractures tend to have less estrogenic activity than the normal postmenopausal population. Thus, as already stated, only 25% to 50% of a normal population have vaginal atrophy, whereas in the authors' fracture population zero maturation values have been recorded in approximately 75% of cases (Table 2).

TABLE 2

Relation Between Vaginal Smear and Fracture History in Postmenopausal Women

		Fractures	No fractures
Vaginal smear	>0	18	67
Vaginal smear	0	60	65

Chi-squared 17.5; $P<0.001$

TABLE 3

Twenty-four Hour Total Urine Estrogens in Crush Fracture and Control Patients

Urine *estrogens* μg/24 hr	Crush fractures	*Postmenopausal* normals
$\geqslant 2$	4	22
<2	14	15

Chi-squared 5.33; $p < 0.05$

III. BIOCHEMICAL CHANGES AT THE MENOPAUSE

Until recently no biochemical differences in-

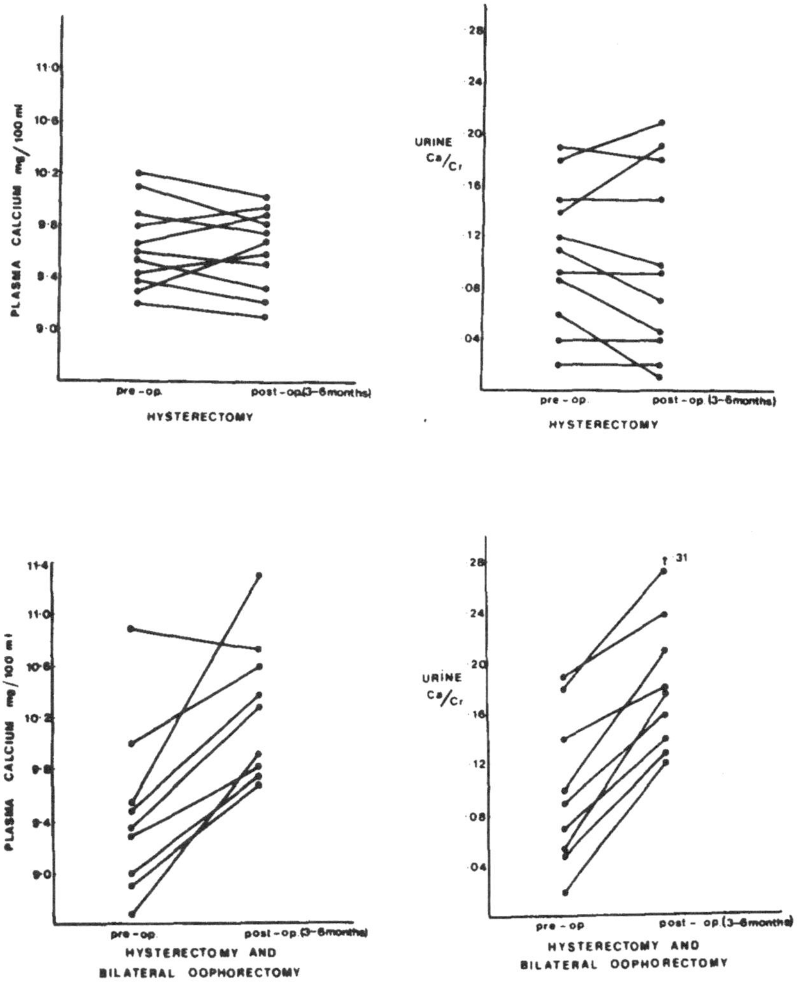

FIGURE 5. The effect of hysterectomy with oophorectomy or hysterectomy only on fasting plasma calcium and urine calcium expressed in calcium/creatinine (Ca/Cr).

volving calcium and phosphorus metabolism had been detected between pre- and postmenopausal women. At first sight this is not surprising since the average skeletal loss in normal females between the ages of 40 and 80 is only about 14% of the skeleton.[16] Since the total bone calcium is approximately 1,000 g, a loss of 14% in 40 years represents only 140 g of calcium, equivalent to 3.5 g per annum, or 10 mg daily. This small increase would be impossible to detect in a 24-hr collection of urine calcium, the value of which is highly dependent on absorbed dietary calcium.

However, in 1967 Young and Nordin[43] in a retrospective study showed that the fasting plasma and urine calcium were significantly higher in postmenopausal than in premenopausal women, and even higher in oophorectomized patients. In a more recent prospective study, Gallagher et al.[44] have shown that oophorectomy is followed by a significant rise in the fasting plasma and urine calcium, whereas simply hysterectomy is not (Figure 5 and Table 4). The measurement of urine calcium in the fasting state is particularly important. After an overnight fast, the bladder is emptied and the patient then drinks half a liter of water. Urine is collected between 8:00 and 9:00 a.m. and analyzed for calcium, phosphorus, hydroxyproline, and creatinine. Since absorption of calcium is complete within about 6 hr of a meal,[45] the morning fasting urine sample is unlikely to contain absorbed dietary calcium. The calcium in this sample can only come from bone and must

TABLE 4

Mean Values of Plasma Calcium and Urine Calcium/Creatinine and Hydroxyproline/Creatinine Ratio in Women Who Underwent Either Hysterectomy Only or Hysterectomy and Oophorectomy. The Pre- and Postoperative Values Are Compared.

	No.	Pre-op observations Mean (±S.E.M.)	Post-op (3 to 6 months) Mean (±S.E.M.)	Paired 't' 'p'
Hysterectomy and bilateral oophorectomy				
Plasma calcium (mg/100 ml)	9	9.44 (±0.22)	10.26 (±0.017)	<0.01
2-hr Urine Ca/Cr	9	0.10 (±0.02)	0.18 (±0.02)	<0.001
24-hr Hydroxyproline/Cr	9	0.013(±0.001)	0.025(±0.006)	<0.025
Hysterectomy only				
Plasma calcium (mg/100 ml)	11	9.64 (±0.09)	9.59 (±0.08)	n.s.
2-hr Urine Ca/Cr	11	0.11 (±0.01)	0.10 (±0.02)	n.s.
24-hr Hydroxyproline/Cr	11	0.013(±0.001)	0.013(±0.001)	n.s.

represent net bone resorption (i.e., resorption minus mineralization). In theory, a rise in the fasting urine calcium could represent a decrease in mineralization or an increase in resorption, but after oophorectomy it almost certainly represents an increase in bone resorption since in our prospective study urinary hydroxyproline (which is a measure of bone resorption) was significantly increased in the oophorectomized patients. Figure 6 shows that the fasting urinary hydroxyproline/creatinine ratio is significantly higher in women who have undergone an artificial menopause than in natural menopause women, and in both groups it is significantly higher than in premenopausal women.

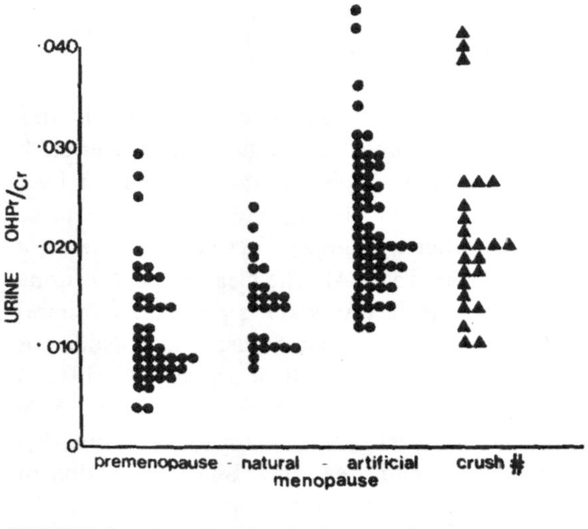

FIGURE 6. Fasting urine hydroxyproline expressed as hydroxyproline/creatinine (OHPr/Cr) is shown in pre-, post-, and artificial menopause patients and in osteoporotic patients with vertebral crush fractures.

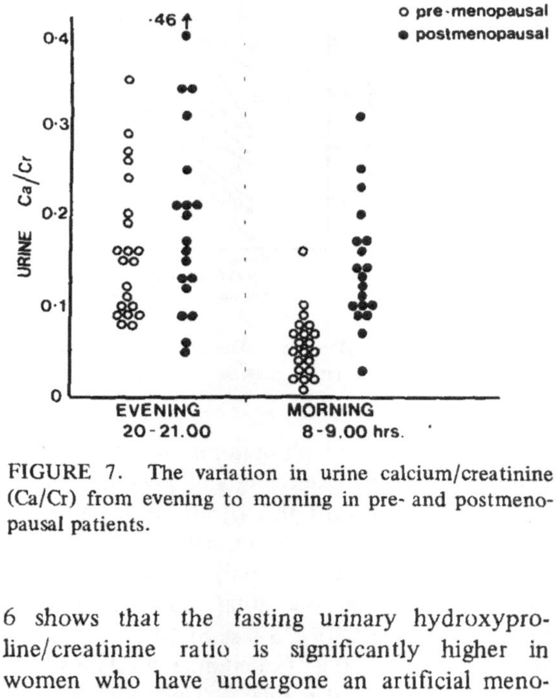

FIGURE 7. The variation in urine calcium/creatinine (Ca/Cr) from evening to morning in pre- and postmenopausal patients.

Thus, it seems that postmenopausal women cannot adapt satisfactorily to an overnight fast. In the daytime, their urine calcium is no different from that of premenopausal women, implying that there is little difference between them in the absorption of dietary calcium, but in the early morning urine there is a clear difference between the groups. Thus, premenopausal women reduce their urine calcium excretion to a lower value after the overnight fast than postmenopausal women are able to do. This is demonstrated in Figure 7, which shows that the evening (8 to 9 p.m.) urine calcium creatinine ratio is similar in pre- and postmenopausal women, whereas the morning (8 to 9 a.m.) fasting urine calcium is lower in the premenopausal than in the postmenopausal women. These studies indicate that after the menopause women are less able to adapt to calcium deprivation such as occurs during an overnight fast or, presumably, on a low-calcium diet.

IV. THE CRUSH FRACTURE SYNDROME

There is considerable confusion about the use of the term osteoporosis. On the one hand, it is widely used to denote nonspecific loss of bone with age, and on the other it is sometimes reserved more particularly for cases with excessive bone loss or with evidence that the bone loss is of clinical significance. However the term is used, it is clear that there is an identifiable syndrome of spinal osteoporosis, as originally described by Albright et al.[5] and characterized by biconcavity and/or collapse of the vertebral body. As already indicated, this syndrome, which may be very disabling, is most commonly seen in postmenopausal women, although it may occur at any age and does also occur in the male sex. It may be associated with hyperthyroidism, hypogonadism, hypoadrenocorticism, or malabsorption syndrome, but in the great majority of cases no underlying disease is discovered and the condition has to be treated as a disease in its own right. The incidence of vertebral crush fractures in a normal population has been defined by Smith and Rizek.[21] Visual assessment of the spine X-ray shows that the frequency of women who develop vertebral fractures increases sharply after the age of 60 years and is associated with an increase in the number of X-ray films of lower optical density with age.

Postmenopausal crush fracture cases differ from other postmenopausal women in their degree of bone loss, particularly of trabecular bone. This is best seen in the iliac crest biopsies where quantitative histological examination generally reveals a bone volume/volume ratio of less than 15%.[46,47] Measurements on peripheral bone may also be abnormal in severe cases, but frequently the metacarpal and femoral cortical thickness is virtually normal. However, sequential measurements of metacarpal cortical thickness of the type described above generally show that these patients are losing bone from the hand at an accelerated rate[47] and it would appear likely that in this state of accelerated bone loss trabecular bone is mobilized preferentially and may lead to vertebral collapse before cortical bone is visibly affected. Although the reason for this preferential resorption of trabecular bone is not entirely clear, it has its counterpart in the trabecular bone loss which characterizes calcium deficiency osteoporosis in animals[48] and it is at least compatible with the concept that it is a response to the demands of plasma calcium homeostasis. It is in fact exactly what would be expected if bone were mobilized by parathyroid hormone to maintain the plasma calcium concentration in the face of a state of negative calcium balance. The pathogenesis of this condition will be considered below in the discussion of the mechanism of postmenopausal bone loss. At this stage it is sufficient to note that although fully developed osteoporosis with crush fractures is a relatively rare condition, it is not impossible that a significant proportion of the backache which affects postmenopausal women represents a subclinical form of the same condition.

V. PATHOGENESIS OF POSTMENOPAUSAL BONE LOSS

The original hypothesis of Albright,[5] subsequently developed by Albright[14] and Reifenstein,[15] was that all osteoporosis was attributable to an impairment of new bone formation resulting from a failure either of osteoblastic activity or of protein synthesis. In support of this concept, they invoked the anti-anabolic action of corticosteroid hormones and the negative nitrogen balance of hyperthyroidism, extrapolating from the osteoporosis of these conditions to all other forms of osteoporosis. This hypothesis was widely accepted until new data produced by new techniques cast increasing doubt upon its validity.

The first such observation was probably that of Krane et al.,[49] who showed that bone turnover in hyperthyroidism, as judged by radiocalcium kinetics, was increased rather than decreased, and that there was a discrepancy between the nitrogen and calcium balances in this condition. Subsequently, Heaney and Whedon[50] and Nordin[51] reported that bone turnover in spinal osteoporosis was comparable to that in normal subjects. Jowsey et al.[52] showed by microradiographic techniques that the loss of bone associated with aging was attributable to an increased rate of bone resorption rather than to a decreased rate of bone formation. This increase in resorption with age, however, was not confirmed by Sissons et al.[53] using quantitative histological measurements on iliac crest biopsies; neither did they find any change in bone formation surfaces with age. More recently, Meunier[46] interprets his quantitative histological data to imply that the loss of bone with age and the excessive loss in spinal osteoporosis are attributable to a reduced rate of new bone formation.

In seeking an explanation for the osteoporotic process, if such it may be called, Nordin[48] drew attention to the extensive literature on calcium deficiency osteoporosis in animals, including a paper by Albright et al.[54] in which trabecular bone loss was produced by low-calcium diets in kittens. Nordin argued that there was a difference between the effect of vitamin D deficiency, which caused osteomalacia or rickets by its effect on plasma calcium and/or phosphate levels, and calcium deficiency, which simply led to mobilization of bone (particularly trabecular bone) for the maintenance of the plasma calcium in the face of continued loss in the urine. These early observations were subsequently confirmed by many other workers but it was Jowsey and Raisz[55] who convincingly demonstrated the role of the parathyroid glands in this process by showing that calcium deficiency osteoporosis could not be induced after parathyroidectomy in cats. Although this work was subsequently questioned,[56] there would probably be general agreement at the present time that calcium deficiency osteoporosis in animals is parathyroid hormone-mediated, although it is likely that some calcium homeostatic mechanism also operates in the absence of the parathyroid glands, enabling plasma calcium to be maintained albeit at a subnormal level, even in the face of a low-calcium diet.

Although it cannot be assumed that experimental calcium deficiency osteoporosis is the correct model to consider in seeking to explain human osteoporosis, it represents at the present time virtually the only plausible model available. Although there are nutritionists, such as Hegsted et al.[57] and Walker[58] who argue that calcium is a "nutrient in search of a disease," it is difficult to believe that calcium deficiency in its widest sense has no effect on the skeleton. If the animal model is a valid one, there can be little doubt that the clinical syndrome of calcium deficiency is osteoporosis, and that clinical osteoporosis is most easily understood in terms of calcium deficiency.

Arguing from these premises, the increase in fasting plasma and urine calcium that accompanies the menopause must represent an increase in bone resorption which is probably parathyroid hormone-mediated. From this it follows *either* that parathyroid activity is increased in the postmenopausal period *or* that the bone has become more sensitive to the action of the hormone. At the present time, evidence on plasma parathyroid hormone levels in postmenopausal osteoporosis is conflicting. Fujita et al.[59] report that plasma parathyroid hormone levels are increased, whereas Riggs et al.[60] report that they are reduced. In the authors' experience, and granted the limitations of existing methodology, plasma parathyroid hormone levels do not change after the menopause.

We are, therefore, forced to the conclusion that there is an alteration in bone sensitivity to parathyroid hormone in the absence of estrogens,[61,62] and the effects of estrogens (to be described below) are entirely compatible with this concept. Experiments with bone in tissue culture[63] have demonstrated inhibition of parathyroid hormone-induced bone resorption by various estrogenic hormones in a dose-related manner. Moreover, it is an interesting fact that primary hyperparathyroidism (particularly the form with bone disease) is more common in postmenopausal women than in any other age group[64,65] possibly because the increase in bone sensitivity to parathyroid hormone leads to the presentation of the condition after the menopause. Finally, it is known that postmenopausal hyperparathyroid women tend to have less cortical bone and hypoparathyroid women more cortical bone than normal controls.[66] This difference is not apparent before the menopause. The fact that low-calcium diets appear to accelerate bone loss in postmenopausal women (see below) is also compatible with the concept that postmenopausal bone loss is parathyroid hormone-mediated.

Nonetheless, the increase in bone resorption

after the menopause could have other explanations. Other agents are known to increase bone resorption, particularly thyroid hormone, and it could be argued that postmenopausal bone resorption reflects an increased sensitivity of bone to thyroid hormone in the absence of estrogens. Although the present authors regard this as a less likely explanation of the facts, it cannot be entirely excluded. It should be noted in this connection that Smith et al.[67] have shown accelerated bone loss in postmenopausal, but not in premenopausal thyrotoxic women. It could also be argued that postmenopausal bone is abnormally sensitive to resorption by corticosteroid hormone. In this connection it should be noted that Gallagher et al.[68] reported significant bone loss in corticosteroid-treated postmenopausal women, but not in corticosteroid-treated males.

Therefore, it seems reasonable to suppose that the crush fracture syndrome develops in postmenopausal women subjected to some additional factor over and above the estrogen insufficiency of the normal population. The only such factor identified by Gallagher et al.[47] was malabsorption of calcium, which characterizes the majority of such cases. Malabsorption of calcium is not normally a feature of the postmenopausal period,[69] but when present appears to produce the accelerated bone loss associated with vertebral compression fractures.

VI. PREVENTION AND TREATMENT

A. Spinal Osteoporosis

Albright et al.,[5] arguing that estrogenic hormones stimulated new bone formation, recommended estrogen therapy in spinal osteoporosis and showed that this therapy improved the calcium balance by lowering fecal and urinary calcium. Subsequently, Henneman and Wallach[70] claimed that estrogen therapy prevented loss of height in patients with osteoporosis, although no increase in spinal density in the patients on estrogens was observed. However, Lafferty et al.,[71] using calcium balance and isotope studies, reported that estrogens acted primarily by inhibiting bone resorption and that their relative therapeutic inefficacy was attributable to a secondary depression of new bone formation, which occurred after several months on estrogens. Estrogens have been shown to reduce both bone formation and bone resorption in crush fracture patients, as measured by histological techniques on iliac crest biopsies.[60]

As far as the treatment of established osteoporosis with crush fractures is concerned, we have already noted that these patients generally suffer from malabsorption of calcium and reduced estrogenic activity. Insofar as a general therapeutic regime can be recommended, the authors consider that it should be a combination of vitamin D, estrogens, and calcium supplements.[47] However, the cases are not all alike and this policy is to tailor the treatment to the individual case. This implies that vitamin D in doses of about 20,000 units daily is administered to those with malabsorption of calcium, and estrogens to those with vaginal atrophy, although we have experienced some cases who cannot tolerate large doses of both at the same time. We are inclined to the view that all the cases should have a calcium supplement. We have no experience of the benefits of fluoride therapy,[60] calcium infusions,[77] or anabolic agents,[78] although claims have been made for all three.

B. Postmenopausal Bone Loss

Now that the fasting plasma and urine calcium are known to rise after oophorectomy (see above), these parameters can be used to monitor the effect of estrogenic hormones on bone resorption. The results of such a study are shown in Figures 8 and 9. The administration of ethinyl estradiol 0.05 mg daily to normal postmenopausal women produces a rapid fall in fasting plasma and urine calcium and in urinary hydroxyproline to premenopausal values, indicating the inhibition of bone resorption by this form of therapy. The beneficial effect of this therapy on calcium balance, mainly attributable to a fall in urinary calcium, is illustrated in Figure 10. We agree, however, with Lafferty et al.[71] and Riggs et al.[60] that there is a tendency for estrogens ultimately to reduce bone formation as well as bone resorption, although the effect on the former is not as marked as on the latter and the overall effect is an improvement in calcium balance.

A similar effect can be seen when postmenopausal patients with primary hyperparathyroidism are given estrogens. In a study carried out in ten patients[72], ethinyl estradiol produced a highly significant fall in plasma and urine calcium and urine hydroxyproline, indicating that bone resorp-

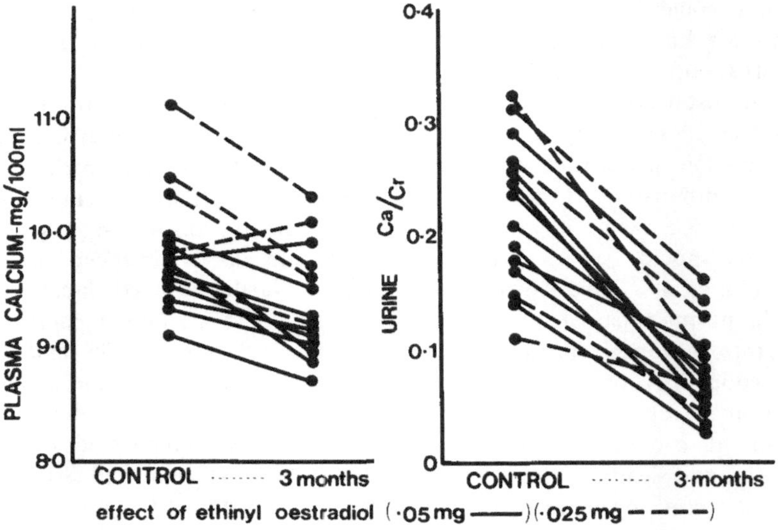

FIGURE 8. The effect of ethinyl estradiol 0.05 mg or 0.025 mg daily on the fasting plasma and urine calcium/creatinine (Ca/Cr) in postmenopausal patients.

tion was inhibited. Further metabolic studies showed that in all the patients treated, there was a significant improvement in calcium balance.[73] Using radiocalcium kinetic techniques, it was demonstrated that in all cases there was a fall in the bone mineralization rate and resorption rate after estrogen administration (Figure 11). This effect is quite reversible, for when ethinyl estradiol was stopped, the plasma and urine calcium and hydroxyproline returned to their previous level during the following 3 months. It is striking that even in the presence of marked parathyroid hormone-induced bone resorption (circulating levels of parathyroid hormone are high), estrogens are capable of restoring bone turnover to a normal rate.

These studies do suggest that if parathyroid hormone is important, as we believe it is, in the pathogenesis of postmenopausal bone loss, then estrogen administration should prevent the development of the bone loss in the postmenopausal state. This is in fact the case. Retrospective studies by various workers[75] have indicated reduced bone loss in postmenopausal women on estrogen replacement therapy; this has been confirmed in a prospective study.[76]

We have also shown reduced bone loss on estrogen therapy in a small prospective study based on sequential measurements of metacarpal cortical thickness. In this study, postmenopausal women observed for periods of up to 5 years lost metacarpal cortical bone at an average rate of 0.044 mm per annum, whereas women on estrogen replacement therapy lost bone at an average rate of 0.007 mm per annum (Figure 12). This difference is highly significant.

Postmenopausal bone loss can also be reduced by calcium supplements or accelerated by calcium deficiency. In another small prospective study, we have observed metacarpal cortical bone loss at a mean rate of 0.085 mm per annum in 6 post menopausal stone-forming women on low-calcium

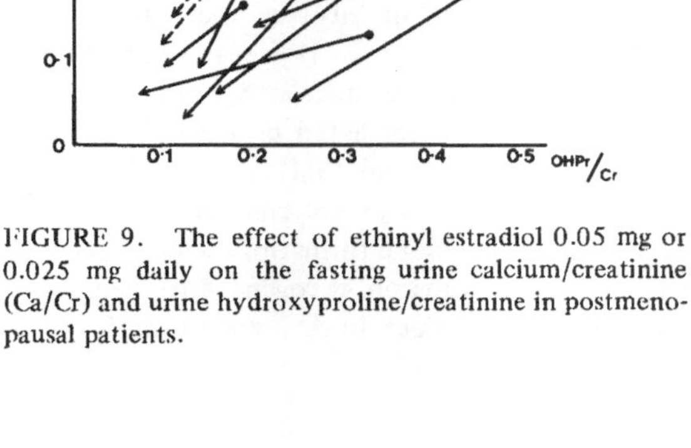

FIGURE 9. The effect of ethinyl estradiol 0.05 mg or 0.025 mg daily on the fasting urine calcium/creatinine (Ca/Cr) and urine hydroxyproline/creatinine in postmenopausal patients.

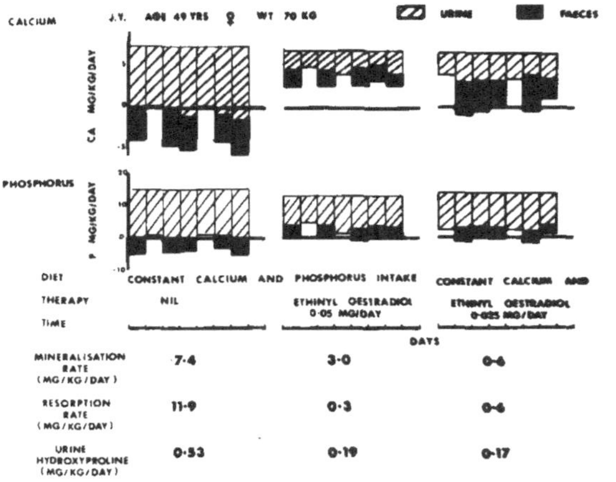

FIGURE 10. An example of the effect of ethinyl estradiol 0.05 mg and 0.025 mg on daily calcium and phosphorus balance in a postmenopausal patient. All values are expressed as mg/kg/day. Ethinyl estradiol had been given for 4 months before each repeat balance was performed. The days on which no fecal values are shown were the days when no fecal samples were obtained.

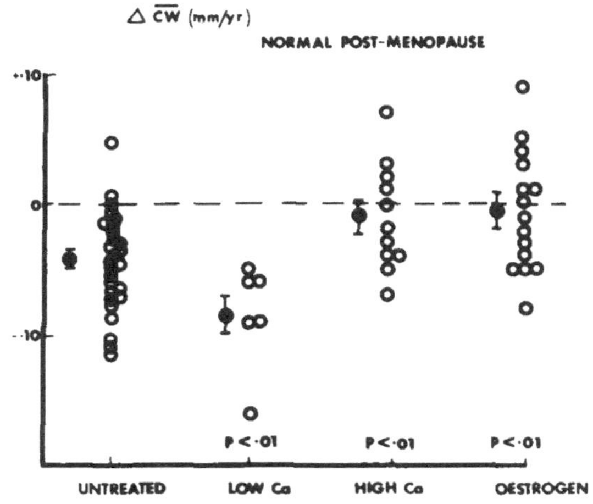

FIGURE 12. The change in mean metacarpal cortical width (mm/year) is shown for 29 normal postmenopausal women, 6 postmenopausal women on a low-calcium diet, 11 postmenopausal women on calcium supplements, and 15 postmenopausal women on estrogens. The patients on the low-calcium diet have significantly greater bone loss than the untreated patients, while the patients on calcium supplements and estrogen therapy have had no significant bone loss during the period of observation which ranges from 3 to 6 years.

diets and a mean loss of 0.011 mm per annum in 11 normal postmenopausal women on calcium supplements (1 g of additional calcium daily as calcium citrate). The rate of loss on the low-calcium diet is significantly greater than normal and the rate of loss on the calcium supplement significantly less than normal (Figure 12). These observations could be interpreted to mean that the body's calcium economy is more efficient in the presence of estrogenic hormones than in their absence, but that if sufficient calcium is given, the effect of estrogen insufficiency can be offset. To draw an analogy, it could be said that the body's sodium economy is less efficient in the absence of aldosterone, but that a patient with aldosterone insufficiency can be maintained on salt supplements if sufficient salt is given.

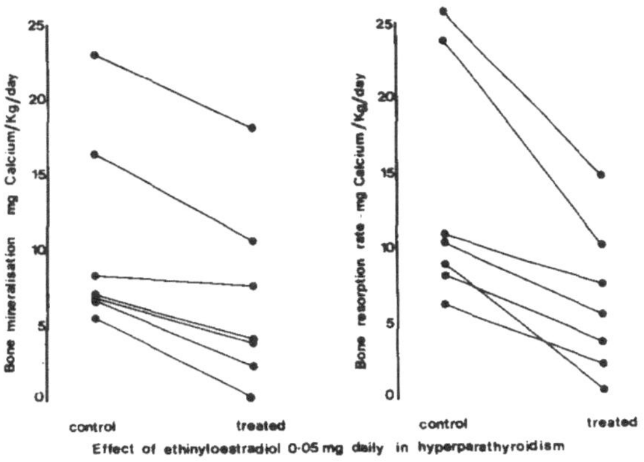

FIGURE 11. The effect of ethinyl estradiol 0.05 mg daily on the bone mineralization rate and resorption rate in seven postmenopausal patients with primary hyperparathyroidism.[73] All values are expressed in mg/kg/day.

In is of interest to note that preliminary studies with two progestogens (norethisterone and 17-hydroxyprogesterone) have shown a similar effect on the fasting urine calcium to that of estrogens (Figure 13) and it is possible that some anabolic agents are in fact acting as progestogens. It is not impossible that the combination of estrogen and a progestogen, as in some contraceptive pills, may prove the most suitable form of replacement therapy for postmenopausal women.

It is not clear at the present time whether all postmenopausal women should be given substitution therapy, or whether is should be confined to selected cases and if so which and for how long. If our preliminary data on the relation between the vaginal smear and the predisposition to fracture (Table 2) are confirmed in larger series, it could be argued that a prima facie case exists for substitution therapy in all patients with vaginal atrophy (i.e., vaginal smear with no estrogenic activity). This could imply administration of estrogens to about 25 to 50% of all postmenopausal women.

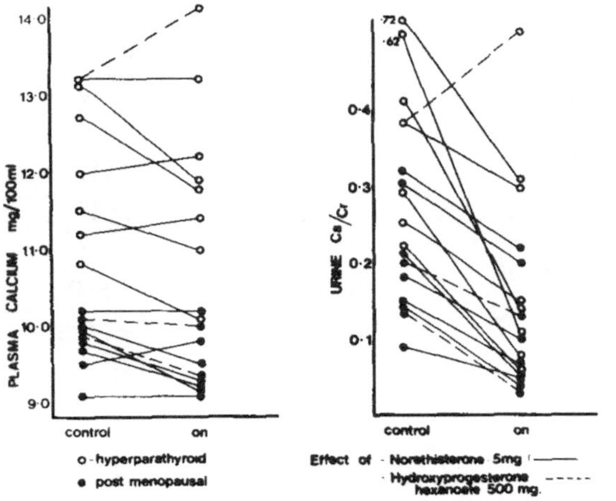

FIGURE 13. The effect of norethisterone or 17-hydroxyprogesterone hexanoate on fasting plasma and urine calcium/creatinine (Ca/Cr) in postmenopausal patients.

The dose we favor would be in the range 25 to 50 μg of ethinyl estradiol or its equivalent in other estrogens, given cyclically for 3 weeks out of every 4 (to all but hysterectomized cases) to allow shedding of the endometrium. (In our experience, most postmenopausal women do not in fact experience regular withdrawal bleeding on this regime.) An alternative regime, recommended by Ferin and Thomas,[79] is to give a 5-day course of a progestogen every 4 to 6 weeks during continuous estrogen therapy. If for any reasons, such as a history of thromboembolic disease, estrogens are contraindicated, a progestogen may be equally effective, but this is speculative at present. It could be argued that postmenopausal women should also have a calcium supplement, possibly in the form of a late night milk drink.

As for the duration of substitution therapy, we are inclined to the view that it should be continued indefinitely. We have observed many old ladies with femoral neck fracture, vaginal atrophy, and a high fasting urine calcium in whom estrogen therapy has reduced the urine calcium and produced a positive vaginal smear. Although we believe that malabsorption of calcium is a major factor in the senile osteoporosis that leads to femoral neck fracture, postmenopausal bone loss must predispose to this fracture as well, and its control by estrogens is probably desirable as a preventive measure. In fact, it has even been suggested that estrogens may improve calcium absorption in senile osteoporosis,[80] but our own data on this point are inconclusive.

VII. CONCLUSIONS

There is little doubt that declining ovarian function is associated with the onset of bone loss in women, although we believe that it may slightly precede the actual cessation of menstruation. The loss of bone appears to be due to an increase in bone resorption rather than to a decline in bone formation, and its biochemical manifestations are reversible with estrogens. Data are now appearing to show that estrogens also prevent or delay the loss of bone.

The underlying mechanisms are only poorly understood, but present data are compatible with the concept that estrogens reduce the sensitivity of bone to the resorbing action of parathyroid hormone. It is possible, however, that the action of estrogens is less specific than this explanation would suggest and that they tend to oppose the bone-resorbing effect of other physiological agents such as thyroid hormone, cortisol, and other as yet unidentified substances.

Thus, from the point of view of skeletal health, a case can be made for the more general use of hormonal replacement therapy, particularly in postmenopausal women with severely reduced estrogenic activity. Preventive therapy is particularly important in relation to the skeletal system since no current therapeutic agents have been shown effectively to restore the bone that has been lost.

REFERENCES

1. **Cooper, Sir Astley,** *Treatise on Dislocations and on Fractures of the Joints,* Longman, London, 1824.
2. **Pommer, G.,** *Untersuchungen Uber Osteomalacie und Rachitis,* Leipzig, 1885.
3. **Howland, J. and Kramer, B.,** Calcium and phosphorus in the serum in relation to rickets, *Am. J. Dis. Child.,* 22, 105, 1921.
4. **Meulengracht, E.,** Osteomalacia of the spinal column from deficient diet or from disease of the digestive tract, *Acta Med. Scand.,* 101, 138, 1939.

5. **Albright, F., Smith, P. H., and Richardson, A. M.,** Postmenopausal osteoporosis; its clinical features, *J.A.M.A.*, 116, 2465, 1941.
6. **Riddle, O. and Rheinhart, W. H.,** Studies on the physiology of reproduction in birds. XXI. Blood calcium changes in the reproductive cycle, *Am. J. Physiol.*, 76, 660, 1926.
7. **Hughes, J. S., Titus, R. W., and Smith, B. L.,** Increase in the calcium of hens' blood accompanying egg production, *Science,* 65, 264, 1927.
8. **Kyes, P. and Potter, T. S.,** Physiological marrow ossification in female pigeons, *Anat. Rec.*, 60, 377, 1934.
9. **Pfeiffer, C. A. and Gardner, W. U.,** Skeletal changes and blood serum calcium level in pigeons receiving estrogens, *Endocrinology,* 23, 485, 1938.
10. **McLean, F. C. and Urist, M. R.,** *Bone. Fundamentals of the Physiology of Skeletal Tissue,* 3rd ed., University of Chicago Press, Chicago, 1968.
11. **Gardner, W. U. and Pfeiffer, C. A.,** Sex hormones and bone changes in mice, *Anat. Rec.*, Suppl. 73, 1939.
12. **Lindquist, B., Budy, A. M., McLean, F. C., and Howard, J. L.,** Skeletal Metabolism in estrogen-treated rats studied by means of Ca^{45}, *Endocrinology,* 66, 100, 1960.
13. **Albright, F., Smith, P. H., and Fraser, R.,** A syndrome characterized by primary ovarian insufficiency and decreased stature. Report of 11 cases with a digression on hormonal control of axillary and pubic hair, *Am. J. Med. Sci.*, 204, 625, 1942.
14. **Albright, F.,** Osteoporosis, *Ann. Intern. Med.*, 27, 861, 1947.
15. **Reifenstein, E. C.,** Relationship of steroid hormone to the development and management of osteoporosis in ageing people, *Clin. Orthop.*, 10, 206, 1957.
16. **Trotter, M., Broman, G. E., and Peterson, R. P.,** Densities of bones of white and Negro skeletons, *J. Bone Joint Surg.*, 42A, 50, 1960.
17. **Barnett, E. and Nordin, B. E. C.,** The radiological diagnosis of osteoporosis: a new approach, *Clin. Radiol.*, 11, 166, 1960.
18. **Meema, H. E.,** Cortical bone atrophy and osteoporosis as a manifestation of ageing, *Am. J. Roentgenol.*, 89, 1287, 1963.
19. **Nordin, B. E. C., MacGregor, J., and Smith, D. A.,** The incidence of osteoporosis in normal women: its relation to age and the menopause, *Quart. J. Med.*, 137, 25, 1966.
20. **Garn, S. M.,** *The Earlier Gain and Later Loss of Cortical Bone,* Charles C Thomas, Springfield, Ill., 1970.
21. **Smith, R. W. and Rizek, J.,** Epidemiological studies of osteoporosis in women of Puerto Rico and South Eastern Michigan with special reference to age, national origin and to other related or associated findings, *Clin. Orthop.*, 45, 31, 1966.
22. **Lindahl, O. and Lindgren, E. H.,** Grading of osteoporosis in autopsy specimens, *Acta Orthop. Scand.*, 32, 85, 1962.
23. **Baylink, D. J., Vose, G. P., Dotter, W. E., and Hurxthal, L. M.,** Two new methods for the study of osteoporosis and other metabolic bone disease. Part 2. Vertebral bone densitometry, *Lahey Clin. Found. Bull.*, 13, 217, 1964.
24. **Arnold, J. S.,** The quantitation of bone mineralisation as an organ and tissue in osteoporosis, in *Dynamic Studies in Metabolic Bone Disease,* Pearson, O. H. and Joplin, G. F., Eds., Blackwell, Oxford, 1964, 59.
25. **Arnold, J. S.,** Amount and quality of trabecular bone in osteoporotic vertebral fractures, *Clin. Endocrinol. Metab.*, 2, 221, 1973.
26. **Caldwell, R. A.,** Observations on the incidence aetiology and pathology of senile osteoporosis, *J. Clin. Pathol.*, 15, 421, 1962.
27. **Nordin, B. E. C., Young, M. M., Bentley, B., Ormondroyd, P., and Sykes, J.,** Lumbar spine densitometry methodology and results in relation to the menopause, *Clin. Radiol.*, 19, 459, 1968.
28. **Adamopoulos, D. A., Loraine, J. A., and Dove, G. A.,** Endocrinological studies in women approaching the menopause, *J. Obstet. Gynecol.*, 78, 62, 1971.
29. **Gallagher, J. C., Bulusu, L., and Nordin, B. E. C.,** Oestrogenic hormones and bone resorption, in *Proceedings of International Symposium on Clinical Aspects of Metabolic Bone Disease,* Henry Ford Hospital, Detroit, Excerpta Medica, Amsterdam, 1972.
30. **Aitkin, J. M., Hart, D. M., Anderson, J. B., Lindsay, R., Smith, D. A., and Speirs, C. F.,** Osteoporosis after oophorectomy for non-malignant disease in premenopausal women, *Br. Med. J.*, 1, 325, 1973.
31. **Horsman, A. and Nordin, B. E. C.,** The quantitative assessment of sequential changes in cortical bone geometry, in *Proceedings of the 9th European Symposium of Calcified Tissues,* Czitzober, H. and Eschberger, J., Eds., Facta Publication, Vienna, 1973.
32. **Alffram, P. A. and Bauer, G. C. H.,** Epidemiology of fractures of the forearm. A biomechanical investigation of bone strength, *J. Bone Joint Surg.*, 44A, 105, 1962.
33. **Knowelden, J., Buhr, A. J., and Dunbar, O.,** Incidence of fractures in persons over 35 years of age, *Br. J. Prev. Soc. Med.*, 18, 130, 1964.
34. **Nordin, B. E. C.,** Clinical significance and pathogenesis of osteoporosis, *Br. Med. J.*, 1, 571, 1971.
35. **Newton-John, H. F. and Morgan, D. B.,** The loss of bone with age, osteoporosis and fractures, *Clin. Orthop.*, 71, 229, 1970.
36. **Stone, D. F., Sedlis, A., Stone, M. L., and Turkel, W. L.,** Oestrogen-like effects in the vaginal smears of post-menopausal women, *Acta Cytol.*, 11, 349, 1967.

37. **Brown, J. B., Klopper, A., and Loraine, J. A.,** The urinary excretion of oestrogens, pregnanediol and gonadotrophins during the menstrual cycle, *J. Endocrinol.*, 17, 401, 1958.
38. **Meisels, A.,** The menopause. A cytohormonal study, *Acta Cytol.*, 10, 49, 1966.
39. **Brown, J. B.,** A new method for the determination of oestrogens in urine and its application to a study of the oestrogen excretion in the menstrual cycle, *Mem. Soc. Endocrinol.*, 3, 1, 1955.
40. **McBride, J. M.,** Oestrogen excretion levels in the normal postmenopausal woman, *J. Clin. Endocrinol.*, 17, 1440, 1957.
41. **Procope, B. J. and Adlercreutz, H.,** Oestrogen production in postmenopausal women, in *Ageing and Oestrogens*, van Keep, P. and Lauritzen, C., Eds., S. Karger, Basel, 1973, 55.
42. **Grodin, J. M., Siiteri, P. K., and MacDonald, P. C.,** Source of estrogen production in postmenopausal women, *J. Clin. Endocrinol.*, 36, 207, 1973.
43. **Young, M. M. and Nordin, B. E. C.,** The effect of the natural and the artificial menopause on plasma and urinary calcium and phosphorus, *Lancet*, ii, 118, 1967.
44. **Gallagher, J. C., Young, M. M., and Nordin, B. E. C.,** Effects of artificial menopause on plasma and urine calcium and phosphate, *Clin. Endocrinol.*, 1, 57, 1972.
45. **Birge, S. J., Peck, W. A., Berman, M., and Whedon, G. D.,** Study of calcium absorption in man: a kinetic analysis and physiologic model, *J. Clin. Invest.*, 48, 1705, 1969.
46. **Meunier, P., Courpron, P., Edouard, C., Bernard, J., Bringuier, J., and Vignon, G.,** Physiological and senile involution and pathological rarefaction of bone, *Clin. Endocrinol. Metab.*, 2, 239, 1973.
47. **Gallagher, J. C., Aaron, J., Horsman, A., Marshall, D. H., Wilkinson, R., Nordin, B. E. C.,** The crush fracture syndrome in postmenopausal women, *Clin. Endocrinol. Metab.*, 2, 293, 1973.
48. **Nordin, B. E. C.,** Osteoporosis, osteomalacia and calcium deficiency, *Clin. Orthop.*, 17, 235, 1960.
49. **Krane, S. M., Brownell, G. L., Stanbury, J. B., and Corrigan, H.,** The effect of thyroid disease on calcium metabolism in man, *J. Clin. Invest.*, 35, 874, 1956.
50. **Heaney, R. P. and Whedon, G. D.,** Radiocalcium studies of bone formation rate in human metabolic bone disease, *J. Clin. Endocrinol.*, 18, 1246, 1958.
51. **Nordin, B. E. C.,** Investigation of bone metabolism with Ca^{47}, *Proc. R. Soc. Med.*, 52, 351, 1959.
52. **Jowsey, J., Kelly, P. J., Riggs, B. L., Bianco, A. J., Scholz, D. A., and Gershon-Cohen, J.,** Quantitative microradiographic studies of normal and osteoporotic bones, *J. Bone Joint Surg.*, 47A, 785, 1965.
53. **Sissons, H. A., Holley, K. J., and Heighway, J.,** Natural bone structure in relation to osteomalacia, in *L'Osteomalacie,* Hioco, D. J., Ed., Masson & Cie, Paris, 1967, 19.
54. **Albright, F., Bauer, W., and Aub, J. C.,** The calcium excretion of normal individuals on a low calcium diet, *J. Clin. Invest.*, 7, 75, 1929.
55. **Jowsey, J. and Raisz, L. G.,** Experimental osteoporosis and parathyroid activity, *Endocrinology*, 82, 384, 1968.
56. **Sevastikoglou, J. A.,** The effect of prolonged calcium deficiency on the skeleton of the thyroparathyroidectomized adult rat, *Clin. Orthop.*, 85, 177, 1972.
57. **Hegsted, D. M., Moscoso, I., and Collazos, C.,** A study of the minimum calcium requirements of adult men, *J. Nutr.*, 46, 181, 1952.
58. **Walker, A. R. P.,** The human requirements of calcium: should low intakes be supplemented, *Am. J. Clin. Nutr.*, 25, 518, 1972.
59. **Fujita, T., Orimo, H., Okano, K., Yoshikawa, M., Shimo, R., Inoue, T., and Itami, Y.,** Radioimmunoassay of serum parathyroid hormone in postmenopausal osteoporosis, *Endocrinol. Jap.*, 19, 571, 1972.
60. **Riggs, B. L., Jowsey, J., Kelly, P. J., Hoffman, D. L., and Arnaud, C. D.,** Studies on pathogenesis and treatment in post-menopausal and senile osteoporosis, *Clin. Endocrinol. Metab.*, 2, 317, 1973.
61. **Heaney, R. P.,** A unified concept of osteoporosis, *Am. J. Med.*, 39, 877, 1965.
62. **Jasani, C., Nordin, B. E. C., Smith, D. A., and Swanson, I.,** Spinal osteoporosis and the menopause, *Proc. R. Soc. Med.*, 58, 441, 1965.
63. **Atkins, D., Zanelli, J. M., Peacock, M., and Nordin, B. E. C.,** The effect of oestrogens on the response of bone to parathyroid hormone *in vitro*, *J. Endocrinol.*, 54, 107, 1972.
64. **Muller, H.,** Sex, age and hyperparathyroidism, *Lancet*, i, 449, 1969.
65. **McGeown, J. G.,** Sex, age and hyperparathyroidism, *Lancet*, i, 887, 1969.
66. **Hossain, M., Smith, D. A., and Nordin, B. E. C.,** Parathyroid activity and post-menopausal osteoporosis, *Lancet*, i, 809, 1970.
67. **Smith, D. A., Fraser, S. A., and Wilson, G. M.,** Hyperthyroidism and calcium metabolism, *Clin. Endocrinol. Metab.*, 2, 333, 1973.
68. **Gallagher, J. C., Aaron, J., Horsman, A., Wilkinson, R., and Nordin, B. E. C.,** Corticosteroid osteoporosis, *Clin. Endocrinol. Metab.*, 2, 355, 1973.
69. **Bullamore, J. R., Gallagher, J. C., Wilkinson, R., Nordin, B. E. C., and Marshall, D. H.,** Effect of age on calcium absorption, *Lancet*, 2, 535, 1970.
70. **Henneman, P. H. and Wallach, S.,** A review of the prolonged use of estrogens and androgens in postmenopausal and senile osteoporosis, *Arch. Intern. Med.*, 100, 715, 1957.

71. **Lafferty, F. W., Spencer, G. E., and Pearson, O. H.,** Effects of androgens, estrogens and high calcium intakes on bone formation and resorption in osteoporosis, *Am. J. Med.,* 36, 514, 1964.
72. **Gallagher, J. C. and Nordin, B. E. C.,** Treatment with oestrogens of primary hyperparathyroidism in post-menopausal women, *Lancet,* i, 503, 1972.
73. **Gallagher, J. C. and Wilkinson, R.,** The effect of ethinyl oestradiol on calcium and phosphorus metabolism of post-menopausal women with primary hyperparathyroidism, *Clin. Sci.,* 45, 785, 1973.
74. **Davis, M. E., Lanzl, L. H., and Cox, A. B.,** Detection, prevention and retardation of post-menopausal osteoporosis, *Obstet. Gynecol.,* 36, 187, 1970.
75. **Meema, H. E. and Meema, S.,** Prevention of postmenopausal osteoporosis by hormone treatment of the menopause, *Can. Med. Assoc. J.,* 99, 248, 1968.
76. **Aitkin, J. M., Hart, D. M., and Lindsay, R.,** Oestrogen replacement therapy for prevention of osteoporosis after oophorectomy, *Br. Med. J.,* 3, 515, 1973.
77. **Pak, C. Y. C. and Bartter, F. C.,** Treatment of osteoporosis with calcium infusions, *Semin. Drug Treat.,* 2, 39, 1972.
78. **Gordan, G. S. and Eisenberg, E.,** The effect of oestrogens, androgens and corticoids on skeletal kinetics in man, *Proc. R. Soc. Med.,* 56, 1022, 1963.
79. **Ferin, J. and Thomas, K.,** Long-term replacement therapy in post-menopausal women. Pituitary gonadotrophin secretion and endometrial studies, in *Ageing and Oestrogens,* van Keep, P. and Lauritzen, C., Eds., S. Karger, Basel, 1973, 134.
80. **Caniggia, A., Gennari, C., Borrello, G., Bencini, M., Cesari, L., Poggi, C., and Escobar, S.,** Intestinal absorption of calcium^{47} after treatment with oral oestrogen-gestogens in senile osteoporosis, *Br. Med. J.,* 4, 30, 1970.

THE BIOCHEMISTRY OF DEPRESSION IN WOMEN

P. A. Toseland

TABLE OF CONTENTS

I. INTRODUCTION

Professor Miller once said,[1] "The cardinal feature of depression of the endogenous type is its totality. It involves the entire person. The patient looks ill, feels ill and is ill."

It must be pointed out immediately, even in these enlightened days of sexual equality, that in the diagnosis of depression resulting from the use of the contraceptive pill, pregnancy, premenstrual tension, or the menopause the description "man" is a very strong negative discriminant. In other words, the male appears to be at less risk. In a recent study of depressed patients,[2] all the persons involved were women, a fact which passed without comment in the paper.

II. THE BIOCHEMICAL BACKGROUND

A. Theories

Five possible theories will be considered, all of them related to the metabolism of L-tryptophan. Although they will be examined under separate headings, it must be apparent that several or, in fact, all of these possibilities could in some way form in the human female a dynamic equilibrium and that depression may occur when a certain balance is disturbed. These headings are the serotonin theory, the tryptophan theory, the vitamin B_6 theory, the tryptamine theory, and the nicotinic acid theory. As can be seen in Figure 1, all these possible explanations are very closely interconnected, thus making it difficult to form separate watertight sections on each idea. They are placed in the order that at the time of writing represents them in the order they are receiving attention. All the theories are concerned with the metabolism of tryptophan and the amount and form in which this amino acid is available for metabolism.

That depression occurs in oral contraceptive therapy, in pregnancy, and during premenstrual tension is in no doubt, although it must be stated that few investigators involved themselves in the biochemical investigation of depression occurring in pregnancy and the menopause until "the pill" arrived as a form of therapy. Furthermore, as the basis of any of the current theories is estrogen administration, or estrogen predominance in

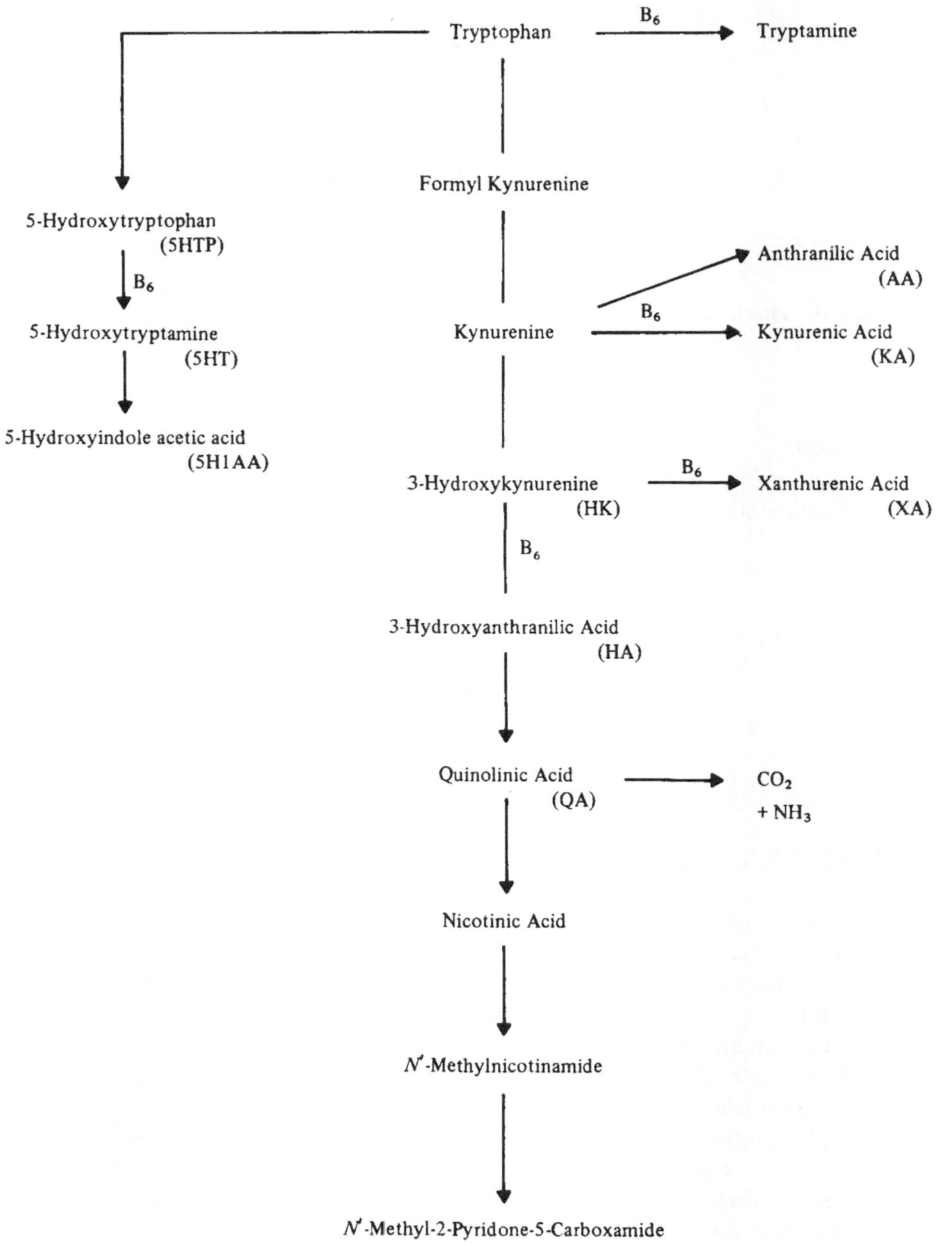

FIGURE 1. The metabolism of L-tryptophan.

hormone imbalance, then the same theory is unacceptable for similar depressive symptoms occurring during the menopause. The metabolic disturbances of tryptophan metabolism resulting from oral contraceptive therapy have been reported on many occasions during the last 7 years. Prior to this, tryptophan metabolism had been reported as abnormal in rheumatoid arthritis[3,4] and disseminated lupus erythematosus.[5] Now the pill is reported to produce signs mimicking rheumatoid arthritis[6,7] and exacerbations of disseminated lupus.[8]

III. DEPRESSION

One of the few facts generally agreed is that

depression results from the contraceptive pill;[9–13] it is also one of the principal reasons for ceasing therapy.[14]

In one study reported in 1970,[15] the incidence of depression was reported to be as high as 6.6%. The symptoms described in this particular study revealed that pessimism, dissatisfaction, crying, and tension were dominant, and sleep disorder and appetite lack were uncommon. At the same time that these reports of depression were appearing in the literature, a chance finding by Rose[16] turned out to be of the greatest significance. He was studying amino acid metabolism in various clinical disorders, when he noticed that some of his controls, normal healthy young women, excreted larger amounts of xanthurenic acid, which was explained by the fact that they were taking oral contraceptives. These findings were confirmed by extended studies by Price and his co-workers in Madison, Wisc.,[17] the center that many consider to be the "home" of "tryptophan metabolism." Subsequently, Rose worked with Brown (Price's successor) and later in the Metabolic Research Laboratory at St. Mary's Hospital in London in company with Wynn.[18–23] The investigations covered 8 years of study from one chance finding and represent one of the most searching examinations of the tryptophan-niacin pathway in man and animals, and in particular how the metabolism of tryptophan is influenced by sex hormones.

The work culminates with the demonstration that in a double-blind cross-over trial, a group of women on the pill, who showed evidence of vitamin B_6 deficiency, responded clinically to the effect of extra pyridoxine hydrochloride.[23]

A. Theories of Depression

1. Serotonin 5-Hydroxytryptamine (5HT)

a. Principle

There is considerable evidence that the synthesis of 5-hydroxytryptamine (5HT) from tryptophan is abnormally low in depressive illness and that this may be related to the origin of the disorder. The 5HT theory is the one that has received most attention; although the theory relating to tryptamine is probably as attractive, it has not received the same support.

According to Dewhurst,[24] "Many papers carry the implicit notion that serotonin (5HT) deficiency is associated with depression, although the hypothesis is rarely made explicit."

5HT was found to be low in the urine of depressed patients.[25] Although 5HT is synthesized in both the liver and kidney, the amine formed in these organs (the same as that appearing in the urine) cannot cross the blood-brain barrier,[26] so the amine affecting the brain must be produced there.

Further work showed that total 5-hydroxyindoles in cerebral spinal fluid was low in a depressed group of patients[27] and brain levels of 5HT were also low.[28,29] 5-Hydroxyindole acetic acid (5HIAA), the principal metabolite of 5HT, was again found low in a similar situation,[30] but oddly enough in the same study it was also found to be low in mania.

5-Hydroxytryptophan (5HTP), the precursor of 5HT, can pass the blood-brain barrier, where it is decarboxylated. Kline and Sacks[31] gave 5HTP to depressed patients and although they first reported it as alleviating some of the symptoms in depression, they were unable to confirm this in a later controlled study.[32] Again the conversion of 5HTP to 5HIAA has been reported to be normal in depression.[33,34]

The best evidence for depletion of brain 5HT resulting from stress comes from the elegant work of Curzon.[35] Curzon's hypothesis is that raised plasma corticosteroid levels (from stress) induce a high activity of tryptophan oxygenase, the enzyme which is the rate-controlling step of the tryptophan-nicotinic acid pathway, thus carrying tryptophan away from the 5HT pathway.

In this section it will be assumed that 5HT is low in depressive mood and the possible hormonal or other influences that bring about its depletion will be examined.

As can be seen in Figure 2, any increase in the activities of steps 1 or 4, or decrease in the activities of 2 or 3, would result in a lowering of brain 5HT. Furthermore, any lowering of the amount of tryptophan passing the blood-brain barrier, i. e., any reduction in the plasma free (or unbound) tryptophan, would produce the same effect.

Step 1. Diversion of Tryptophan to Other Pathways

What may be described as the Curzon theory[35] is that induced tryptophan oxygenase enzyme diverts tryptophan to the nicotinic acid pathway, thus lowering 5HT.

Pregnancy or the administration of estrogens increases tryptophan oxygenase in rat liver;[36,37] in the human pregnancy, treatment with estrogens

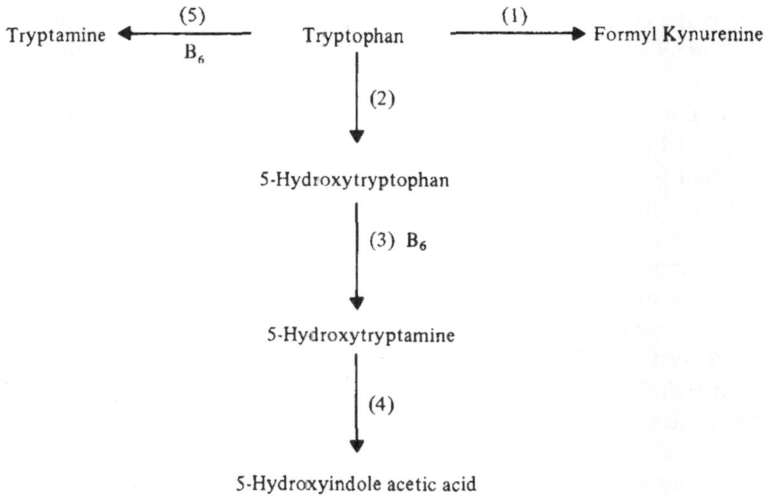

1. Tryptophan oxygenase
2. Tryptophan hydroxylase
3. 5-Hydroxytryptophan decarboxylase (B_6-dependent)
4. Monoamine oxidase
5. Tryptophan decarboxylase (B_6-dependent)

FIGURE 2. Serotonin (5HT) metabolic routes.

and the use of estrogen-containing oral contraceptives all produce greatly elevated levels of urinary excretions of tryptophan metabolites.[16,17,38]

The effect of estrogens appears to operate exclusively by action on the adrenal glands, for high-enzyme levels do not occur when estrogens are given to adrenolectomized rats.[39] In pregnancy there are probably two causes for abnormal tryptophan metabolism:

i. Induction of tryptophan oxygenase

ii. Pyridoxal deficiency, which is known to occur in pregnancy, and which has been described as being the end result of fetal demands for the vitamin.[40–42] This latter finding is further confirmed by the recently reported finding[43] that in 58 pregnant women all had progressive falls in vitamin B_6.

Estrogens appear to induce the oxygenase enzyme via glucocorticoids. It has long been known that estrogens increase total plasma cortisol but do not increase urinary excretion of cortisol metabolites.[44] This somewhat anomalous result is explained by the fact that estrogen-treated subjects show increased cortisol-binding to plasma proteins.[45] The increase occurs in the corticosteroid-binding globulin (CBG), depends on the estrogen level,[46] and is similar to that of other liver enzymes induced by estrogen.[47]

Another enzyme induced is alanine aminotransferase, which is elevated in both the liver and pancreas by cortisol. Keller et al.[48] have shown that in the rat when plasma corticosteroids are elevated by estrogens there is an increase in the activity of the liver enzyme, but not of that in the pancreas. Their explanation was that CBG can gain access to the enzyme-producing systems of the liver by pinocytosis, followed by liberation of the free steroid, whereas the thick basement membrane of the pancreatic cell prevents entrance of the protein-bound steroid in that organ. The possibility exists that estrogens increase tryptophan oxygenase in the same way and that the effect is dose related.

The dose-dependent thesis is very important within the context of the trend toward a low-dose estrogen pill. For example, urinary free cortisol is normal in women taking the low-dose estrogen or progestogen-only contraceptives. It is raised to a minor extent in the standard pill, but is increased threefold in the high-dose estrogen therapy for prostatic cancer.[49]

The significance of this is emphasized by the finding that the progestogen content of the pill is not implicated in causing depression,[15,23] although Grant and Mears[50] have suggested that a high progestogen content pill is the one responsible for the depressive symptoms.

Most of the early investigators of tryptophan metabolism and the pill were concerned with determining metabolites of the nicotinic acid pathway following a tryptophan load of either 2 or 5 g. This means that there exists the possibility that the hormonal effect plus the large load of tryptophan produced its own pyridoxine deficiency. Indeed, Price et al.,[17] having analyzed urine samples from a group of women all taking Enovid-E®, concluded that the abnormality was only detectable following a load of tryptophan; however, when the basal *N'*-methylnicotinamide results were examined from the same group of women, it was found that the Enovid-E group had significantly higher levels.[18] Later, in a longitudinal study of 18 women starting the pill, Price[51] showed that 3-hydroxyanthranilic acid (3HA) excretion measured without prior tryptophan load had a significant rise after only 1 month, and the high level was maintained throughout the 6 months study. Pyridoxine administration reduced the 3HA excretion to pretreatment levels. It was demonstrated in the same study that the 3HA excretion was increased in a male subject made pyridoxine deficient by diet.

The conclusion of this work must mean that an abnormality exists along the tryptophan-nicotinic acid pathway, and that it is not a biochemical abnormality produced by the unphysiological tryptophan load.

Step 2. Lowering the Activity of 5-Hydroxytryptophan Hydroxylase

The administration of L-tryptophan as a form of therapy in the relief of depression[52] may have an effect on the hydroxylase, which is not pyridoxine dependent. As tryptophan can pass the blood-brain barrier, it may restore 5HT synthesis to normal. The only evidence available is that the 5 hydroxytryptophan decarboxylase is the rate-limiting enzyme for the synthesis of brain 5HT in man; thus, it is unlikely that the hydroxylase is a factor in the lowering of 5HT.[53]

Step 3. The Activity of 5-Hydroxytryptophan Decarboxylase

As the decarboxylase is pyridoxine dependent, it is possible that the clinical improvement[23] of endogenous depression was confined to the group of women with an absolute pyridoxine deficiency and that tryptophan diversion to the nicotinic acid pathway (step 1) was also involved; if this is so, there are grounds for suggesting that the decarboxylase was increased by pyridoxine therapy. As the diversion of the tryptophan is not reversed with extra pyridoxine, then the improvement must have resulted from the stimulated decarboxylase. Lack of coenzyme is associated with reduced activity and decreased amounts of latent apoenzyme. Thus, 5-hydroxytryptophan decarboxylase activities in pyridoxine-deficient rats were increased with extra pyridoxine; however, the levels were lower than those prior to the deficiency.[54] These effects are particularly striking in the liver and kidney, but less so in the brain. It is possible that the clinical improvement noted in the study of Adams et al.[23] was due to induction of synthesis of apoenzyme as well as stimulation of "latent apoenzyme."

Step 4. Monoamine Oxidase

The catabolism of 5HT to 5HIAA is controlled by the enzyme monoamine oxidase. Any impairment in the activity of this oxidase enzyme would lower the catabolism of 5HT and inevitably lead to higher brain levels. The rationale of the use of the drugs, the monoamine oxidase inhibitors (MAOI), for the control of endogenous depression is based on the concept of restoring normal brain 5HT levels to normal, although there is much dispute over their value.[55] Not only are there differences of opinion about the type of patient likely to respond, but there is also a lack of agreement concerning the correctness of their use in clinical psychiatry. Most of the known MAOI's in current use are derivatives of hydrazine, e.g., $R{\cdot}NH{\cdot}NH_2$. What is surprising is that one of the earliest forms of therapy was isoniazid, the antituberculostatic drug, a compound that led to the introduction of MAOI's, the surprise being that isoniazid is a pyridoxine antagonist. It has an action virtually equivalent to deoxypyridoxine, the compound with the most powerful antivitamin action.[20,57]

This must virtually rule out pyridoxine deficiency per se as the cause of endogenous depression in those women taking oral contraceptives. On the one hand, we have a relative deficient state allegedly producing depression, both of which are corrected in exactly half the patients studied[23] when pyridoxine is compared to placebo and, on the other hand, pyridoxine deficiency is made more severe by isoniazid, a compound which has been used to relieve depression.

Although it may be heaping on the confusion, it must also be stated that the finding that isoniazid produced euphoria in some patients was noted in 1959;[58] the double-blind controlled clinical trial was not undertaken by psychiatrists until 1960.[59]

If we accept that MAOI's result in relief of depression and that this operates by increasing brain 5HT levels, we must examine what happens to MAO's with estrogen treatment.

Any increase in brain MAO or decrease in its inhibition would be reflected in a decrease of brain 5HT. MAO is considered to be involved in the in vivo inactivation of catecholamines, tyramine, and 5HT. The enzyme is widely distributed in man and there is considerable speculation on the possible role of placental MAO activity in toxemia of pregnancy.[60] Grant and Pryse-Davies[61] examined endometrial MAO activity in women taking the pill and discovered that with the strong progestogen pill the increased MAO was detectable on the 12th day of the menstrual cycle, that is, extra to the normal cyclical variations, whereas with strongly estrogenic pills endometrial enzyme levels remained low throughout the cycle. This suggests that endometrial MAO is related to hormone balance. This is the thesis of Grant and Mears,[50] who implicate a high-progestogen pill in the production of depression, although they are alone in this opinion.[13,15,23,62] Estrogens alone appear unlikely to modify the rate of 5HT production via MAO activity.

2. The Tryptamine Theory

As tryptamine is produced via the decarboxylation of tryptophan, a pyridoxine-dependent step, and its production is dependent on the amount of tryptophan available, a deficiency of the vitamin could produce the same effect as 5-hydroxytryptamine. However, there is one important difference between tryptamine and its 5-hydroxy analogue. The former can pass the blood-brain barrier and large amounts of tryptamine occurring in extracerebral tissue will also contribute to cerebral effects.[24] Dewhurst points out that if the antidepressant action of MAOI's is accepted, one should take into account the finding that after MAOI treatment the brain tryptamine quadrupled, the 5HT was not quite trebled, and the catecholamines showed less increase. The amines showing least change can also be metabolized by *O*-methylation, whereas tryptamine cannot; hence, it is much more sensitive to interference with MAO. Although this theory has not received very much support, its principal deficiency being that tryptamine has no effect when administered intravenously,[63] it does not necessarily establish that it can be ignored entirely.

a. The Binding of Tryptophan

Tryptophan is an essential amino acid and thus the plasma level must, in some way, control the brain level. In fact the total plasma level and the brain level are not significantly correlated,[64] but the brain level is controlled by the free (or unbound) tryptophan.[65] Tryptophan is largely bound to the plasma albumin.[66] Any disturbance in the binding would lead to greatly increased unbound and brain levels, respectively. Curzon et al. have shown[67] recently that plasma unesterified fatty acids have an effect on the binding of tryptophan. They suggest that changes in the fatty acid fraction have a role in determining the availability of tryptophan to the brain. Tryptophan hydroxylase, the rate-limiting enzyme for 5HT synthesis, is normally saturated with respect to tryptophan.[68,69] This means that any effect on the binding of tryptophan will be immediately reflected in changes in 5HT metabolism.

Although a few women taking oral contraceptives have been noted with much higher free fatty acid levels, produced in response to a glucose load,[70] there is no general increase and the mean fatty acid levels do not differ from the mean of the controls.

A decrease in plasma albumin would probably result in an increased free tryptophan fraction in plasma. Similarly, an increase in the albumin concentration or appearance of a further protein fraction, resulting from pregnancy or "the pill," that binds tryptophan strongly could lead to a lowering of brain tryptophan. A decrease in total plasma protein has been well documented in pregnancy and a similar but less marked effect is noted in estrogen therapy.[71] A further protein migrating with the α_2-globulin is seen in pregnancy[72] and this "pregnancy zone" can be seen also in pill-treated subjects.[47,73] However, there is no evidence that the binding of tryptophan is impaired.

Coppen et al.[74] have shown that the concentration of free tryptophan in plasma was significantly lower in female depressive patients and that this was reflected in lower CSF concentrations.

In a study involving 80 depressed patients who had classical or definite rheumatoid arthritis, Aylward and Maddock[75] have demonstrated that a correlation existed between plasma tryptophan levels and the exhibition of symptoms of mental depression.

It had been demonstrated previously that most of the clinically effective antirheumatic drugs displace tryptophan from human plasma proteins in vitro[76] and that this effect does not occur with other drugs lacking clinical anti-inflammatory action, but which are protein bound, e.g., paracetamol.[77] There is evidence that the percentage of tryptophan bound to plasma proteins is reduced in patients with rheumatoid arthritis who are receiving antirheumatic drugs.[78] Since aspirin increases brain tryptophan levels[79] and the latter depend on the plasma nonprotein-bound fraction, then the anti-inflammatory drugs which depress the protein-bound tryptophan may have a beneficial effect on the endogenous depression which plays an important part in rheumatoid arthritis.[80]

When Aylward and Maddock investigated the relationship between clinically anti-inflammatory action and a drug's ability to displace tryptophan from plasma proteins in vivo, they found that amelioration in depression scores (using the Hamilton Depressive Scale[81]) was greatest in those patients with the highest plasma free tryptophan. The highest concentrations of free tryptophan were observed in those patients receiving alclofenac, which group incidentally manifested the greatest improvement.

Another point of interest concerning this study and a finding not entirely unexpected is that there is an altered tryptophan metabolism in rheumatoid arthritis.[3,4]

3. The Nicotinic Acid Therapy

The administration of nicotinic acid as a form of therapy has been reported to produce endogenous depression.[82,83]

Although direct measurements of nicotinic acid itself have not been made in oral contraceptive therapy and pregnancy, the measurements of N'-methylnicotinamide and its metabolite N'-methyl-2-pyridone-5-carboxamide have both been made in spontaneous urines (without prior tryptophan loading) and following a 2-g load.[18] Women receiving Enovid-E (2.5 mg norethynodrel and 0.1 mg mestranol) excreted more N'-methylnicotinamide and the same or lower levels of the pyridone. As quinolinic acid, the precursor of nicotinic acid, is also raised in oral contraceptive therapy,[57] it is likely that nicotinic acid levels will also be increased.

Reasons for a causal relationship between excess nicotinic acid and depression are purely speculative.

Nicotinic acid activates cyclic AMP phosphordiesterase and decreased cyclic AMP excretion has been reported in depression.[2] Nicotinic acid administration also lowers plasma free fatty acid and triglyceride levels, and partially blocked the stress-induced free fatty acid rise.[84] Free fatty acids and plasma free tryptophan are correlated.[67] Thus, nicotinic acid may lower brain 5HT.

IV. BIOCHEMICAL METHODS

No consideration will be given here to the psychiatric assessment of the patients; attention will be focused on the biochemical changes resulting from pregnancy and oral contraceptive therapy.

Several biochemical criteria can be used to assess absolute vitamin B_6 deficiency, although it is still a matter of debate,[23] if the psychiatric state is a direct result of this deficiency.

Of 22 depressed women receiving oral contraceptive therapy, 11 showed evidence of an absolute vitamin B_6 deficiency, and these were the women who responded clinically in a double-blind cross-over trial to the administration of pyridoxine hydrochloride. The remaining 11 women showed no response and placebo administration was without effect.

Using the same biochemical criteria, 80% of women taking oral contraceptives had abnormal tryptophan metabolism indicating a relative vitamin B_6 deficiency.[85] The criteria were

i. A low urinary 4-pyridoxic acid[86] (4-pyridoxic acid is the major urinary metabolite of vitamin B_6)

ii. An increased HK/HA ratio[87]

iii. Lower activities of pyridoxal phosphate-dependent enzymes, e.g., erythrocyte aspartate and alanine aminotransferases[88]

The best indices appear to be the last two, especially when used in combination. 4-Pyridoxic acid was found to be a poor index of pyridoxine status in the studies of Adams et al.;[23] this is hardly surprising when one considers that these women are not receiving less vitamin B_6; they simply require more.

All these patients were assessed by employing the oral administration of a 2-g loading dose of L-tryptophan, as this load is the one used for most of the baseline studies.[20]

The erythrocyte enzymes alanine and aspartate aminotransferases were usually assayed by the method of Woodring and Storvick[89] with and without the addition of 100 μg of pyridoxal phosphate. The assay without added coenzyme is a measure of basal activity of the enzyme in vivo; the assay with added coenzyme enabled calculation of the percentage stimulation of the enzyme in vitro. In one study,[23] the mean alanine aminotransferase activity of 11 vitamin B_6-deficient women was lower than that of women not taking the pill. More significantly, the alanine aminotransferase percent stimulation in vitro, which is increased in pill-taking women, increases further in absolute vitamin B_6 deficiency.[85] It appears that two parameters can give clear evidence of absolute vitamin B_6 deficiency; the best two indices are the HK/HA ratio and the erythrocyte alanine aminotransferase with its percentage stimulation in vitro.

Undoubtedly one of the best methods of establishing pyridoxal deficiency is the measure of the plasma pyridoxal level itself. No direct measurements have so far been reported, although there is the implication in the work of Anderson et al.,[42] that this finding may explain some of their results.

V. SPONTANEOUS EXCRETION MEASUREMENTS

An important step in the investigation of vitamin B_6 lack would accrue if methods were available to assess accurately the spontaneous urinary excretion of HK and HA in order to measure the HK/HA ratio without giving the patient an unphysiological load of tryptophan.

Methods for HA measurement in urine without loading appear to be reasonably simple and successful. In fact, the measurement of this metabolite without prior loading has been used to demonstrate elevated excretions of HA in 30 women[51] taking oral contraceptives. A quick, reliable method for the spontaneous measurement of urinary HK would prove a valuable asset in this field of study.

VI. TREATMENT

Undoubtedly if oral contraceptives and pregnancy result in endogenous depression in women of childbearing age, then one is presented with a "heads-you lose, tails-you-lose" situation.

It has been suggested that pyridoxine supplements should be given to all women taking oral contraceptives.[13,91,92] In view of the 50% success rate with administered pyridoxine hydrochloride in the controlled trial of Adams and his co-workers,[23] this is a very attractive proposition. However, these authors as well as Rose et al.[88] point out that administration of large doses of pyridoxine hydrochloride may lead to further difficulties. It is known, for example, that estrogens stimulate both alanine – and tyrosine – aminotransferase enzymes in rat liver.[39,90] Furthermore, it has been shown that plasma concentrations of several amino acids are decreased in women taking oral contraceptives.[93,94] Thus, if coenzyme induction is superimposed on liver amino acid catabolizing enzymes, further lowering of plasma amino acids and the amino acid pool would be undesirable in populations where protein lack exists.

A high-protein diet and pyridoxine supplementation is one course of suggested treatment and undoubtedly the best course of action for depression when this occurs in pregnancy. For oral contraceptive users the choice would lie between this course of action and changing to a progestogen-only pill. If we accept the findings of Aylward and Maddock,[75] then aspirin is as good as anything.

REFERENCES

1. **Miller, H.**, Depression, *Br. Med. J.*, i, 257, 1967.
2. **Abdulla, Y. H. and Hamadah, K.**, 3′, 5′ Cyclic adenosine monophosphate in depression and mania, *Lancet*, i, 378, 1970.
3. **McMillan, M.**, The identification of a fluorescent reducing substance in the urine of patients with rheumatoid arthritis: the excretion of 3-hydroxyanthranilic acid in this and other conditions, *J. Clin. Pathol.*, 13, 140, 1960.
4. **Flinn, J. H., Price, J. M., Yess, N., and Brown, R. R.**, Excretion of tryptophan metabolites by patients with rheumatoid arthritis, *Arch. Rheum.*, 7, 201, 1964.
5. **Bett, I.**, Urinary tryptophan metabolites in rheumatoid arthritis and other diseases, *Ann. Rheum. Dis.*, 25, 556, 1966.
6. **Bole, G. G., Friedlander, M. M., and Smith, C. K.**, Rheumatic symptoms and serological abnormalities induced by oral contraceptives, *Lancet*, i, 323, 1969.
7. **Speira, H. and Plotz, C.**, Rheumatic symptoms and oral contraceptives, *Lancet*, i, 571, 1969.
8. **Pimstone, B. L.**, L. E. cells after oral contraceptives, *Lancet*, i, 1153, 1968.
9. **Kaye, B. M.**, Oral contraceptives and depression, *J.A.M.A.*, 186, 522, 1963.
10. **Tharkudas, H.**, Oral contraceptives and depression, *Lancet*, i, 1164, 1967.
11. **McGregor, C. M.**, Oral contraceptives and depression, *Lancet*, i, 1231, 1967.
12. **Daly, R. J., Kane, F. J., Jr., and Ewing, J. A.**, Psychosis associated with the use of a sequential oral contraceptive, *Lancet*, ii, 444, 1967.
13. **Winston, F.**, Oral contraceptives and depression, *Lancet*, i, 1209, 1969.
14. **Dennis, K. J. and Jeffery, J. d'A.**, Depression and oral contraceptives, *Lancet*, ii, 454, 1968.
15. **Herzberg, B. N., Johnson, A. L., and Brown, S.**, Depressive symptoms and oral contraceptives, *Br. Med. J.*, iv, 142, 1970.
16. **Rose, D. P.**, Excretion of xanthurenic acid in the urine of women taking progestogen-oestrogen preparations, *Nature*, 210, 196, 1966.
17. **Price, J. M., Thorton, J. M., and Mueller, L. M.**, Tryptophan metabolism in woman using steroid hormones for ovulation control, *Am. J. Clin. Nutr.*, 20, 452, 1967.
18. **Rose, D. P., Brown, R. R., and Price, J. M.**, Metabolism of tryptophan to nicotinic acid derivatives by women taking oestrogen-progestogen preparations, *Nature*, 219, 1259, 1968.
19. **Rose, D. P. and Brown, R. R.**, The influence of sex and oestrogens on liver kynureninase and kynurenine aminotransferase in the rat, *Biochim. Biophys. Acta*, 184, 412, 1969.
20. **Brown, R. R., Rose, D. P., Price, J. M., and Wolf, H.**, Tryptophan metabolism as affected by anovulatory agents, *Ann. N.Y. Acad. Sci.*, 166, 44, 1969.
21. **Rose, D. P.**, Effect of ovarian hormones and oral contraceptives on tryptophan metabolism, *J. Clin. Pathol.*, 23 (Suppl. 3), 37, 1969.
22. **Rose, D. P.**, Aspects of tryptophan metabolism in health and disease: a review, *J. Clin. Pathol.*, 25, 17, 1972.
23. **Adams, P. W., Wynn, V., Rose, D. P., Seed, M., Folkard, J., and Strong, R.**, Effect of pyridoxine hydrochloride (vitamin B_6) upon depression associated with oral contraceptives, *Lancet*, i, 897, 1973.
24. **Dewhurst, W. G.**, Amines and abnormal mood, *Proc. R. Soc. Med.*, 62, 1102, 1969.
25. **Rodnight, R.**, Body fluid indoles in mental illness, *Int. Rev. Neurobiol.*, 3, 251, 1961.
26. **Schanberg, S. M.**, A study of the transport of 5-hydroxy tryptophan and 5-hydroxytryptamine into brain, *J. Pharmacol. Exp. Ther.*, 139, 191, 1963.
27. **Ashcroft, G. W., Crawford, T. B. B., Eccleston, D., Sharman, D. F., MacDougal, E. J., Stanton, J. B., and Binns, J. K.**, 5-Hydroxyindole compounds in the CSF of patients with psychiatric or neurological diseases, *Lancet*, ii, 1049, 1966.
28. **Bourne, H. R., Bunney, W. E., Colburn, R. W., Davis, J. M., Davis, J. N., Shaw, D. M., and Coppen, A. J.**, Noradrenaline, 5-hydroxytryptamine and 5-hydroxyindole acetic acid in hind brain of suicidal patients, *Lancet*, ii, 805, 1968.
29. **Shaw, D. M., Camps, F. E., and Eccleston, E. G.**, 5-Hydroxytryptamine in the hind-brain of depressive suicides, *Br. J. Psychiatry*, 113, 1407, 1967.
30. **Dencker, S. J., Malm, U., Roos, B. E., and Werdinius, B.**, Acid monoamine metabolites of cerebrospinal fluid in mental depression and mania, *J. Neurochem.*, 13, 1545, 1966.
31. **Kline, N. S. and Sacks, W.**, Relief of depression within one day using an M.A.O. inhibitor and intravenous 5-HTP, *Am. J. Psychiatry*, 120, 274, 1963.
32. **Kline, N. S., Sacks, W., and Simpson, G. M.**, Further studies on one day treatment of depression with 5-HTP, *Am. J. Psychiatry*, 121, 379, 1964.
33. **Dewhurst, W. G. and Pare, C. M. B.**, A clinical and biochemical study of M.A.O. inhibition in depressed patients, *J. Ment. Sci.*, 107, 239, 1961.
34. **Feldstein, A., Hoagland, H., Wong, K. K., Oktem, M. R., and Freeman, H.**, M.A.O. activity in relation to depression, *Am. J. Psychiatry*, 120, 1192, 1965.

35. **Curzon, G.**, Tryptophan pyrrolase – a biochemical factor in depressive illness? *Br. J. Psychiatry,* 115, 1367, 1969.
36. **Auricchio, S., Rigillo, N., and Di Toro, R.**, Tryptophan pyrrolase and 3-hydroxyanthranilic oxidase activity of the rat liver, *Minerva Pediatr.,* 12, 1463, 1960.
37. **Rose, D. P. and Braidman, I. P.**, Oral contraceptives, depression and amino acid metabolism, *Lancet,* i, 1117, 1970.
38. **Brown, R. R., Thornton, M. J., and Price, J. M.**, The effect of vitamin supplementation of the urinary excretion of tryptophan metabolites by pregnant women, *J. Clin. Invest.,* 40, 617, 1961.
39. **Braidman, I. P. and Rose, D. P.**, Effects of sex hormones on three glucocorticoid-inducible enzymes concerned with amino acid metabolism in rat liver, *Endocrinology,* 89, 1250, 1971.
40. **Wachstein, M. and Gouditis, A.**, Disturbances of vitamin B_6 metabolism in pregnancy, *J. Lab. Clin. Med.,* 42, 98, 1953.
41. **Hamfelt, A. and Hahn, L.**, Pyridoxal phosphate concentration in plasma and tryptophan load test during pregnancy, *Clin. Chim. Acta,* 25, 91, 1969.
42. **Anderson, B. B., Peart, M. B., and Fulford-Jones, C. E.**, The measurement of serum pyridoxal by a microbiological assay using *Lactobacillus casei, J. Clin. Pathol.,* 23, 232, 1970.
43. **Hamfelt, A. and Tuvemo, T.**, Pyridoxal phosphate and folic acid concentration in blood and erythrocyte aspartate aminotransferase activity during pregnancy, *Clin. Chim. Acta,* 41, 287, 1972.
44. **Petesson, R. E., Nokes, G., Chen, P. S., Jr., and Black, R. L.**, Estrogens and adrenocortical function in man, *J. Clin. Endocrinol.,* 20, 495, 1960.
45. **Sandberg, A. A. and Slaunwhite, W. R., Jr.**, Transcortin: a corticosteroid-binding protein of plasma. II. Levels in various conditions and the effects of oestrogens, *J. Clin. Invest.,* 38, 1290, 1959.
46. **Doe, R. P., Mellinger, G. T., Swaim, W. R., and Seal, U. S.**, Estrogen dosage effects on serum proteins: a longitudinal study, *J. Clin. Endocrinol.,* 27, 1081, 1967.
47. **Musa, B. U., Doe, R. P., and Seal, U. S.**, Serum protein alterations produced in women by synthetic oestrogens, *J. Clin. Endocrinol.,* 27, 1463, 1967.
48. **Keller, N., Richardson, U. I., and Yates, F. E.**, Protein binding and the biological activity of corticosteroids: in vivo induction of hepatic and pancreatic alanine aminotransferases by corticosteroids in normal and estrogen-treated rats, *Endocrinology,* 84, 49, 1969.
49. **Burke, C. W.**, The effects of oral contraceptives on cortisol metabolism, *J. Clin. Pathol.,* 23, (Suppl. 3), 11, 1969.
50. **Grant, E. C. G. and Mears, E.**, Mental effects of oral contraceptives, *Lancet,* ii, 945, 1967.
51. **Price, S. A., Rose, D. P., and Toseland, P. A.**, Effects of dietary vitamin B_6 deficiency and oral contraceptives on the spontaneous excretion of 3-hydroxyanthranilic acid, *Am. J. Clin. Nutr.,* 25, 494, 1972.
52. **Coppen, A., Shaw, D. M., Herzberg, B., and Maggs, R.**, Tryptophan in the treatment of depression, *Lancet,* ii, 1178, 1967.
53. **Robins, E., Robins, J. M., Croninger, A. B., Moses, S. G., Spencer, S. J., and Hudgens, R. W.**, The low level of 5-hydroxytryptophan decarboxylase in human brain, *Biochem. Med.,* 1, 240, 1967.
54. **Davis, V. E.**, Effect of cortisone and thyroxine on aromatic acid decarboxylation, *Endocrinology,* 72, 33, 1963.
55. **Tyrer, P.**, Current status of monoamine oxidase inhibitors in psychiatry, *Br. J. Hosp. Med.,* 9, 795, 1973.
56. **Zeller, E. A. and Barsky, J.**, In vivo inhibition of liver and brain monoamine oxidase by 1-isonicotinyl-2-isopropyl hydrazine, *Proc. Soc. Exp. Biol. Med.,* 81, 459, 1952.
57. **Rose, D. P. and Toseland, P. A.**, Urinary excretion of quinolinic acid and other tryptophan metabolites after deoxypyridoxine or oral contraceptive administration, *Metabolism,* 22, 165, 1973.
58. **Bosworth, D. M.**, Iproniazid: a brief review of its introduction and clinical use, *Ann. N. Y. Acad. Sci.,* 80, 809, 1959.
59. **Rees, L.**, Treatment of depression by drugs and other means, *Nature,* 186, 114, 1960.
60. **Southgate, J., Grant, E. C. G., Pollard, W., Pryse-Davies, J., and Sandler, M.**, Cyclical variations in endometrial monamine oxidase, *Biochem. Pharmacol.,* 17, 721, 1968.
61. **Grant, E. C. G. and Pryse-Davies, J.**, Effect of oral contraceptives on depressive mood changes and on endometrial monoamine oxidase and phosphatases, *Br. Med. J.,* 3, 777, 1968.
62. **Baumblatt, M. J. and Winston, F.**, Pyridoxine and the pill, *Lancet,* i, 832, 1970.
63. **Coppen, A., Shaw, D. M., Malleson, A., Eccleston, E., and Gundy, G.**, Tryptamine metabolism in depression, *Br. J. Psychiatry,* 111, 993, 1965.
64. **Curzon, G., Joseph, M. H., and Knott, P. J.**, Effects of immobilisation and food deprivation on rat brain metabolism, *J. Neurochem.,* 19, 1967, 1972.
65. **Knott, P. J. and Curzon, G.**, Free tryptophan in plasma and brain tryptophan metabolism, *Nature,* 239, 452, 1972.
66. **McMenamy, R. M. and Oncley, J. L.**, The specific binding of L-tryptophan to serum albumin, *J. Biol. Chem.,* 233, 1436, 1958.
67. **Curzon, G., Friedel, J., and Knott, P. J.**, The effect of fatty acids on the binding of tryptophan to plasma protein, *Nature,* 242, 198, 1973.
68. **Eccleston, D., Ashcroft, G. W., and Crawford, T. B. B.**, 5-Hydroxyindole metabolism in rat brain. A study of intermediate metabolism using the technique of tryptophan loading, *J. Neurochem.,* 13, 493, 1965.
69. **Friedman, P. A., Kappelman, A. H., and Daufman, S.**, Partial purification and coactivisation of tryptophan hydroxylase from rabbit hind-brain, *J. Biol. Chem.,* 247, 4165, 1972.

70. **Wynn, V. and Doar, J. W. H.,** Effects of oral contraceptives on carbohydrate metabolism, *J. Clin. Pathol.*, 23 (Suppl. 3), 19, 1969.
71. **Musa, B. U., Doe, R. P., and Seal, U. S.,** Elevation of certain plasma proteins in man following estrogen administration: a dose-response relationship, *J. Clin. Endocrinol.*, 25, 1163, 1965.
72. **Robinson, J. C., London, W. T., and Pierce, J. E.,** Observations on the origin of pregnancy-associated plasma proteins, *Am. J. Obstet. Gynecol.*, 96, 226, 1966.
73. **Alfonso, J. F. and de Alvarez, R. R.,** Further starch gel fractionation of new protein zones in pregnancy, *Am. J. Obstet. Gynecol.*, 86, 815, 1963.
74. **Coppen, A., Brooksbank, B. W. L., and Peet, M.,** Tryptophan concentration in the CSF of depressive patients, *Lancet*, i, 1393, 1972.
75. **Aylward, M. and Maddock, J.,** Plasma tryptophan levels in depression, *Lancet*, i, 936, 1973.
76. **McArthur, J. N. and Dawkins, P. D.,** The effect of sodium salicylate on the binding of L-tryptophan to proteins, *J. Pharm. Pharmacol.*, 21, 744, 1969.
77. **McArthur, J. N., Dawkins, P. D., and Smith, M. J. H.,** The displacement of L-tryptophan and dipeptides from bovine albumin in vitro and from human plasma in vivo by antirheumatic drugs, *J. Pharm. Pharmacol.*, 23, 393, 1971.
78. **McArthur, J. N., Dawkins, P. D., Smith, M. J. H., and Hamilton, E. B. D.,** Mode of action of antirheumatic drugs, *Br. Med. J.*, ii, 677, 1971.
79. **Tagliamonte, A., Biggio, G., and Gessa, G. L.,** Aspirin and brain tryptophan, *Riv. Farm. Ther.*, 11, 251, 1971.
80. **Glyn, J. H.** Rheumatic pains: some concepts and hypotheses, *Proc. R. Soc. Med.*, 64, 354, 1971.
81. **Hamilton, M.,** A rating scale for depression, *J. Neurol. Neurosurg. Psychiatry*, 23, 56, 1960.
82. **Zolner, N. and Gudenzi, N.,** Behandlung der Hypercholesterinämie mit Beta-Pyridyl Karbinol. II. Mekenwirkungen und Diskussion, *Med. Klin.*, 61, 2036, 1966.
83. **Danto, B. L.,** Triflupromazine versus pentylenetetrazolnicotinic acid for treatment of chronic brain disease on a general hospital psychiatric service, *J. Am. Geriatr. Soc.*, 17, 414, 1969.
84. **Carlson, L. A., Levi, L., and Orö, L.,** Plasma lipids and urinary excretion of catecholamines during acute emotional stress in man and their modification by nicotinic acid, *Forsvarsmedicin*, 3, 129, 1967.
85. **Rose, D. P., Strong, R., Adams, P. W., and Harding, P. E.,** Experimental vitamin B_6 deficiency and the effect of oestrogen containing oral contraceptives on tryptophan metabolism and vitamin B_6 requirements, *Clin. Sci.*, 42, 465, 1972.
86. **Reddy, S. K., Reynolds, M. S., and Price, J. M.,** The determination of 4-pyridoxic acid in human urine, *J. Biol. Chem.*, 233, 691, 1958.
87. **Heeley, A. F.,** The effect of pyridoxine on tryptophan metabolism in phenylketonuria, *Clin. Sci.*, 29, 465, 1965.
88. **Rose, D. P., Strong, R., Folkard, J., and Adams, P. W.,** Erythrocyte aminotransferase activities in women using oral contraceptives and the effect of vitamin B_6 supplementation, *Am. J. Clin. Nutr.*, 26, 48, 1973.
89. **Woodring, M. and Storvick, C.,** Effect of pyridoxine supplementation on glutamic-pyruvic transaminase and *in vitro* stimulation in erythrocytes of normal women, *Am. J. Clin. Nutr.*, 23, 1385, 1970.
90. **Doberenz, A. R., Van Miller, J. P., Green, J. R., and Beaton, J. R.,** Vitamin B_6 depletion in women using oral contraceptives as determined by erythrocyte glutamic-pyruvic transaminase activities, *Proc. Soc. Biol. Med.*, 137, 1100, 1971.
91. **Price, S. A. and Toseland, P. A.,** Oral contraceptives and depression, *Lancet*, ii, 158, 1969.
92. **Luhby, A. L., Brin, M., Gordon, M., Davis, P., Murphy, M., and Spiegel, H.,** Vitamin B_6 metabolism in users of oral contraceptives, *Am. J. Clin. Nutr.*, 24, 284, 1971.
93. **Rose, D. P. and Cramp, D. G.,** Reduction of plasma tyrosine by oral contraceptives and oestrogens: a possible consequence of tyrosine aminotransferase induction, *Clin. Chim. Acta*, 29, 49, 1970.
94. **Craft, I. L. and Peters, T. J.,** Quantitative changes in plasma amino acids induced by oral contraceptives, *Clin. Sci.*, 41, 301, 1971.

INDEX

F

G

H

I

J

L

M

N